PHILOSOPHICAL ANALYSIS IN LATIN AMERICA

SYNTHESE LIBRARY

STUDIES IN EPISTEMOLOGY,

LOGIC, METHODOLOGY, AND PHILOSOPHY OF SCIENCE

VOLUME 172

PHILOSOPHICAL ANALYSIS IN LATIN AMERICA

Edited by

JORGE J. E. GRACIA
State University of New York at Buffalo

EDUARDO RABOSSI
Universidad de la Plata

ENRIQUE VILLANUEVA
Univerdidad Nacional Autónoma de México

and

MARCELO DASCAL
Universidade Estadual de Campinas and Tel-Aviv University

D. REIDEL PUBLISHING COMPANY

A MEMBER OF THE KLUWER ACADEMIC PUBLISHERS GROUP

DORDRECHT / BOSTON / LANCASTER

Library of Congress Cataloging in Publication Data

CIP

Main entry under title:

Philosophical analysis in Latin America.

(Synthese library; v. 172)
Includes index.
1. Philosophy–Addresses, essays, lectures. 2. Philosophy–Latin America–Addresses, essays, lectures. I. Gracia, Jorge J. E. II. Series.
B29.P454 1984 199'.8 84-3342
ISBN 90-277-1749-4

Published by D. Reidel Publishing Company,
P.O. Box 17, 3300 AA Dordrecht, Holland.

Sold and distributed in the U.S.A. and Canada
by Kluwer Academic Publishers,
190 Old Derby Street, Hingham, MA 02043, U.S.A.

In all other countries, sold and distributed
by Kluwer Academic Publishers Group,
P.O. Box 322, 3300 AH Dordrecht, Holland.

Printed in The Netherlands

For if any one composes and joins by the understanding that which nature does not suffer to be joined, no one is unaware that that is false, as would be the case should one join by the imagination horse and man and construct a centaur. But if it be done by division and by abstraction, the thing would not be constituted as the idea is, yet for all that, the idea is still not in the least false.

Boethius, _Second Commentary on Porphyry's "Isagoge"_, Bk. I; _PL_ 64, 84.

TABLE OF CONTENTS

PREFACE

Historians of Latin American philosophy have paid relatively little attention to the development of philosophical analysis in Latin America. There are two reasons for this neglect: First, they have been primarily concerned with the formative period of philosophical development, in particular with the so called "founders" of Latin American philosophy. And second, philosophical analysis did not become a noticeable philosophical trend in Latin America until recent years. True, a number of Latin American philosophers took notice of Moore, Russell, the members of the Vienna Circle and other important figures in the analytic movement quite early. But these were isolated instances that lacked the sustained effort and broad base indispensible to make a serious impact in the development of Latin American philosophy. That has changed now. There are not only good numbers of philosophers who work within the analytic tradition, but also some journals and institutes dedicated to the analytic mode of philosophizing. It is, therefore, most appropriate to publish a collection of articles which would introduce the reader of philosophy to the most representative analytic material produced so far in Latin America. Indeed, it is not only appropriate, but also necessary, since most of the published analytic literature to date is scattered in various journals, sometimes of difficult access. Moreover, not all that has been published is representative of the best already produced and of the potential that the movement has in Latin America. Much of it is simply the result of an effort to absorb ideas already proposed elsewhere, lacking as a result sufficient originality to be taken as serious contributions to philosophy. The aim of this collection, then, is not only to introduce the reader to a vital philosophical current in contemporary Latin American philosophy, but also to present some of what the editors think is the best material that has been produced by philosophical analysis in Latin America to date.

The criteria used in the selection had little to do with time or themes. Philosophical analysis is a new current and, therefore, most of those who have adopted it are young or at most mature philosophers; only a handful are past the age of sixty. Therefore, instead of setting

J. J. E. Gracia et al. (eds.), Philosophical Analysis in Latin America, ix–xii.

artificial thematic or temporal boundaries at the start, we decided to begin with tentative lists of authors who had consistently published and are still actively publishing material which could be classified as part of the analytic tradition, and then select from their writings what seemed best and most appropriate for a collection of this size and intent. Not every author included, however, can or would accept to be described as an "analytic philosopher." Some would argue, and rightly so, that the philosophical orientation of some or even many of their published works is not analytic. Nevertheless, all those included here have to some extent contributed to the development and dissemination of philosophical analysis in Latin America. In general, long or very technical papers were excluded as well as those whose main intent was the mere exposition or negative criticism of someone else's views. We looked for clear style, accessibility to a general philosophical audience, and originality.

The selection was done under the general editorship of Jorge J. E. Gracia. He launched the idea, coordinated the efforts of the editors, kept the channels of communication open, and formulated and maintained the general perspective of the volume. Each editor had primary responsibility for a predetermined geographical area, but also worked with the others in the selection of the various articles. Eduardo Rabossi, in particular, worked closely with the general editor in the selection of the articles and the formulation of criteria.

In order to make clear the interrelationships among the selected authors, and to provide a setting where their work may be understood, they were gathered into four groups, according to the country or geographic area where they reside. Philosophers from Argentina, Mexico, and Brazil constitute the first three groups, since these are the countries that have produced the largest number of analytic philosophers; the fourth group comprises authors from the rest of Latin America. The articles have been arranged in alphabetical order according to the author's last name. Each area is preceded by an introduction which discusses the development of philosophical analysis in that location. A general introduction at the beginning of the collection presents a brief historical account of philosophical analysis and summarizes its present state in Latin America.

Concurrently with this English edition, a Spanish edition is being published by the Fondo de Cultura Económica. It is a much longer work, including a number of papers from the various geographical areas left out in the English edition as well as an entire section on Latin American philosophers who work within analytic parameters in the United States and Canada. These authors were excluded because of limitations of space imposed by the publisher and/or because of difficulties related to the translation of their works. Since the work of authors who live in the United States and Canada is well known in English speaking countries, it was a natural choice to leave them out in a work whose main purpose is to make the work of Latin American philosophers known in the English speaking world. The authors and papers left out in this edition are the following: "On the Fregean Solution of Morton White to the Paradox of Analysis," Thomas M. Simpson (Argentina); "Political Assasination and Tyrannicide," Javier Esquivel (Mexico); "Validity, Inference and Implications," Hugo Margáin (Mexico); "Truth and Practice," Mario H. Otero (Mexico); "The General Theory of Law and Juridical Science," Roberto Vernengo (Argentina and Mexico); "The Difference Between 'to be certain' and 'to know'," Luis Villoro (Mexico); "Ontology and Semantics," Luis Enrique Lopes dos Santos (Brazil); "On the Concept of Object in the *Tractatus*," Rafael Burgos (Venezuela); "Black and Tarski and the Philosophical Problem of Truth," Juan A. Nuño (Venezuela); "The Difficulty of Choosing," Augusto Salazar Bondy (Peru); "Wittgenstein's *Investigations* and the Possibility of Philosophy," Roberto Torretti (Chile-Puerto Rico); "The Fregean Concept of 'Bedeutung'," Ignacio Angelelli (Argentina, USA); "Are There Propositions?" Mario Bunge (Argentina, Canada); "Perceptual Propositions," Héctor-Neri Castañeda (Guatemala, USA); "Notes on the Framework of Contemporary Philosophy of Science," J. Alberto Coffa (Argentina, USA); "Individuals as Instances," Jorge J. E. Gracia (Cuba, Canada, USA); "A Theory of Knowledge," Ernesto Sosa (Cuba, USA). Also it should be noted that in a couple of cases the articles from particular authors printed in the Spanish and English edition are not the same. The English edition contains a single jointly authored article by Carlos E. Alchourrón and Eugenio Bulygin instead of the two found in the Spanish version: "Is There a Real Disagreement Between Strawson and Russell with Respect to

Definite Descriptions?" and "Omnipotence, Omniscience and Freedom." Eduardo Rabossi's article in the Spanish version is "Concerning a Possible Proof of Ethical First Principles"; Carlos Santiago Nino's article is "Is to Act the Same as to Omit?"; and Fernando Salmerón's article is "Philosophy and Moral Attitudes." It should also be added that some articles have been abridged in the English edition. Moreover, although the papers from authors working in North America and Canada have been left out in this edition, we have added an appendix with some information about their work.

The preparation of a volume like this involves substantial consultation not only among those who put it together and those whose work is included, but also with many others who, because of their interest in Latin American philosophy or their expertise in philosophical analysis, can give advice to the editors; we must acknowledge their contribution. Among those whom we would like to single out for special mention are Risieri Frondizi, Alberto Cordero, Teresa Salas, Iván Jaksić, Eloy Neira, Oscar Martí, Alberto Coffa, John Kearns and Ezequiel de Olaso. We are also grateful to Joel Tierno, James Griffis and Christopher Pines for helping with the copy editing and the proofreading of the manuscript. And we would like to thank the various authors of the articles for making the appropriate arrangements for their translation and the journals and other publications where they appeared for their permission to reprint them here. Finally, we are particularly grateful to Dr. Edward Dudley, Director of the Council of International Studies at the State University of New York at Buffalo, and to Dr. Ross MacKinnon, Dean of the Faculty of the Social Sciences of the same institution, for the financial assistance they provided to make possible the preparation of the camera-ready copy of the manuscript.

The Editors
1983

Jorge J. E. Gracia

INTRODUCTION

1. PHILOSOPHICAL ANALYSIS

One of the most difficult problems which faces the analytic philosopher, or even the historian of the analytic movement, is the characterization of analysis. For once this issue is raised it becomes increasingly difficult to find any satisfactory answer, since most analytic philosophers vary widely in their views not only about substantive philosophical issues but even about what analysis is or is supposed to be. Indeed, from a non-historical point of view, it is difficult to find anything that is uniformly accepted by or common to all the so called "analytic philosophers." They seem to have no more in common than a family resemblance, to use Wittgenstein's terminology.[1] But then, a problem arises: if we do not find a set of common features, not even one feature that they all share, how is analysis to be distinguished from other philosophical traditions?

The complex philosophical and historical issues involved in this question preclude its adequate treatment in an introductory essay such as this. We must content ourselves with a considerably more modest task, namely, identifying some of the ideas and attitudes that gave rise to the movement. Hopefully this will reveal some of the basic assumptions of philosophical analysis and will be sufficient to point in the direction of a more complete characterization of the movement. It will also help to uncover the similarities and differences between the conditions which gave rise to it in Europe and the conditions which made some philosophers turn to analysis in Latin America. However, given the limitations of space imposed here, the task of uncovering these similarities and differences will have to be left to other historians of the movement.[2]

J. J. E. Gracia et al. (eds.), Philosophical Analysis in Latin America, 1–15.

What is known today as "philosophical analysis" is a movement which arose from the philosophical activity and reflections of Bertrand Russell and G. E. Moore in England and the members of the Vienna Circle in Austria. The two English philosophers in particular were reacting against the excesses of post-Hegelian philosophy, although they did so in different ways, and the members of the Vienna Circle were reacting against the metaphysical views of some of their contemporaries. In fact, from their explicit aims there is little that one could expect these philosophers to have in common, and their views differed considerably. Nevertheless, in spite of these differences, the manner in which they discussed philosophy and the problems that interested them created a common ground on which a dialogue developed and a tradition was founded.

The relative value of the ideas of the various participants in the movement is a matter of debate; determining these values is marginal to our present task. What is important for us here is to note the basic contribution which each individual or group made to the movement. Nor is it necessary for us to trace the development of this movement from its beginning to the present or to name all the philosophers who have participated in it. For our purpose it will suffice to point to some of the basic ideas and attitudes which contributed to the creation of this tradition, even if some of those ideas and attitudes have since been abandoned.

1.1 G. E. Moore

If one sought to single out the most characteristic philosopher of the movement in its early stages, most would point to G. E. Moore. His influence on the course of the analytic movement was enormous. Nevertheless, this influence seems to have been more an influence of personality than of views, for it was not so much the conclusions he reached,--some of which he adopted as tentative and later discarded,--but the way he went about reaching them that changed the course of philosophy. Moreover, concerning the very notion of analysis, it is only after it was well under way as a movement that Moore attempted a clarification of the philosophical method implied by it.

Perhaps the best way to characterize the aim which Moore pursued through analysis is as a <u>search for clarification</u>. The search for clarification is to be understood in the context of his puzzlement about what other philosophers said. He was not particularly concerned with the commonsensical or the scientific ways of looking at the world; he found these perfectly clear and acceptable. What bothered him were the statements which philosophers made when dealing with these. As he put it:

> I do not think that the world or the sciences would ever have suggested to me any philosophical problems. What has suggested philosophical problems to me is things which other philosophers have said about the world or the sciences.[3]

One of the main problems with what philosophers have said is that their statements often contradict generally accepted propositions which, as Moore puts it, "I <u>know</u>, with certainty, to be true." Take for example the Kantian view that we are unable to counter doubts or provide any satisfactory proof of the claim that there are things outside us. According to Moore, it is perfectly true that there are some things about which it is doubtful whether they should be regarded as "existing outside of us," but certainly there are many which, provided they exist, would exist outside of us. Moore's proof: "Here is one hand; here is another: therefore at least two things outside us exist."[4]

The best way to unravel the mystery which surrounds many philosophical statements, according to Moore, is through analysis. What this method is, however, and how it is to be employed became a source of puzzlement for Moore; with his characteristically honest approach, he was one of the first to point out its shortcomings. Nonetheless, there are at least two things certain about the method. The first is the assumption that analysis works on something complex whose parts or elements are to be distinguished and whose relation to those parts is to be made clear. The second is the fact that, although the complex object of analysis is not a linguistic entity, analysis has to do with language.

Late in his career, Moore attempted too clarify in greater detail what he meant by analysis by pointing out three necessary conditions of it:

> If you are to "give an analysis" of a given _concept_, which is the _analysandum_, you must mention, as your _analysans_, a _concept_ such that (a) nobody can know that the _analysandum_ applies to an object without knowing that the _analysans_ applies to it, (b) nobody can verify that the _analysandum_ applies without verifying that the _analysans_ applies, (c) any expression which expresses the _analysandum_ must be synonymous with any expression which expresses the _analysans_.[5]

According to this explanation, there are, for example, four ways of analyzing the concept of "brother": (1) "The concept 'being a brother' is identical with the concept 'being a male sibling.'" (2) "The propositional function 'X is a brother' is identical with the propositional function 'X is a male sibling.'" (3) "To say that a person is a brother is the same thing as to say that that person is a male sibling." And, (4) "To be a brother is the same thing as to be a male sibling."

It is clear, then, that the aim of analysis is not the discovery of facts about the world. That seems to be the function of science. Analysis is concerned only with the clarification of concepts or propositions. Moreover, this clarification involves a breaking down of these concepts or propopsitions into simpler elements, and it is to be applied to the linguistic expressions of these concepts or propositions.

1.2 _Bertrand Russell_

Russell, unlike Moore, never attempted to articulate what he meant by analysis, but nonetheless he is regarded by some as "the foremost practitioner of the method."[6] Moreover, the aim of analysis for Russell was different from that of Moore. The latter was searching for clarification; Russell was _searching for the ultimate constituents of reality_. As such his search was metaphysical, although he rejected what he considered to be the inflated claims of other metaphysicians, such as Bradley and the idealists. Moore, on the other hand, was not particularly interested in formulating a general view of reality. True, he did not hold that the only function of philosophy is clarification. He accepted as a legitimate aim of other philosophers to "give a general description of the _whole_ of the Universe,"[7] but certainly he neither did nor attempted to do so. Russell's efforts, on the contrary, were fundamentally directed toward that aim. What Moore and Russell had in common, on the other hand, was that both regarded analysis as the way of achieving their respective goals: for Moore, clarification of philosophical statements; for Russell, the discovery of the ultimate constituents of the universe.

What is to be analyzed in the search for simples, according to Russell, are facts:

> The things in the world have properties, and stand in various relations to each other. That they have these properties and relations are _facts_, and the things and their qualities or relations are quite clearly in some sense or other components of the facts that they have those qualities or relations. The analysis of apparently complex _things_... can be reduced by various means, to the analysis of facts which are apparently about those things. Therefore, it is with the analysis of _facts_ that one's consideration of the problem of complexity must begin.[8]

Facts, then, are the object of analysis, but how does one get at them? Russell's procedure indicates that it is through propositions. Facts are stated in propositions and it is facts that make propositions true or false. Moreover, it is somehow the structure of propositions that reveals the structure of facts. This is most evident in what Russell called "atomic propositions," i.e., propositions which assert the possession of a quality or relation by some thing, for example, 'This is red.' To each atomic proposition corresponds an atomic fact. It is Russell's emphasis on atomic facts and on the logical analysis of the ultimate structures of language and the world that gave the view its name: _logical atomism_.

Within this framework it naturally became very important to uncover the real structure of language. This was partly done by creating what Russell called a "logically perfect language" which could serve as a model to unravel the problems created by the language ordinarily used.[9] The perfect language took the form of a notation which was truth-functional, that is, complex formulae were constructed on the bases of simple ones, so that the truth value of the complex would depend solely on the truth value of the simple. It was assumed by Russell, as well as the others who favored this approach, that this notation embodied the essence of language, and that where languages differed or common language appeared to diverge, it was merely that "the essential skeleton was concealed."[10]

Perhaps the best example of Russell's analysis occurs in his theory of descriptions. Russell became very concerned about certain propositions and expressions that deny existence or that deal with unreal or self-contradictory objects, such as 'The round square does not exist,' 'the present King of France,' and similar ones. The problem with these propositions and expressions is that in order to deny the existence of an object, they seem to posit the object as existing: to state that the present King of France does not exist seems to require the existence in some sense of the present King of France. Otherwise, how could we talk about him? Yet, if he exists, we fall into a contradiction--the object exists and does not exist at the same time.

Russell's solution to this problem is to introduce a distinction between two types of symbols. Proper names are one type of symbol. A proper name is a simple symbol such as 'Scott,' which designates an individual.[11] The other are called "descriptions." Descriptions are complex symbols such as 'the author of *Waverley*'; they do not designate individuals directly and are therefore incomplete. The meaning of descriptions is given only in context. This means that they do not refer to an actual entity as 'Scott' and other proper names do. Analysis then is to be applied to complex symbols that are complete, namely, propositions such as 'Scott is the author of *Waverley*' and not to incomplete complex symbols such as 'the author of *Waverley*.' The analysis of this particular proposition in Russellian terms would be as follows:

(1) "X wrote *Waverley*" is not always false;
(2) "if X and Y wrote *Waverley*, X and Y are identical" is always true;

(3) "if X wrote _Waverley_, X was Scott" is always true.

Those three propositions, translated into ordinary language, state:

(1) at least one person wrote _Waverley_;
(2) at most one person wrote _Waverley_;
(3) whoever wrote _Waverley_ was Scott.[12]

This analysis reveals, according to Russell, that the simplicity of the proposition 'Scott is the author of _Waverley_' is apparent, and that the description 'the author of _Waverley_' does not name any real thing. The same is true of other problematic expressions such as 'the present King of France' or 'the square circle.' From this it becomes clear, according to Russell, that what has misled philosophers into positing all sorts of non-existing entities is their lack of understanding language; they have been "misled by grammar."[13] Careful analysis of language makes many of these philosophical problems disappear.

1.3 Vienna Circle

While Moore left open the question as to the nature of philosophy and Russell clearly engaged in metaphysical speculation, the members of the Vienna Circle explicitly repudiated metaphysics and identified philosophy with analysis. They described themselves as "logical positivists"—positivists because they rejected metaphysics as spurious and emphasized science; logical because the method of philosophy they favored was formal. The most important members of the group were Moritz Schlick, Rudolph Carnap, Kurt Gödel, Herbert Feigl and Friedrich Waismann. Ludwig Wittgenstein was not a member of the group, but he lived nearby and had frequent conversations with them. His _Tractatus logico-philosophicus_ (1919) was widely discussed by the members of the Vienna Circle.

The cornerstone of the Circle's views was the so called Verification Principle. This was seen as a criterion of meaning: The meaning of a sentence depends on its verifiability. If it is verifiable, the sentence has cognitive meaning; if it is not, then it is either a tautology or it is meaningless. Verifiability was generally assumed to be related to empirical observation or sense data.

The rigorous application of this criterion neatly does away with most sentences contained in books of traditional philosophy. Only scientific statements, which are, of course, empirically verifiable, and tautological statements contained in formal sciences, such as mathematics and logic, remain. The job of the philosopher, then, is greatly simplified. The function of philosophy, as Carnap sees it, for example, has nothing to do with the meaning of terms and sentences and their relation to the world, but rather with the analysis of syntactical relations.[13] Philosophy deals primarily with the logical analysis of the sentences, terms and concepts used in science. It also exposes the misleading grammatical structure of traditional metaphysical sentences. For instance, it can show that the sentence "The rose is a thing" is misleading. For, _prima facie_ this sentence is like the empirical sentence, "The rose is red" in form. But in fact, it is like the syntactical sentence "The word 'rose' is a thing-word" in content. It is its similarity with empirical sentences that is the source of difficulty. In fact, the sentence is syntactical and should be translated accordingly.

The task of the Vienna Circle can be seen, then, as the _search for the conditions of certainty through analysis_. The only sentences about whose truth value we can be certain are tautologies, whose truth value depends on the meaning of their terms, and scientific sentences, whose truth value depends on empirical facts. It is this strong empirical and scientific emphasis and the rejection of all metaphysical speculation that constitutes the Circle's major contribution to the analytic movement.

1.4 Summary

Moore, Russell and the members of the Vienna Circle all agreed (1) that analysis plays an important part in philosophy and (2) that it is somehow related to language, although they did not necessarily agree on the exact nature of the method, the reasons why it should be used, or the results it yields. It is perhaps for the last three reasons that the validity of the method itself came into question quite early. Moore himself formulated the so called "paradox of analysis," which was going to do much to convince others that analysis as a method was at least inadequate in some respects and at most useless. And Wittgenstein, after he returned to Cambridge in 1929, engaged in a process of self-criticism that was going to shake the very foundations of his previous views as outlined in the Tractatus and accepted to a great extent by the members of the Vienna Circle. Since then, many have tried to modify analysis and present it in various garbs as "linguistic analysis," "ordinary language analysis," etc., while others have simply abandoned the term and substituted for it others such as 'elucidation.' Nevertheless, the fact is that these changes have occurred within the philosophical tradition established by Moore, Russell, the Circle's members, and Wittgenstein, and that it is their students and disciples that brought them about. There is, therefore, a historical continuity that justifies the use of the term 'philosophical analysis' and its application to include such divergent figures as the later Wittgenstein, Austin, Ryle, Quine, Chisholm, Strawson, and many others. There are also some general basic assumptions which seem to run through most of the members of the movement: the concern with language and the clarification of its meaning, the interest in logic and its use in philosophical discourse, a positive attitude toward science, and the conviction that non-empirical claims of a non-syntactical type are to be subjected to careful scrutiny. These and other features, rooted in a recent but nonetheless well established tradition, distinguish philosophical analysis from other traditional or contemporary movements. They also make the movement appear as a turn toward the development of a more scientific and rigorous way of doing philosophy. In the words of Max Black: "the label of 'analysis,'... serves well enough to identify philosophers who share a common intellectual heritage and are committed to the clarification of basic philosohical concepts."[14]

In Latin America, analytic presuppositions and the close relations between analysis and logical positivism at the beginning of the century have produced many confusions concerning the nature and method of analysis, leading some to identify mistakenly this philosophical current with logical positivism, mathematical logic, or even with the philosophy of science. In what follows, I shall present a broad outline of the development and present situation of philosophical analysis in Latin America.

2. PHILOSOPHICAL ANALYSIS IN LATIN AMERICA

Prior to 1940 little notice was taken in Latin America of philosophical analysis. Except for Spanish translations of Moore's Ethics, Russell's Problems of Philosophy (Argentina, 1928) and of the second into Portuguese (São Paulo, 1939), there is little evidence of serious interest on analysis. Latin America was still rejecting positivism. This is the time when vitalism and the influence of Bergson were at their height, although German spiritualism was making serious inroads owing to the influence of Ortega and his Revista de Occidente. This vitalistic and spiritualistic reaction of Franco-German origin neutralizes the scientistic influence of positivism which could have served as basis for the analytic current of Anglo-Saxon origin, creating an obstacle for the introduction and development of analysis.

In the 1940's there was increased acquaintance among a few Latin America philosophers with some of the basic ideas and figures of philosophical analysis, but still their knowledge of the movement was very limited. As late as 1947, Risieri Frondizi, referring to the current philosophical situation in Latin America at the time, wrote in the Handbook of Latin American Studies: "The complete lack of interest and information about logical positivism and semantics [and we could add: about analysis in general, not just about these two aspects of the movement], matters of so much concern in North America, stand out."[15] He had already said the same about logic and epistemology in 1943.[16] The two congresses that took place in this decade, the Primera Conferencia Interamericana de Filosofía and the Segundo Congreso Interamericano de Filosofía, paid practically no attention to analysis. Even most logic textbooks and works on epistemology published or used at this time ignored the work of contemporary logicians who belonged to the analytic movement. Carlos Vaz Ferreira's Lógica viva, published at the beginning of the century and reprinted as late as 1945, has nothing to do with mathematical logic, ignoring the most important discoveries in this field made by members of the analytic movement. And the same can be said about the Lógica y nociones de teoría del conocimiento written jointly by Francisco Romero and

Eugenio Pucciarelli. Most of the logical texts written at this period follow scholastic models. Moreover, what is written about contemporary developments in philosophical analysis is mainly negative. Frondizi was responsible for one of the first serious criticisms of logical empiricism published in Latin America; it is contained in his book, *El punto de partida del filosofar* (1945).[17] Moreover, Vaz Ferreira attacked modern logic in *Trascendentalizaciones matemáticas ilegítimas y falacias correlacionadas* (1940).

There are exceptions to this attitude, however. Vicente Ferreira da Silva published *Elementos de lógica matemática* as early as 1940 in Brazil. This is the first effort of its kind in Latin America. Francisco Miró Quesada accomplished the same task for Spanish speakers a few years later in his *Lógica* (1946), in which he took into account theoretical Anglo-American developments. These texts, however, deal primarily with logic and did not necessarily imply a corresponding interest in analysis as such. Moreover, they were exceptions in a generally non-analytic philosophical climate. Nonetheless, and perhaps as a result of the early interest in symbolic logic in Brazil, Quine was invited to the country at the beginning of the decade and, although he did not find or left behind any followers of analysis, from then on Brazilian philosophers took some note of the work of analytic philosophers. Part of his legacy was a book, *O sentido da nova lógica*, published in São Paulo in 1944. Moreover, in Argentina, Hans A. Lindemann, a former member of the Vienna Circle, maintained interest in analytic philosophy for many years.[17] Around this time works by Russell and others were studied and discussed by a group of philosophers in Buenos Aires and some of them were translated (for example, Russell's *Our Knowledge of the External World*, 1946, and *An Inquiry into Meaning and Truth*, 1946). The group included Gregorio Klimovsky, Julio Rey Pastor, and others. Indeed, it is in Buenos Aires, as it will become clear in the introduction to the Argentinian section, that the impact of philosophical analysis was first and most profoundly felt in Latin America. Still, the force of the impact was limited. The predominant Latin American philosophical loyalties still went to phenomenology, existentialism and Thomism. The first two in particular were very strong. It must be remembered that this was the decade when the Spanish emigrés, fugitives of the Spanish Civil War, arrived in Latin America. Among them, a distinguished group of philosophers, such as Xirau, García Bacca, Gaos, Nicol, Echevarría, Imaz, and others, brought with them the Germanic ideas of Ortega and his followers.[19] Some of these, however, like García Bacca, were aware of developments in analytic philosophy and therefore contributed to the dissemination of these ideas in Latin America. In fact, Ferrater Mora's well known _Diccionario de Filosofía_ was one of the few sources on analytic philosophy available at this time in the continent.

Philosophical activity in the fifties increased considerably. In 1941 Frondizi complained in the _Handbook of Latin American Studies_ that at the time there was no single journal dedicated entirely to philosophy in the whole of Latin America. By the mid-fifties, however, there were more than a dozen. Moreover, many regional gatherings took place during this period in addition to four Inter-American congresses, while the Sociedad Interamericana de Filosofía was founded in 1954. With the growth of philosophical activity, the interest in philosophical analysis also grew. A number of works by Cohen, Hans Reichenbach, Russell, Nagel and others were translated into Spanish during this period. Russell's _The Analysis of Mind_ was translated in 1950 and G. E. Moore's _Principia ethica_ appeared in Mexico in 1959. And there was renewed interest in the British empiricist tradition, particularly on Berkeley, Locke and Mill, and in American pragmatism. Still existentialism and Thomism dominated. Most of the articles that dealt with ideas and figures of the analytic movement were critical, although the cricitism was generally restricted to logical positivism, which some Latin Americans confuse with philosophical analysis to this day. Interest in contemporary logic of Anglo-American origin and in philosophical analysis in general was restricted to small nuclei in Argentina, Brazil, Chile and Mexico or to isolated figures in the rest of Latin America (Miró Quesada in Peru, Castañeda in Guatemala and later in the U.S.A.). In the Quinto Congreso Interamericano de Filosofía (1959) there were only four Latin Americans who adopted an analytic posture or discussed ideas proposed by analysts: Bunge (Argentina), Castañeda (Guatemala-U.S.A.), Miró Quesada (Peru), and Torretti (Chile).

The situation changed considerably in the sixties. This decade cannot be characterized any longer as one of _exposure_; it is a period of _consolidation_. In 1967,

Villoro, Rossi and Salmerón founded _Crítica_, the most important organ of the analytic trend in Latin America to date. In addition, the interest in philosophical analysis widened. In the middle and late sixties it extended to most countries that had normal philosophical activity. And although in a few, such as Colombia, the reaction was still primarily negative, most countries had at least two or three productive analytic philosophers at this time. Important works began to come out, such as Bunge's _Causalidad_ (1961), Thomas Moro Simpson's _Formas lógicas, realidad y significado_, both of which appeared in Argentina, and _La filosofía y las matemáticas_, which was published in Mexico in 1968 by Fernando Salmerón and José Adem. In the meantime the process of translation and propagation continued. The Universidad de Buenos Aires published translations of Ayer, Black, Carnap, Cohen, Goodman, Kaplan, Ryle, Strawson, and others to facilitate their study, and several works, such as Leônidas Hegenberg's text, _Lógica simbólica_ (Brazil, 1966) appeared in other places.

It is not until the decade of the seventies, however, that philosophical analysis can be considered an influential philosophical trend in Latin America. For it is only during this period that the quality of publications reaches a satisfactory degree of philosophical sophistication and their number achieves what might be called "critical mass." The caliber of articles included in this volume attests sufficiently to the degree of quality achieved. Moreover, the introductions to each geographical area will show how the analytic perspective and the problems usually associated with it receive increasing attention in philosophical journals and the university curriculum in spite of the resistance of some reactionary and/or ideological factions. While ten years before it was difficult to find evidence of philosophical analysis in the periodical literature, in the seventies it is difficult not to notice it.

From all this it should be clear that philosophical analysis is now established in Latin America. After periods of _exposure_ in the fifties and _consolidation_ in the sixties, it has achieved a certain _stability_ in the seventies. As far as quantity is concerned, there is considerable material not only in _Crítica_, but also in almost any philosophical journal published in Latin America during this period. To this it can also be added that the interest in analysis has contributed and in turn has fed upon an increasing interest in linguistics. This was particularly true in the sixties and seventies with the development of transformational grammar and other related subjects. In Brazil especially, there is an important group of linguists who regularly contribute articles in the analytic vein to international journals of philosophy. The center of analytic activity in the country is the Centro de Lógica, Epistemologia e História da Ciência in Campinas. The Center publishes a largely analytic journal, Manuscrito, founded in 1977, and a series entitled "Cadernos de Filosofia da Ciência." In Mexico it is the Instituto de Investigaciones Filosóficas that carries on most of the analytical activity. In addition to Crítica it publishes a series of translations entitled "Cuadernos de Crítica," founded in 1977. In Argentina the Sociedad Argentina de Análisis Filosófico began the publication of a journal, Análisis Filosófico, in 1981, and the Revista Latinoamericana de Filosofía gives considerable attention to analysis, as do Diálogos, published in Puerto Rico, and Discurso, published in Sao Paulo. Moreover, elsewhere in Latin America other centers and journals dedicated to analytic philosophy are also springing up.

What has been said should suffice as a general outline of the development of philosophical analysis in Latin America. The introductory essays to the various sections of this book contain the details relevant to the situation in each country.

NOTES

1. I am not the first to describe the common characteristics among analytic philosophers as a family resemblance. See, E. Rabossi, La filosofía analítica y la actividad filosófica (Buenos Aires: CNICT, 1972), p. 4.
2. I have tried to identify some of these in the Latin American context in "Philosophical Analysis in Latin America," History of Philosophy Quarterly 2(1984).
3. The Philosophy of G. E. Moore, ed. Paul A. Schilpp (Chicago: Northwestern University Press, 1942), p. 14.
4. Proof of the External World, Hertz Annual Philosophical Lecture (London: Milford). Cited by G. J. Warnock in English Philosophy Since 1900 (London: Oxford

University Press, 1958), p. 19.
5. "Reply to My Critics," in The Philosophy..., p. 663.
6. Morris Weitz, "Analysis, Philosophical," in The Encyclopedia of Philosophy, vol. 1-2 (New York: Macmillan, 1967), p. 97a.
7. Some Main Problems of Philosophy (London: Allen and Unwin, 1953; conferences from 1910 to 1911), p. 1.
8. Logic and Knowledge (London: Allen and Unwin, 1956), p. 192.
9. Ibid., p. 198.
10. Warnock, op. cit., p. 38.
11. Introduction to Mathematical Philosophy (New York: Macmillan, 1930), p. 177.
12. Ibid., p. 168.
13. See, Philosophy and Logical Syntax (London: Kegan and Paul, 1935).
14. Philosophical Analysis: A Collection of Essays (Englewood Cliffs: Prentice-Hall, 1963), p. V. For more information concerning analysis, see A. J. Ayer, Russell and Moore: The Analytical Heritage (Cambridge: Harvard Univ. Press, 1971); M. J. Charlesworth, Philosophy and Linguistic Analysis (Pittsburgh: Duquesne University Press, 1961); J. O. Urmson, Philosophical Analysis (Oxford: Clarendon Press, 1956); E. Rabossi, op. cit., pp. 14-16; and Javier Mugüerza, La concepción analítica de la filosofía (Madrid: Alianza Editorial , 1981, 2nd. ed.).
15. Vol. 14 (1947), p. 204.
16. Ibid., vol. 9 (1943), p. 463.
17. See my "Philosophical Analysis in Latin America," cited in n. 2 above.
18. See Victor Kraft, The Vienna Circle (New York: Philosophical Library, 1953), p. 9.
19. For a study of the Spanish emigrés, see J. L. Abellán, Filosofía española en América (1936-1966) (Madrid: Guadarrama, 1967).

Eduardo Rabossi

I. PHILOSOPHICAL ANALYSIS IN ARGENTINA

The conceptual weapons used by Latin American philosophers against nineteenth-century positivism were drawn from arsenals provided by German and French philosophers. The so called "founding period" of Latin American philosophy must be identified with the "discovery" and adoption of neo-Hegelianism, vitalism, historicism and some minor "isms" somehow accepted in German and French philosophical circles. Later on, Heideggerianism, phenomenology, neo-Thomism and neo-Marxism were added to the list. This search for new philosophical horizons was valuable in itself as far as it was related to a strict professional concern. But it had little value in the long run. The new philosophical wave was mainly responsible for some of the negative features that, somehow, characterize Latin American philosophical thought: lack of inventiveness, parochialism, disregard for standards of rigor, clarity and criticism, and disregard for logic and science.

Argentina was no exception in this respect, and the origins of philosophical analysis in Argentina may be traced back to the critical stance adopted by some philosophers and philosophically minded scientists in a _milieu_ dominated by the negative effects of some fashionable German and French philosophies.

In what follows I do not intend to present a history of the analytical tradition in Argentina. I only shall point out some aspects of its development and shall highlight some trends of interest, personalities, institutions and periodicals.

1. FIRST STAGE: INCEPTION

The first signals of a certain interest in philosophical analysis may be traced back to the forties, immediately after the second World War. Mario Bunge started a

J. J. E. Gracia et al. (eds.), Philosophical Analysis in Latin America, 17–23.

periodical, Minerva: Revista Continental de Filosofía, which, in spite of a short life (1944-45), was very influential in setting up standards of professionalism. Although Bunge himself regarded the Vienna Circle as a hybrid school, Minerva published some interesting information concerning logical positivism.[1] From 1945 onwards, some members of the Argentine Group of the International Academy of History and Philosophy of Science started discussions of some of the works of Russell and Carnap. Later, in 1952, Vicente Fatone, Julio Rey Pastor, Carlos Prelat, Eberto Fuentes, and Gregorio Klimovsky, among others, attended these meetings. In the same year, Vicente Fatone, Gregorio Klimovsky and Rolando García began a series of meetings to discuss some topics in logic and philosophy of science. It was also in 1952 that the Philosophical Circle of Buenos Aires began its activities. Mario Bunge acted as chairman and was very active in the organization of discussion groups. In 1954, he handed over to the members of the Circle a draft of his book on causality. Enrique Mathov, Hernán Rodríguez, Manuel Sadosky and José Westerkamp belonged to the Circle. In the Faculty of Law and Social Sciences of the University of Buenos Aires--traditionally an active philosophical center--Carlos Cossio held a seminar on von Wright's "Essay on Modal Logic." The seminar took place in 1954 and gave the students who attended it the first opportunity to study symbolic logic.

These references are surely very incomplete. Nevertheless, they give an idea of the way in which philosophical analysis was received in Argentina. The negative attitude prevailing in the faculties of philosophy toward philosophical trends which could affect the "theoretical status quo," moved some people to look for new trends of thought which favored a critical attitude and a scientific outlook. It goes without saying that participation in such groups did not mean a positive attitude toward philosophical analysis. Fatone and Cossio, for example, belonged to quite different philosophical traditions. What was important, even in those cases, was the receptive attitude, the interest in discussing with others the new ideas proposed some time before by the Vienna Circle, Bertrand Russell, Tarski, etc.

2. SECOND PERIOD: DEVELOPMENT

The real development of philosophical analysis in Argentina started in 1956. At that moment a new stage began which lasted until 1966.

In 1956, Bunge was appointed full professor of Philosophy of Science at the Faculty of Philosophy of the University of Buenos Aires. In 1957, Klimovsky took over the professorship of Logic. At the School of Law, Ambrosio Gioja was appointed successor to Cossio. Although Gioja was influenced by Husserl's approach, his openness of mind and fondness for discussion favored the development of analytical views.

It is interesting to stress here the activity that Bunge and Klimovsky performed meanwhile in the Faculty of Philosophy. In a short time they gathered a group of students who were able to assist them, quite soon, in their educational task. They presided over seminars and promoted groups of study and discussion. They carried out their duties, imposing upon the members of their staff strict professional standards. Bunge himself started a collection of booklets for the use of students, through which they were able to become acquainted with an up-to-date bibliography on epistemology, semantics and related areas. Bunge played also a leading role as a leader of the opposition against the "philosophy of the _status quo_".

During this period some books by leading analytical philosophers were translated into Spanish, and the first signals of originality among their followers in Latin America became apparent. In 1961 Bunge published _Causalidad_ and in 1964 Tomás Simpson published _Formas lógicas, realidad y significado_. Both books were edited by the University Press of Buenos Aires.

In 1956, the River Plate Group of Logic and Scientific Philosophy (Agrupación Rioplatense de Lógica y Filosofía Científica--ARLYF) started activities. Its main purpose was to gather philosophers and scientists from Argentina and Uruguay interested in a critical and rational philosophical attitude. The ARLYF organized many symposia on a wide variety of topics. Among the active members were Alchourrón, Bunge, Germani, Klimovsky, Rabossi, Raggio,

Reis, Sadovsky.

Outside Buenos Aires, philosophical analysis attracted also the attention of several philosophers. In Cordoba, for instance, Andrés Raggio was very keen on topics related to philosophy of logic, philosophy of science and the foundations of mathematics. In Tucuman, Roberto Rojo discussed Carnap, Quine, and Wittgenstein.

In the stage of inception, it also is important to stress that some non-analytical philosophers helped in the development of the analytical groups. I have already mentioned Gioja's case. I must also mention now Vicente Fatone and Risieri Frondizi. Acting as chairmen of the Philosophy Department of the Faculty of Philosophy, they showed an extremely positive attitude toward analytical philosophers. Frondizi used to criticize Moore and the emotivists in his lectures on ethics. For many students this was the first opportunity to acquaint themselves with an unknown philosophical world.

Unfortunately, the stage of development stopped suddenly. In 1965, Bunge decided to continue his philosophical career abroad. In 1966, control of the national universities was taken by the Armed Forces, after having overthrown the constitutional order. A wave of resignations--as a signal of protest--, extended over the university community. All the members of the analytical group of the Faculty of Philosophy resigned their positions. Raggio remained in Cordoba, but in 1968 he too left the country. In 1968, Alberto Coffa established himself in the United States. From 1968 to 1971, Eugenio Bulygin, Genaro Carrió, Tomás Simpson and Eduardo Rabossi carried on research in Oxford. For several years, Klimovsky was the only one who kept the analytical spirit alive in Argentina.

3. THIRD PERIOD: SETTLEMENT

The decade of the seventies is the most unfortunate period of contemporary Argentina. Political instability and frustration, fanaticism and economic crises joined together to create an extremely dark picture. Outrage and violence were normal features in Argentinian life and, as so often happens in such circumstances, philosophy became an ideological instrument. From 1970 to 1974 a wave of Marxism of catechistic overtones invaded the philosophical circles. After 1977, a similar phenomenon took place: this time the circumstances favored a sort of catechistic Thomism. In spite of difficulties of the most disparate kind, philosophical analysis gained a force which had been unknown up to that moment. It is possible to speak, thus, of a stage of settlement, beginning in the seventies and continuing up to the present.

A crucial factor in the definite affirmation of the analytical tradition in our country was the founding of the Argentinian Society for Philosophical Analysis (Sociedad Argentina de Análisis Filosófico--SADAF). It was started in 1972 by a group of philosophers, on the model of the Aristotelian Society. Membership was quite open and did not require a particular philosophical affiliation. The only requirement was to accept the rules of rational discussion and a critical standpoint. In a short time, SADAF counted with a fairly large group of members whose dues (the only financial source of the Society) serve to cover basic expenses. With the renting of a flat in downtown Buenos Aires, SADAF started activities such as courses of specialization, seminars, research groups, etc. Nowadays, SADAF counts with an Institute for Philosophical Research, a post-graduate seminar, a post-graduate course in philosophy leading to a Master Degree, and a library. It maintains close ties with leading analytical philosophers. SADAF is a member of the Inter-American Society of Philosophy and recently it was accepted as a member of the International Federation of Societies of Philosophy. In 1981 it began to edit _Análisis Filosófico_, which is published twice a year.

I will not dwell too much on this period. I would suggest that the reader go over the papers included in this volume and become acquainted with the _vitae_ of their authors. I think that these papers give an idea of the present situation of philosophical analysis in Argentina. Only fifteen years ago the possibility of gathering such group of papers was unthinkable as was unthinkable the diversity of topics and of technical approaches which they address and use. During the stage of inception, being interested in philosophical analysis meant being interested, in one way or another, in logic, philosophy of logic, or philosophy of science. During the stage of development, philosophical analysis was seen as a battlefield for formalists and non-formalists. Semantics (formal or otherwise) dominated the scene. In this third stage the diversification of topics and the liberation from methodological orthodoxies is apparent.

But let me add this. During this stage, philosophical analysis in Argentina has been under fire in a very peculiar way. Tags such as 'positivism,' 'empiricism,' 'logical positivism,' etc. have been and are stubbornly being used as synonyms for 'philosophical analysis' and/or 'analytic philosophy.' From 1970 to 1974 the analysts were accused of being "scientificists". What was meant, actually, was that the analyst did not care very much about "social reality," "concrete problems," etc. It suggested, also, a certain dependence (not necessarily intentional) on the supposed ideological involvement of the Anglo-Saxon mentors. In 1976, the criticism changed (the political situation was different). The rationalist vein of philosophical analysis brought about some distrust. Pluralism and criticism led some people to accuse analysts of favoring somehow international Marxism. The First World Christian Congress of Philosophy, which took place in Cordoba in 1979, with official support, launched a strong attack following these lines, against philosophical analysis.

In normal circumstances this sort of criticism does not deserve even a passing reference. But in the shaken Argentina of the seventies the criticism had a different import. It has to be taken into account because it helps to explain the systematic lack of acknowledgment of merits, the rejection from academic positions, the postponements of promotions, etc. Paradoxically (or not) it must be mentioned because it also explains, to some extent, the lively interest of philosophical analysis in Argentina in the last few years. Practicing analysis in Argentina nowadays brings out the feeling not only that we are doing philosophy in earnest, but that one takes part in a sort of enlightening crusade, producing models of rational thinking and showing how conceptual analysis is a requirement for the development of a rational attitude, since the understanding and modification of reality requires conceptual tools that philosophical analysis can provide, at least in part.

NOTES

* I wish to thank Mario Bunge and Gregorio Klimovsky for their help in the reconstruction of the beginnings of philosophical analysis in Argentina. José Babibni was also kind in letting me consult past issues of Minerva.

1. CF., for instance, H. A. Lindemann, "El Círculo de Viena y la filosofía crítica," Minerva (1944), 123-151. Minerva's aim was to "defend" reason and was pluralistic. Rodolfo Mondolfo, Risieri Frondizi, Luis Farré, Julio Rey Pastor and Francisco Romero contributed to it. As to the attitude of Bunge concerning the Vienna Circle, see his article "¿Qué es la epistemología?," Minerva 1 (1944), 27-43.
2. Almost fifteen years later Bunge described the ARLYF as a society which strove to back up an enlightened frame of mind in an environment where any obscure passage translated from the German or the French acquired a prestige denied to most valuable contributions from analytical philosophers.
3. In my book, Análisis filosófico, lenguaje y metafísica (Caracas: Monte Avila, 1977), I offer a personal evaluation of Analytical Philosophy and the stereotypes associated with it.
4. I wish to mention, among others, Tomás M. Simpson, Félix Schuster, Ernesto Garzón Valdés, Martin Farrell, Juan C. D'Alessio, Roberto Vernengo, Ricardo Guibourg, Oscar Nudler, Juan R. Larreta, Alberto Moretti, Gladys Palau y Telma Barreiro as active philosophers practicing philosophical analysis.

Carlos E. Alchourrón and Eugenio Bulygin

NORMATIVE KNOWLEDGE AND TRUTH

1. INTRODUCTION

One of the most important (though not the only) role of norms is to guide human behavior. Obviously only future actions can be guided; this reference to the future is essential for what may be called the _regulative function_ of norms.[1] (Norms can also be used to evaluate or interpret past actions or events, but our attention will be directed toward the regulation of actions.)

The regulative function of norms or deontic sentences[2] consists in giving an answer to the practical question "What should I do?", that is, deontic sentences provide a guide for the decision an agent must make on a given occasion concerning his actions in some future moment. Such a situation will be called "practical" and the decision of the agent a "practical decision." A practical decision presupposes the existence of at least two alternatives for actions; if there are no alternatives, there is no room for a choice and hence no decision can take place.

In order to give an answer to a practical question a deontic sentence must be _normative_ or _practical_ and not merely descriptive.[3] Any information to the effect that a certain action is obligatory according to a given norm or set of norms (normative system) is compatible with the information that the same action is prohibited by another norm or normative system, because both may be true. But such an information does not solve a practical problem, for it does not tell what an agent should actually do. Only a practical deontic sentence is able to play the required role.

A practical decision will be justified if it is based on a true deontic sentence. But have practical deontic sentences truth values? And if so, under what conditions are they true?

J. J. E. Gracia et al. (eds.), Philosophical Analysis in Latin America, 25–45.

The answers to these questions vary according to the philosophical position adopted. Broadly speaking one can distinquish two types of answers, which correspond to two traditions in the field of practical philosophy (ethics, axiology, philosophy of law).

For the realist[4] tradition there are certain facts that are part of reality and make norms true. Hence, norms are true or false in the sense of the correspondence to reality. These facts are normative or evaluative. This means that values are objective; they are part of the universe and they do not depend on human actions or attitudes. Something is valuable not because men approve, desire or prefer it; on the contrary, men have reasons to approve, desire or prefer certain things because they are valuable (good).

For the conceptualist tradition values are not given in reality, but are a result of human activity; they are constituted by man. Norms lack truth values, because there is no correspondence to reality, as there are no normative or evaluative facts. We can only speak of truth in relation to already constituted values; hence deontic sentences can only be true or false in a framework of valuations and their truth is relative to this framework.

These are two incompatible conceptions of normative knowledge and of deontic truth. The aim of this paper is to outline in some detail the two conceptions in order to make clear their presuppositions and to locate the central points of disagreement.

In what follows we shall take into account the investigations of deontic logicians and especially the semantic interpretations of deontic logic in terms of possible worlds. Logicians have been concerned with the truth conditions of deontic sentences, without being particularly interested in taking a position regarding the controversy between realism and conceptualism. They succeeded in showing that in order to give an account of deontic truth one must enrich the notion of reality; it is not enough to consider the real world, but one must also make references to possible worlds, comparing them with the real one. A reasonable view on possible worlds is to interpret them as future, not yet realized situations. In this way we can give a very plausible meaning to possible worlds, even if it implies a certain ontological commitment.

The disagreement between realism and conceptualism is purely theoretical; our spontaneous reaction is in practice realistic. (This is true both of our attitude regarding the external world and regarding values.) And what is even more important, our language is realistic.[5] But these facts do not prove that realism as a philosophical theory is justified. Conceptualism tries to question this attitude from a philosophical point of view and realism is an attempt at its theoretical justification. It is a philosophical disagreement that cannot be settled by empirical observation, nor by linguistic or conceptual analysis, for it presupposes the adoption of a metaphysical position.

2. TRUTH CONDITIONS FOR DEONTIC SENTENCES

Our problem is to determine under what conditions a sentence that declares a certain action obligatory, prohibited or permitted in a given situation can be regarded as true or false. The truth values of practical deontic sentences depend on several factors that may be classified into three groups.

a) The actual situation of the agent

We need a complete description of the actual situation in which the agent must make a decision regarding his future actions. This description must include all relevant information about such past and present facts as, e.g., actions already performed by the agent himself, his attitudes (preferences and desires), as well as actions and attitudes of other agents and "natural" facts or events. Of course, not all these facts are relevant, but in order to find out which of them are, one must have a complete list of all facts that characterize the state of the world at the moment in question.

b) Future situations and alternatives

We must also take into account the consequences of all possible actions, that is, all future situations that may arise as a consequence of the actions the agent might perform and all alternative actions that will be accessible to him in each of these future situations.[6] This reference to future situations and actions is independent of a consequentialist approach to ethics.[7]

Future situations and alternatives may be represented by means of a topological tree, in which the different situations are symbolized by circles and alternative actions by lines (see Diagram I).

Each branch of the tree that picks up one situation corresponding to each temporal point will be called a <u>history</u>.[8] In Diagram I six histories (h^1-h^6) are depicted, which embrace three successive temporal points.

c) <u>The hierarchical ordering of histories</u>

In order to determine what an agent ought to do in each situation a hierarchical ordering of all histories is needed, that would fix the relative value of each history. In other words, starting from the actual situation in which the agent is placed, we must compare all possible histories to which this situation belongs, i.e. the histories that are accessible to the agent in the situation in question.

The hierarchy divides these histories into two sets: the <u>optimal</u> and the <u>rejected</u> set. The optimal set may contain one or several histories and all other histories are rejected. This division is <u>relative to the situation</u>: a history rejected in one situation may be optimal in a later situation, if the histories that were optimal are no longer accessible to the agent in this later situation. This fact entails the relativity of the truth of deontic sentences regarding the situation in which an agent is placed. The deontic qualifications of actions as obligatory, prohibited and permitted can be defined in terms of the hierarchy.

An action is <u>obligatory</u> if and only if it belongs to all histories of the optimal set; it is <u>facultative</u> (i.e. it is permitted to perform and to omit it) if and only if it belongs to some, but not to all histories of the optimal set; and it is <u>prohibited</u> if and only if it does not belong to any history of the optimal set (i.e. if it belongs only to rejected histories). As the division of histories into optimal and rejected is relative to a situation, so is the truth of deontic practical sentences.

Diagram I

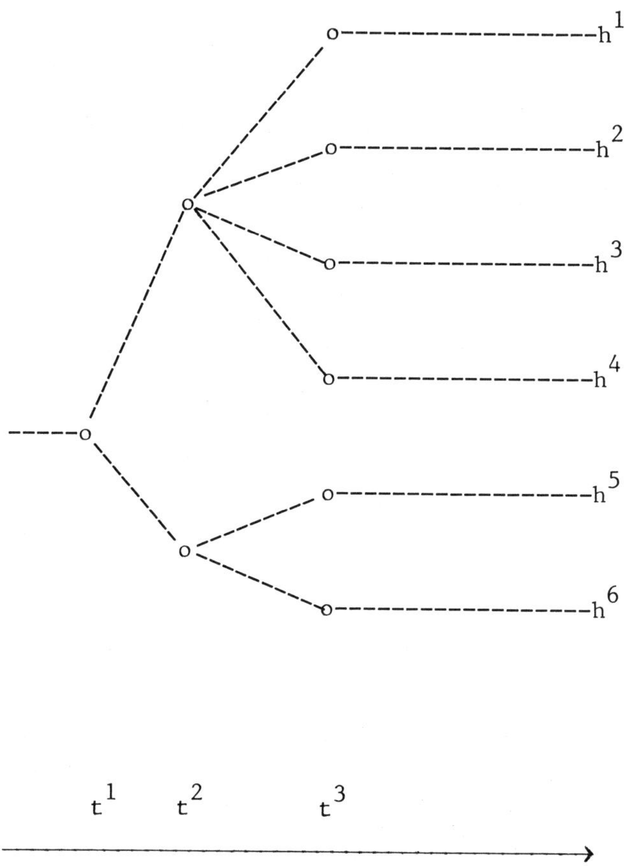

Let us illustrate this with an example. Suppose an agent A has been asked by his nephew B to give him a gun for his birthday (situation 1). A has to decide whether to promise it or not; his decision will lead to the situation 2 (if he refuses to promise) and 3 (if he promises). In situation 2 there are two alternatives; not to give the gun (situation 4) or to give it (situation 5). In situation 3 there are also two possibilities; it may happen at t^3 (i.e. after the promisse is made and before B's birthday) a new law prohibiting the possession of guns is enacted (situation 7), or not (situation 6). Both in 7 and 6 A can either give B the gun or not give it (situations 8-11). Diagram II, where (p) represents the sentence "A gives B the gun," Prom(p) stands for "A promises to give B the gun" and N(p) for "The norm prohibiting the possession of guns is enacted," depicts A's practical problem (see Diagram II).

Suppose now that we adopt the following hierarchical order of histories:

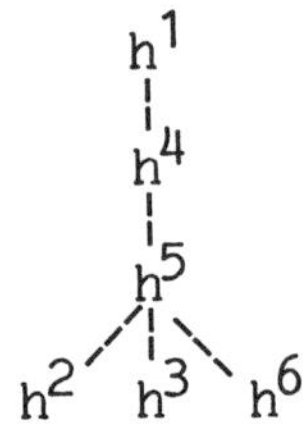

For situation 1, the optimal set is h^1; but in situation 3 it is h^4, for h^1 is no longer accessible to the agent. In situation 7 the optimal set is h^5. This means that A should not promise at t^2 to give B the gun and consequently he ought not to give it at t^4. But if he made the promise at t^2, then he ought to give the gun at t^4. On the other hand, the legislator should not enact a law prohibiting the possession of guns, but if he does enact it (situation 7), then A ought not to give the gun in spite of his promise, because h^5 is better than h^6. So the same action p (giving the gun at t^4) may be obligatory or forbidden, according to what happens before that moment, without any change of the hierarchy.

Diagram II

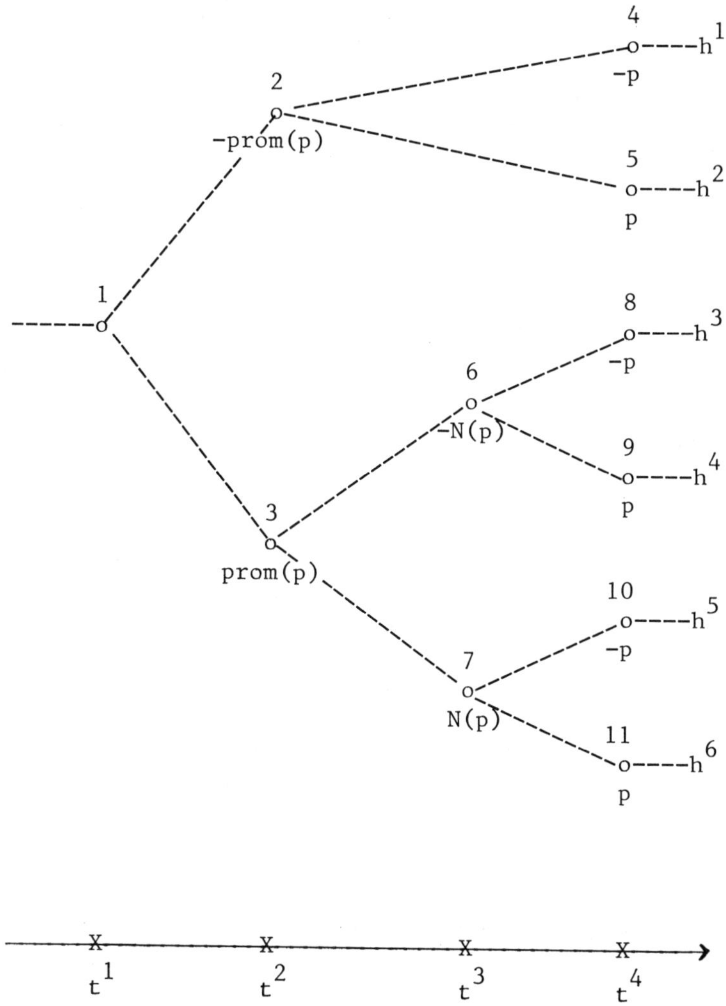
1
2
-prom(p)
3
prom(p)
4
-p
h1
5
p
h2
6
-N(p)
7
N(p)
8
-p
h3
9
p
h4
10
-p
h5
11
p
h6
t1
t2
t3
t4

It is possible in principle to construe a model of this type for every practical problem. The example we have been considering shows certain features of such deontic models that deserve closer attention.

(1) Each model presupposes a complex structure with only one hierarchy of possible histories, which does not change in the course of time. This hierarchy reflects the assumed valuations; all other valuations are considered in the model as mere facts, entering into the constitution of histories. Among them may occur such facts as the existence of legal norms, promises, contracts and other norm creating facts. But none of these valuations is used as such, though the fact of their existence can be relevant for the truth of a deontic sentence. The only valuation used is the one that gives rise to the hierarchical order of histories.

(2) The fact that the hierarchy does not change does not entail however that deontic qualifications remain constant. The truth of deontic sentences is a function of two factors: values and facts. A modification of factual components may produce changes in deontic qualifications of actions, although the values, i.e. the hierarchy, remain the same.

(3) The model shows the holistic nature of deontic concepts. In order to determine what actions are obligatory we must take into account the relative value of total histories, because past facts as well as future facts may have influence on the truth of a deontic sentence. The same type of action may be obligatory, prohibited or facultative in different moments, because it is inserted in a history the properties of which contribute to a deontic qualification of an action. An action that is prohibited in a history in which there is no legal rule concerning this action may become obligatory if a rule to the effect that it ought to be done is enacted by legal authorities.

(4) In view of the fact that deontic truth is relative to the hierarchy adopted, i.e. a certain valuation of histories assumed as such, the contention that practical sentences have truth values is relatively harmless and acceptable even for those who deny that value judgments can correspond to reality. The big problem refers to these and

hence to the criteria for accepting a given hierarchy of histories, giving rise to the two opposite conceptions already mentioned: realism and conceptualism.

3. DEONTIC DISAGREEMENTS

A set of true deontic sentences that qualify deontically certain actions of an agent must fulfill two _adequacy conditions_ in order to be able to justify a practical decision.

In the first place, the deontic qualifications must be _complete_ in the sense that they must qualify all actions (acts and omissions) that an agent can perform in all situations of all possible histories. If some actions were not deontically qualified, the agent could not know what to do regarding these actions; hence a rational decision would be impossible.

In the second place, the deontic qualifications must be _consistent_, in the sense that all actions accessible to an agent in a given situation cannot be prohibited, for in that case deontic sentences could not be used as a guide for actions and a rational decision would be equally impossible.

These requirements mean that the hierarchy must be such that (i) all histories can be compared (there should be no incomparable histories), and (ii) it selects an optimal set of histories for each situation. They are _rational_ requirements, for their fulfilment is a necessary condition for the rationality of a practical decision.

But these two conditions are not yet sufficient to ensure an agreement on the truth of deontic sentences. A situation in which the same deontic sentences are regarded as true by some members of a society and false by others would bring considerable difficulties from the point of view of social cohesion. The ideal situation would be that all members of a social group used the same hierarchy, for then the deontic truth would be at least intersubjective, that is, common to all them. This requirement will be called the _ideal of systemic unity_.

How can this ideal be satisfied? How can one hierarchy be privileged in view of the fact that there may be, and as a matter of fact are, many differences about the acceptable deontic sentences?

In order to satisfy the ideal of systemic unity two further conditions must be met: (i) the histories to be compared must be the same, i.e. there must be one and only one set of histories for each problem; and (ii) all members of a society must use one and the same hierarchy. That is to say, the same hierarchy ranging over the same set of possible histories must be used by all members of a social group for each practical problem.

This shows that disagreements about the deontic truth can stem from two different sources: (i) factual disagreements (in a broad sense of the term "factual") either concerning some actual facts referring to the situation in which an agent is placed at the present moment, or regarding some possible facts that refer to future situations and alternatives which can arise as a consequence of actions the agent can choose or that are due to "external" causes. In other words, factual disagreement concerns the possible histories: these will be different if different facts are taken into account. As we have already pointed out, as long as there are differences about histories that must be compared no agreement about deontic truth is possible. (ii) Evaluative or axiological disagreements, i.e. differences about the comparative value of histories, that is the hierarchical ordering of the same set of histories.

Therefore, in order to reach an agreement about deontic truth, it is necessary to have a unique set of histories to be compared, and to adopt one and the same hierarchical order.

4. PRACTICAL ARGUMENTS

A great part of practical reasonings (both in the field of law and of morals) pursue the end of attaining agreement about relevant (actual and future) facts, i.e. about the question what possible histories should be compared and of achieving acceptance of a given hierarchy as a criterion for judging the truth of deontic sentences. These reasonings are neither deductive nor inductive, for they do not try to justify the truth of certain sentences by means of the truth of others. What they aim at is to present a perspective of reality that should be adopted as a criterion for deontic truth. The acceptance of a hierarchy is constitutive of normative knowledge: only after having adopted a hierarchy are we in a condition to judge about the truth of deontic practical sentences. Of course, the adoption of a hierarchy can be questioned, but this is like questioning a criterion of knowledge and not the truth or falsehood of what results from the application of such a criterion.

Two types of practical arguments can be distinquished; they correspond to the two sources of deontic disagreements: the material and the formal argument.

The purpose of the <u>material argument</u> is to overcome disagreements about the set of histories that should be considered; and the aim of the <u>formal argument</u> is to achieve consensus about the hierarchical order of histories considered and so to overcome the axiological disagreements.

The material argument consists of an enrichment or embellishment of the description of (actual and possible) facts and actions occurring in each history, and in this way leads to new branches of the tree, i.e. to new alternatives that must be considered. This procedure can show that certain circumstances that would be accepted as relevant if considered, have as a matter of fact not been taken into account, and so it can lead eventually toward unifying the factual component of a deontic model, viz. the histories to compare. Thus this argument tends to produce agreement about deontic truth insofar as disagreements are due to the fact that the parties have not been considering the same set of histories, because evaluative agreement presuppposes that the same set of histories is made object of evaluations. This is a necessary, though not a sufficient, condition for the satisfaction of the ideal of systemic unity.

A further step is needed for it: consensus about the hierarchy of histories. This is what the formal argument aims at, pointing to certain formal conditions which are necessary in order to achieve an adequate perspective, suitable for formulating value judgments. If we can show to a person that disagrees with us about the hierarchical order of histories that he is not in a position to judge impartially, according to criteria that he himself accepts, we can induce him to withdraw his valuations and eventually to accept a different hierarchy. The formal argument is based on the conviction that only a person that satisfies certain formal conditions (those of an impartial and rational observer) can issue value judgments which are not infected by spurious elements like self interest, prejudice, passions and the like. It seems not to be difficult to attain agreement about certain basic conditions of this kind; among them will certainly be some requirements of a minimal mental development (excluding children and cases of insanity) and also some degree of cultural development (though here it will no longer be easy to trace a borderline). Further conditions will be designed to eliminate or at least to minimize self interest and other egoistic motivations. Authors like Frankena, Rawls or K. Baier have put special emphasis on these formal conditions for formulating moral judgments and formal requirements that judgments must meet in order to count as ones.

Both arguments, the material and the formal, serve the same purpose, viz. to achieve agreement about the ingredients necessary for deontic truth values, which amounts to the adoption of one and the same hierarchy ranging over the same set of histories, that is, the satisfaction of the ideal of systemic unity. But such an agreement plays a different role in the different philosophical positions that we have distinguished and that must be characterized in more detail.

5. THE REALIST PARADIGM

The realist answer to the problem of satisfaction of the ideal of systemic unity is based on the thesis that it is reality itself that determines the truth or falsehood of deontic sentences. Both elements on which the deontic truth is based, i.e. the factual (sameness of histories) and the axiological (identity of the hierarchy) are given by reality itself. As there is only one real world, there is only one set of histories relevant for each practical problem; and as values are part of the world, value judgments are true or false in the sense of correspondence to reality, i.e. to values inserted in it, with independence of human actions and attitudes. So the priviledged hierarchy is the one resulting from the use of true axiological criteria; the truth of such criteria is a condition of possibility of deontic truth. So it is reality which guarentees the satisfaction of the ideal of systemic unity. In fact, the problem of how to satisfy it is not even posed: the structure of reality is such that the satisfaction of the ideal is ontologically warranted.

Moreover, a hierarchy based on true evaluative criteria is complete and consistent, because all actions are deontically qualified and there is no situation such that in it all actions accessible to an agent would be prohibited. In other words, all deontic sentences are either true or false and none of them is at the same time true and false.

It seems that the thesis that values are part of reality does not justify by itself the contention of completeness (though probably it does justify the thesis of consistency: it is difficult to imagine an inconsistent reality). It could very well happen that the ordering relations would not allow for the selection of an optimal set for certain situations, if some histories were not comparable. Therefore the thesis of completeness seems to be based on independent considerations.

Though for realism values, and hence the hierarchy of histories, are objective, i.e. independent of human attitudes and immutable, this does not preclude the possibility of a change of deontic truth according to the circumstances that characterize the situation. The variability of deontic truth is perfectly compatible with the immutability of values and of the hierarchical order of histories, as our example discussed in s. 2 clearly shows.

Another interesting observation is that realism is capable of giving a satisfactory answer to the question how to combine the objectivity and the immutability of values and the uniqueness of the hierarchy with the obvious fact that certain human actions or attitudes, like acts of legislation or promises, have an influence on the deontic qualification of actions, being capable of modifying it. The answer is to point out that such acts are considered in the model as relevant facts for deontic truth. In this way, the fact of the existence of a legal norm or of facts that give rise to obligations can influence deontic qualifications of actions inside a framework of a hierarchy that rests on objective values.

Disagreements about deontic sentences and hierarchies which in fact occur are explained away by realism as due to lack of knowledge, either of values, or of (actual or possible) facts. And it is precisely here where the most serious difficulties for realism arise. Problems of knowledge appear at two levels: knowledge of the hierarchy and knowledge of histories.

Knowledge of values and hence of the hierarchy requires some kind of ethical intuition, for values cannot be known by empirical observation, nor by a deductive or inductive reasoning. Realism is compelled to postulate some sort of special intuition of values.[9] We shall not discuss the difficulties of an ethical intuitionism. But there is another problem that has to do with knowledge of the hierarchy and therefore with knowledge of deontic truth. This difficulty is of a linguistic kind.

The ways of describing the histories have a special influence on the process that culminates in the adoption of a hierarchy. The description depends on the language. A richer language would allow one to make more distinctions, some of which being considered might be accepted as relevant. Therefore, the acceptance of a hierarchy depends on the language used and in this sense it is relative to it. This fact poses a very serious problem to realism, for which not only deontic sentences have truth values, but also the hierarchies upon which they are founded correspond to reality. But as the acceptance of a hierarchy depends on the relevant facts that occur in each history, the metaphysical position of a realist requires that the expressive capacity of the language in which the histories are described should be such that all relevant facts and properties would be expressible in it. This entails that the language in question must be so rich, that it cannot admit any enrichment. It is not easy even to make intelligible the idea of a perfect language of this kind, and it is fairly certain that no empirical language is perfect in this sense. And if language is not perfect there is no guarentee that all possible aspects that are relevant are expressible in it. This puts a limit if not to knowledge of deontic truth, at least to the communication of such knowledge.

There is another and similar problem regarding knowledge of relevant facts, i.e. knowledge of histories. In virtue of the holistic character of deontic concepts knowledge of the truth of a deontic sentence requires knowledge of all relevant facts concerning all possible situations of all possible alternatives. Again in order to know which are the relevant facts, one must know all the facts. It follows that factual knowledge required for knowledge of deontic truth is an ideal hardly attainable in practice. It is empirically impossible for a human agent to know all past facts and to foresee all future contingencies. Hence a human agent cannot attain full knowledge of deontic truth.

Therefore all judgments about our duties are always fallible; they can always be revised in view of new data. This problem led Sir David Ross to his famous doctrine of _prima facie_ duties. We have a _prima facie_ duty when an action has some property that makes it obligatory, but only if all its properties are considered can we say that it is a duty proper. In other words, an action is obligatory or not, but given the limitation of our knowledge, we can only say that is _prima facie_ obligatory, leaving open the possibility of rectifying our judgment if new data appear. This is why Ross stresses the fact that _prima facie_ duty is not a kind of duty, but something related in a special way to duty.[10]

This argument shows that every non-dogmatic realist must admit that his deontic (unconditional) judgments are in principle fallible. But this does not mean, of course, that deontic sentences are neither true nor false; it is a problem concerning knowledge, not truth.

For these reasons an actual agreement about hierarchy does not guarentee the truth of deontic sentences that results from an application of this hierarchy, for deontic truth depends on reality, not on human beliefs. There is no absolute certainty about deontic truth, because an error of knowledge is never excluded, even if there is consensus about valuations. The satisfaction of formal conditions for issuing value judgments and the consensus about a hierarchy can at most play the role of a criterion of truth, but not that of its definition, for truth is defined in terms of correspondence to reality. But this criterion is never definitive, because our knowledge both of facts and values is essentially fallible. So even if realism is capable of giving satisfactory answers to the problems of unifying histories and hierarchies at the ontological level, these problems reappear at the gnoseological level. And what is decisive for the solution of practical problems is not so much the truth of a deontic sentence, but the knowledge of this truth. Insofar as the latter is fallible, no answer a realist can give to a practical problem is fully satisfactory. This has to do with a peculiar feature of practical problems, viz. their relation to time. An agent must make a decision concerning his future actions within a certain temporal interval, which may be very short. He cannot postpone his decision, for whatever he does, he chooses a certain course of action. The situation is different in the case of a theoretical problem; here we can postpone indefinitely our judgment, in the hope of getting more data for the solution of the problem.

6. THE CONCEPTUALIST PARADIGM

Conceptualists share with the realists the ideal of systemic unity, but they distinguish sharply between the ideal and the real. According to them the main mistake of realists is to confuse an ideal with what happens in fact; what is reasonable as a desideratum becomes a mere illusion when it is presented as a description of what is really the case.

The decisive difference between the two paradigms lies in the fact that conceptualism rejects the transcendence of values postulated by realism. Values are not given, are not part of the fabric of the universe,[11] but constituted by man. The criteria for establishing hierarchy and the ordering of a set of histories are neither true nor false in the sense of correspondence, because there is nothing in the world that would correspond to them. Deontic sentences are true or false with regards to a hierarchy, but the evaluative ordering of histories must be justified in a different way. According to Wittgenstein: "If the true is what is grounded, then the ground is not true, nor yet false."[12] The fact that the hierarchical order lacks true values does not mean, however, that the acceptance of a certain hierarchy is the result of an arbitrary decision or an artificial convention. It is not more arbitrary than the adoption of a certain criterion for appreciating the logical correctness of an inference or for determining the epistemic weight of factual beliefs.

In this conception normative knowledge is a dynamic process of coherent articulation of our beliefs and convictions concerning questions of value. Coherence and comprehensiveness of the set of evaluative and normative beliefs becomes thus a criterion of acceptability. As Rorty puts it: "There is no way to get outside our beliefs and our language, so as to find some test other than coherence."[13] The same idea we find in Wittgenstein: "Our knowledge forms an enormous system. And only within this system has a particular bit the value we give it."[14]

If there is no ontologically privileged hierarchy, then the only way of satisfying the ideal of systemic unity is to reach agreement among the members of society. So the search for agreement becomes the most important and vital task for conceptualism.

The role played by the formal argument is particularly relevant in this context. As evaluations have no truth values, the satisfaction of formal conditions is the only way for privileging some of the judgments over others. Hence writers of the conceptualist trend have put special emphasis on the conditions that must be met in order to render rational value judgments.

The appeal to counterfactual conditions will certainly allow us to discard some evaluations and therefore some of the hierarchies, but will it lead to a consensus about one and only one hierarchy for each problem? Even if we reach agreement about the formal conditions it seems that this would not guarantee by itself that all those that satify these conditions will choose the same hierarchy. In order to warrant this result an additional premise must be accepted, that will be called "the principle of correlation." According to this principle all individuals in ideal conditions for any given set of histories would accept the same hierarchy.

It is here where the opinions diverge, giving rise to two varities of conceptualism: the <u>absolutist</u> and the <u>relativist</u> branch. The acceptance of the principle of correlation is the core of the absolutist position, for admitting the fact that persons satisfying the formal conditions could, nevertheless, choose different hierarchies for the same set of histories would be tantamount to accepting the standpoint of a relativist. So the main disagreement between absolutists and relativists consists in the fact that the former accept whereas the latter reject the principle of correlation.

It is interesting to confront this point with Frankena's remarks about relativism. He points out that in order to prove relativism it is not enough to show that the ethical and evaluative judgments of different people and societies are different, for such disagreements "might be due to differences and incompleteness in their factual beliefs." What one must prove is that such judgments "would differ and conflict even if they were fully enlightened and share all the same factual beliefs." And as this proof is extremely difficult to produce, Frankena concludes that "metaethical relativism has not been proved."[15]

Frankena is certainly right, except in his characterization of relativism. A relativist does not need to maintain that value judgments formulated in ideal conditions would still be different and incompatible. It is sufficient for him to state that one cannot rule out this possibility. In other words, all a relativist must do is deny the principle of correlation, i.e. that the satisfaction of certain formal conditions leads necessarilly to a consensus regarding the same hierarchy. And it is the absolutist who has the burden of proof.

The consensus that the absolutist is interested in is not an agreement of a social group, based on actual preferences of its members. It is a hypothetical agreement of rational agents. But the hierarchy that rational agents in ideal (counterfactual) condition (like Rawl's original position) would accept is the privileged hierarchy. So the position of an absolutist resembles that of a realist; both maintain that for every set of histories there is one objective hierarchy that satisfies the adequacy conditions: consistency and completeness. And both must face the problem of knowledge of the privileged hierarchy. For it is not clear how the principle of correlation could be proved. To maintain that a disagreement about the hierarchy proves that the ideal conditions are not satisfied makes analytically vacuous the principle of correlation: to be in ideal conditions would in that case mean the acceptance of the same hierarchy. If an absolutist wants to avoid the mousetrap of making his position true by convention, all that he can say is that a disagreement about a hierarchy is a sign or symptom of the non-satisfaction of formal conditions. But then an agreement about a hierarchy does not guarantee either the satisfaction of such conditions, but is only a symptom of it. Hence it is always possible to argue that such consensus is not based after all on the satisfaction of the ideal conditions and so the corresponding deontic judgment can in principle be amended.

There is still another difficulty that must be mentioned in this connection. Most absolutists have emphasized the possibility of agreement regarding some very general moral principles.[16] Unfortunately this is not sufficient for the solution of all practical problems; what we need is the acceptance of a hierarchical order of histories that takes into account all the contingencies of a specific situation. It may be the case that even having agreed about some general principles, disagreement persists about the ordering of histories for a given particular problem, and so we do not know which deontic sentences are true. It is for the absolutist to prove that it is always possible to infer from his general principles a concrete hierarchical order of histories that we need for a particular practical problem.

For a relativist the situation is different. What he looks for is an effective agreement of the social group, based upon the actual preferences of its members. The search of consensus becomes the final goal: it is the only way for a peaceful solution of social conflicts. There is no guarantee, of couse, that such a consensus can be reached at all and it will certainly be impossible where there are disagreements about basic values.[17] But many disagreements can be eliminated by showing that they are due to factual components or that the value judgment on which they are based are not basic, or that such judgments do not satisfy the formal conditions already accepted.[18] If, after having exhausted all possible arguments, no agreement is achieved, the problem remains without solution. Of course, this does not prove that such consensus is impossible. The expectations that lie behind the satisfaction of the ideal of systemic unity postulated by realists and absolutists, are not ruled out by relativists.

NOTES

1. Cf. Richard H. Thomason, "Deontic Logic as Founded on Tense Logic" and "Deontic Logic and the Role of Freedom in Moral Deliberation," in Risto Hilpinen (ed.), New Studies in Deontic Logic (Dordrecht-Boston-London: Reidel, 1981).
2. As the term "norm" has connotations of generality, we shall use the more neutral expression "deontic sentence" to cover individual normative sentences as well.
3. Cf. Eugenio Bulygin, "Norms, Normative Propositions, and Legal Statements," in G. Fløistad (ed.), Philosophy of Action, "Chronicles of Philosophy," vol. III (The Hague: M. Nijhoff, 1982).
4. In what follows we use the terms "realism" and "conceptualism" in the medieval sense, in which it has been used in the well known discussion about universals.
5. John L. Mackie, Ethics (London, 1977), pp. 35 ff.
6. Situations and actions correspond to the Universe of Cases and the Universe of Actions: cf. C. E. Alchourrón and E. Bulygin, Normative Systems (Wien-New York, 1971).
7. Cf. Risto Hilpinen, "Normative Conflicts and Legal Reasoning," in "Primer Congreso de Filosofía del Derecho de La Plata," (La Plata, 1982): "If the agent does A, then his behavior can be described in terms of the consequences of A as well as by A itself, and therefore the agent's behavior cannot be evaluated retrospectively (i.e. judged) without considering the consequences of A and the possible consequences of the acts that the agent could have performed instead of A, as well as the normative principles which may require or exclude these consequences. For this reason the traditional distinction between deontological and teleological (or consequentialist) approaches to ethics seems partly illusory: both approaches require the investigation of the (possible) consequences of our actions. Cf. also Andrew Oldenquist, "Rules and Consequences," Mind 75 (1966), pp. 180-192: The "traditional teleological-deontological distinction does not mark two fundamentally different theories about what is relevant to the rightness of an action, but instead only

distinguishes two alternative ways of saying the same thing."

8. Cf. L. Aqvist and J. Hoepelman, "Some Theorems about a Tree System of Deontic Tense Logic," in R. Hilpinen (ed.) New Studies in Deontic Logic, (Dordrecht, 1981).
9. Cf. Mackie, Ethics, pp. 38 ff.
10. D. Ross, The Right and The Good (Oxford, 1930), pp. 19-20.
11. Cf. Mackie, Ethics, p. 15.
12. L. Wittgenstein, On Certainty (Oxford, 1969), No. 205.
13. R. Rorty, Philosophy and the Mirror of Nature (Oxford, 1980), p. 178.
14. Wittgenstein, L. On Certainty, No. 410.
15. Cf. W. K. Frankena, Ethics (Englewood Cliffs, N.Y.: Prentice Hall, 1963), pp. 93-94.
16. Cf. Carlos A. Nino, Los límites de la responsabilidad penal (Buenos Aires, 1980).
17. Cf. A. Sen, Collective Choice and Social Welfare (San Francisco, 1970).
18. Cf. A. Aarnio, On Legal Reasoning (Turku, 1977), p. 107: "Often the differences of opinion are due to the fact that the standpoints are superficially argued. A closer inspection would show that at least some of the differences can be eliminated simply by analyzing more deeply the reasonings and by pointing out the 'actual' grounds for the statements. When an analysis like this has been made, only those differences will remain that are founded on real epistemic and/or valuational differences between the auditories."

Genaro R. Carrió

LEGAL PRINCIPLES AND LEGAL POSITIVISM*

1. INTRODUCTION

The main purpose of this article is to question the soundness of a recent attack on legal positivism launched by Professor Ronald M. Dworkin in his essay "Is law a system of rules?"[1] I do not think any preliminary explanation of the strategy of my argument is necessary except for my use of a model taken from the rules of a game. I am aware of the important differences existing between legal norms and rules of games. But in spite of these differences law, as a normative structure, shares some problematic features with other normative structures, those of games included, and the following non-legal model does, I think, throw light on some of the main issues discussed later.

2. A NON-LEGAL MODEL

I assume that most of my readers are familiar enough with Association football (soccer) to know that in this game there are different types of rules. Let me draw attention to two of them:

1) Rules which forbid definite actions, specifically determined, e.g. the rule against handling, which prohibits all players, except the goalkeeper when within the penalty area, from touching the ball with the hands intentionally. The penalty attached to the violation of this rule is a direct free kick, from the place of the offence, awarded to the non-offending side.

2) Rules which forbid a physically heterogeneous variety of actions defined not in any precise and specific manner but through a general criterion. To this type belongs the rule which forbids players to play "in a dangerous way," the penalty attached to the violation being an indirect free kick awarded to the non-offending side.

J. J. E. Gracia et al. (eds.), Philosophical Analysis in Latin America, 47–71.

It is easy to find legal analogies for these two types of rules. Those belonging to the first type, which may be called specific rules, are analogous to laws prohibiting homicide or theft; those belonging to the second one, which may be called standards, are analogous to laws prohibiting injury to others through lack of due care.

There is at least one rule of soccer which performs a different function from the other rules which I have distinguished so far, and diverges from them with regards to its addressees and has different normative consequences. At the present time referees the world over apply unhesitatingly, albeit with varying emphasis, a rule which may be stated in the following terms: an offense against the rules (handling the ball, a foul, an off-side, dangerous play) shall not be penalized if, as a consequence of the penalty, the offending side would gain an advantage and the non-offending side would be adversely affected. In common parlance this rule is referred to in Argentina as the <u>ley de ventaja</u>. Accordingly I shall call it the "advantage rule."

This rule has multiple applications. A player of team A moves forward. In order to get the ball away from him or to make him lose control of it a player of team B commits a clear foul. The player of team A merely stumbles and, keeping control of the ball, continues his run towards B's goal-line. In such a case the referee must not penalize the foul for if he did he would actually be prejudicing the non-offending side.

To the best of my knowledge this rule was not originally one of the official rules of soccer. It became formally accepted as such only after referees started to use it frequently and consistently. It is, so to speak, a "referee-made" rule.

This rule has four basic features:

(i) It deals with the application of other rules of the game (the rules penalizing handling the ball, fouls, off-side, dangerous play, etc.). It may be considered, therefore, as a second-order rule;

(ii) It is primarily addressed to referees and not to the players as such, insofar as it imposes a duty on the former--not to penalize offenses--but does not alter the normative status of the players' actions. Much more could be said about this but surely it would be not only false but also absurd—nonsensical--to maintain that because under the "advantage rule" the unsuccessful foul of the player in our example is not penalized this rule confers upon him a right to indulge in unsuccessful fouls and only forbids him to commit successful ones;

(iii) It serves to restrict the scope of first-order rules (the rules penalizing handling the ball, fouls, off side, dangerous play, etc.); it justifies the referee in not penalizing acts which accordingly to other rules are plain transgressions;

(iv) It presents a certain measure of "topic-neutrality" in the sense that it is used to justify not penalizing a heterogeneous variety of acts (e.g. handling the ball, tripping or pushing an opponent, dangerously raising your leg in the neighborhood of a player of the opposite team, etc.).

Nowadays it is undeniable that the "advantage rule" is one of the rules of soccer. It was a rule of the game even before it was formally incorporated in the official regulations, insofar as it was consistently and consciously followed by referees as a desirable form of official conduct and thus warranted the assertion that it was followed as a rule and was not a mere habit or just a widespread abuse.

3. ONE MEANING OF THE EXPRESSION "LEGAL PRINCIPLE"

We have already seen that it is easy to find legal analogies for specific rules such as that against handling and for standards such as the one embodied in the rule forbidding dangerous play. Is this also true of the "advantage rule"? Are there in the field of law patterns or criteria which present features similar to this rule? The answer is affirmative.

I shall use an example drawn from American law: the case Riggs vs. Palmer (115 N.Y. 506; 2 N.E. 188 (1889)). X had executed his will and named his grandson Y as his heir. The latter, appraised of the fact, murdered X in order to obtain the inheritance. There ensued a litigation between

Y, who claimed X's estate, and X's other relatives, who naturally opposed Y's claim. The court pointed out that if the statutes regulating the effects of wills and the devolution of property were literally construed in such a way that their scope could under no circumstances be restricted, the murderer was entitled to receive the estate. None of those rules contained any provision that explicitly precluded the murderer's claim. Nevertheless it was rejected by the court. Among other supporting reasons the court advanced the following: "All laws as well as all contracts may be controlled in their operation and effects by general fundamental maxims of the common law. No one shall be permitted to profit by his own fraud or to take advantage of his own wrong, or to found any claim upon his own iniquity, or to acquire property by his own crime... These maxims are dictated by public policy, have their foundation in universal law administered in all civilized countries, and have nowhere been superseded by statute." The court then proceeded to cite several cases which testified to the application of the "maxim" which states that no one should be permitted to benefit from his own transgression or to profit from his own wrong. It went on to say that "these maxims, without any statute giving them force and operation, frequently control the effects and nullify the language of wills."

Leaving aside the particular phraseology in which the court cloaked its effective use of the "maxim" in question, let us first inquire whether there is any functional similarity between the role it was made to play in Riggs vs. Palmer and that which the "advantage rule" performs in soccer. Somewhat widening our perspective, I want to say that that "maxim" belongs to a large family of patterns or criteria which are invoked in the every day practice of law and present the following characteristics:

(i) They presuppose the existence of other rules with which they are concerned. They are, therefore, second-order rules. The one used in Riggs vs. Palmer, for instance, is concerned with norms dealing with the validity and binding force of wills, contracts and other acts-in-the-law.

(ii) They are addressed to those who are in a position to justify decisions, claims, defenses, etc. on the ground of the first-order rules. With many reservations and qualifications we might say they are addressed primarily to the judges and only indirectly to ordinary citizens.

(In this paper I cannot elaborate any further all the arguments which it would be necessary to introduce in order to elucidate this aspect of the problem. I shall only point out that whereas the "advantage rule" is much more naturally conceived of as being addressed not to the players as such but to the officials of the game, some analogy with this aspect of the "advantage rule" appears in a somewhat diluted form in the provision of the maxim that no one shall be _permitted_ to take advantage of its own wrong. Indeed we have seen that we cannot without absurdity argue that the "advantage rule" restricts the obligations imposed on players by the rule against fouls by allowing unsuccessful fouls. It is likewise absurd to argue that the "advantage rule" forbids the players to commit successful fouls. It is not, however, equally absurd to claim that the "maxim" used in _Riggs vs. Palmer_ forbid the wrongdoers to derive benefits from their wrongdoing.)

(iii) They supply guidance on how and when to use the rules with which they are concerned; what scope to give them; how to combine them; when to give precedence to one of them, and so on. Moreover, and under given circumstances, they show how to fill the gaps left by certain groups of first-order rules or, in other words, how to justify, in the decision-making process, the use of _new_ rules for the purpose of solving—_lato sensu_—unprovided cases.

(iv) They exhibit a certain degree of "topic-neutrality" in the sense that they cross the boundaries of different fields of legal regulation or, if more restricted in scope, are related to the construction and use of a multiplicity of rules covering a variety of facts and situations.

Mutatis mutandis we meet here the same four functional features which enabled us, with regard to soccer, to distinguish the "advantage rule" from specific rules such as the rule against handling the ball and from variable standards such as the rule against dangerous play.

Although not always in a clear cut fashion, and often with significant variations of shade and emphasis in different jurisdictions, similar characteristics can be found in a vast family of legal criteria.[2] Let us recall, for instance, that which enables judges to control the "abusive" exercise of rights; that which states that social security laws must be given an extensive interpretation; that which denies the protection of courts to some one who pleads his own wrong; that which proscribes the analogical extension of criminal laws; that which, with regard to the custody of children, guardianship and adoption, requires that the interest of the minors be the overriding consideration; that which establishes the presumptive legitimacy of official acts; that which, with regard to the enforcement of tax laws, provides that the underlying economic reality should be taken into account in preference to the outward legal forms; that which, in matter of public services, requires that the continuity of the service be assured; that which limits the court's powers to declare the unconstitutionality of statutes to the cases where such declaration is found to be unavoidable; that which, with regard to negotiable instruments, provides that rapidity and security of transactions must be given wide recognition; that which states that acts in the law may not be annulled unless there is some interest in declaring them void; that which prescribes a broad view on the admission of evidence; that which says that contracts are made in order to be performed; that which provides that in case of doubt as to the applicability of a tax, the tax-payer must be regarded as exempt; that which states that insurance contracts are contracts of indemnity and should not become a source of undue profit; that which, regarding the interpretation of the rules of maritime law, establishes as an overriding goal that the ship should continue in navigation; that which with regard to social security laws, provides that the underlying purpose must prevail over a rigidly logical construction; that which says that liens are of exceptional nature and subject to restrictive interpretation and that therefore they cannot be granted unless specifically provided for by the law, notwithstanding the consent of the debtor; that which says that, unless it entails a violation of either their letter of spirit, statutes must be construed in such a way that their provisions are compatible with the Constitution; that which says that the interpretation of laws governing labor contracts should favor the continuity of the employment; and many others.

Patterns such as these may be labelled "legal principles" and distinguished from specific rules and variable standards. In so doing we are not deviating from current usage. I should add, however, that the expression "legal principle" has many other meanings to which we shall return later. At any rate here we have one of them.

In what sense of "legal" are these patterns--the ones I have tried to characterize in this section--to be considered legal principles? X may be said to be legal: (1) because the law of a community refers to X and attributes to it a given position or status or attaches certain consequences to it (so we speak of legal acts, of legal conditions, etc); (2) because X refers to the law of a community (so we speak of a legal essay, of a legal doctrine); and (3) because X is a part of the law of a community (so we speak of a legal norm).

As we have previously remarked, our principles are second-order criteria or patterns which are concerned with first-order rules. Therefore, they are clearly legal in the second of the senses of "legal" we have just distinguished. Would it not be possible to argue with good reasons that they are also legal in the third of those senses, that is, as parts of the law? Are we to define the law of a community in such a way as to include, along with the specific rules and variable standards that unquestionably belong to it, the criteria which serve to indicate how those other patterns are to be construed, manipulated, elaborated and completed? Are these legal principles part of the law in the same way that the "advantage rule" is part of rules of soccer?

I shall leave these questions unanswered for the time being and only come back to them after having taken several preliminary steps, one of them being to examine other ways in which the expression "legal principle" is used.

4. OTHER USES OF THE EXPRESSION "LEGAL PRINCIPLE"

In legal contexts the word "principle" is not always used in the sense that I have just pointed out. Far from it. If I were to claim that I have characterized the meaning of the expression "legal principle," I would in fact be basing my claim on a stipulatory definition which would sanction only one of the current usages of the expression or, at any rate, would be assigning it a position of privilege. This is certainly not the only way in which this expression is used and the other uses are well worth examining. Only in so doing shall we have a clearer perspective and be able to make some meaningful contribution to the problem of the relationship between legal principles and legal positivism.

In common parlance the concept of principle is associated with at least seven clusters or foci of meaning. We may say that it is related to:

(I) the notions of "important part," "prime ingredient," "nucleus," "characteristic constituent";

(II) the notion of "a rule or guide very general in character";

(III) the notions of "source," "cause," "origin," "mainspring";

(IV) the notions of "fundamental motive or reason of action," "purpose," "objective," "aim," "goal," "main idea";

(V) the notions of "premise," "postulate," "basis of reasoning," "starting point," "axiom," "fundamental truth";

(VI) the notions of "self-evident rule," "unquestionable ethical truth"; and

(VII) the notions of "maxim," "aphorism," "proverb," "received saying."

The seven clusters or foci of meaning are and have been relevant within the field of law. To a greater or lesser extent jurists rely on them or on combinations derived from them. In legal contexts the word "principle" is used in various senses which mirror, so to speak, those foci, and make up a complex family bound together by complex relationships. This happens with the various activities which, in some way or other, are related to the law; using it; expounding it; criticizing, justifying and amending it. The subject well deserves a careful and detailed analysis; I shall here only sketch a few rough distinctions.

In addition to the use of "legal principle" accounted for in the preceding section, which I shall call use (1), that expression is employed:

(2) to isolate important features or aspects of a legal order which could not be overlooked in a comprehensive account of it (e.g. the so called principles of separation of powers, of the sovereignty of Parliament, of the indissolubility of marriage, of the socialization of means of production). This use is related to the focus of meaning (I).

(3) to refer to illuminating generalizations derived from the rules of the system (e.g. the principle that there is no liability without fault, or that legal transactions must be conducted in good faith). This use is related to the foci of meaning (I) and (II).

(4) to refer to the <u>ratio legis</u> or <u>mens legis</u> of a given rule or set of rules, that is to say, to their purpose, objective, aim, goals. This use is related to the foci of meaning (IV).

(5) to identify certain formal or external requirements which—it is maintained--every legal order must fulfill to be a legal order: for example, that its rules should be general, not retroactive, sufficiently clear, free of contradictions, duly promulgated and that they should not make impossible demands. These are the requirements listed by Professor Lon L. Fuller under the heading of "the morality of law."[3] This use is related to the foci of meaning (V) and (VI).

(6) to refer to rules deemed to be intrinsically and patently just (e.g. the principle which condemns racial or religious discrimination or that which proscribes slavery). This use is related to the focus of meaning (VI).

(7) to refer to guidelines addressed to the legislator and which are of a mere hortatory nature. This use is related to the focus of meaning (II).

(8) to refer to certain value judgments which express some basic demands of positive justice and morality and which are sometimes alledgedly grounded in the "popular

legal conscience." This use is related to the foci of meaning (II) and (III).

(9) to refer to maxims drawn from legal tradition. This use is related to the focus of meaning (VII).

Finally, in legal contexts we find two peculiar uses of the word "principle" which have been the target of severe criticism. These uses are related--consciously or unconsciously--to two schools of thought which even though they have been repeatedly proclaimed dead are still very much alive.

(10) The first of these two uses testifies to the influence of the historical school of law. The word "principle" is sometimes used to refer to a mysterious mainspring or source supposedly underlying a given set of rules and which is conceived of as continuously engendering new rules as it once engendered those. This use is related to the focus of meaning (III).

(11) The second one testifies to the influence of the jurisprudence of conceptions. The word "principle" is sometimes used to isolate statements held to be derived from some inscrutable essence of legal concepts viewed as entities. This use is related to the focus of meaning (V).

At this point I should like to anticipate an objection that is likely to be raised. There may be some overlapping and crisscrossing among these eleven uses because the list I have drawn does not pretend to bring order and systematic precision to the use of the expression "legal principle." It merely attempts to convey it faithfully.

5. LEGAL PRINCIPLES AND LEGAL POSITIVISM: FIRST APPROXIMATION TO THE SUBJECT

In the preceding section I have presented an extensive list of different uses of the word "principle" as found in legal contexts. I do not claim it to be exhaustive yet it would not at all displease me to find that it is.

To what extent are these uses consistent with the approach to law vaguely labelled "legal positivism"? In order to answer this question we must first clarify what we mean by "legal positivism." This is not an unequivocal expression. Let us begin to clear the ground with the aid of two negative characterizations of it.

(1) In its broadest and least precise sense, the expression "legal positivism" is used to exclude that approach to law which requires that the description and, in general, the manipulation of law should take into account certain rules which are held to be necessarily part of every legal order, either because they are intrinsically just or because they are derived from the very meaning of the expression "legal order." If by legal positivism we understand the approach to law which confines itself to the denial of this, that is, to the rejection of Natural Law theory in all its forms, then it is only incompatible with uses (5) and (6) of "legal principles" and is compatible with all others.

(2) The second sense of "legal positivism" excludes not only the various forms of Natural Law theory but also all references to metaphysical entities such as underlying mainsprings, hidden essences and similar non-empirical paraphernalia. If this is what we understand by "legal positivism," then legal positivism is incompatible not only with uses (5) and (6) but likewise with uses (10) and (11), and is consistent with all others. In other words it is compatible:

(i) with the use of "legal principle" to refer to important characteristics of the legal order, to illuminating generalizations derived from its rules and to non-operative rules addressed primarily to the legislator (uses 2, 3 and 7);

(ii) with the use of that expression to refer to second-order rules performing the function we have discussed with the aid of the model of the "advantage rule" in soccer (use 1); and

(iii) with the use of that expression to refer to: a) goals, aims and policies which can be ascribed to the rules of the system, b) demands of positive justice and morality, and c) maxims originated in legal tradition (uses 4, 8 and 9).

Having thus used the filter of a negative characterization of "legal positivism" we have been left with three groups of meaning for the expression "legal principle" which are compatible with the positivist approach so understood. The first one is devoid of interest. Let us focus our attention on the other two. They constitute important cores of meaning of the word "principle" as found in legal contexts. It is worth our while to go over them and attempt a further clarification.

According to one of them, legal principles are second order patterns or criteria which indicate how other rules are to be understood, used and, occasionally, completed. Henceforth I shall call them "principles^{1}."

According to the other, legal principles are the goals, aims or policies of a rule or group of rules of the system; certain basic demands of positive justice and morality; and maxims drawn from legal tradition. As an aid to clear thinking I shall stipulate provisionally that these principles be not principles^{1}. Henceforth I shall call them "principles^{2}."

If we confine ourselves to the negative characterization of "legal positivism" made earlier, it is apparent that either way of understanding the expression "legal principle" is consistent with it. There is nothing in legal positivism, as we have so far characterized it, that precludes speaking of legal principles in any of those senses. Both "principles^{1}" and "principles^{2}" may be used as tools for the description, criticism and application of law; one may admit this without ceasing to adhere to positivism in its negative characterization. Difficulties begin when we attempt to give a positive content to the concept of "legal positivism."

As we undertake this task we come across approaches such as those of both Kelsen and Hart. These two jurists are held to be the contemporary champions of a form of legal positivism supposedly related to the approach found in Austin's school of analytical jurisprudence. That is the reason why it is generally considered that these two thinkers belong to the most orthodox positivist tradition and that in their contributions we find the legal positivism _par excellence_.

The moment has come to raise the crucial question suggested by the title of this paper. We have restricted the meaning of "legal principles" to two nuclei: "principles^1" and "principles^2". For the sake of the clarity of the presentation we have provisionally introduced the artificial component that both categories are mutually exclusive. Moreover, we have accepted the widespread notion that Kelsen's and Hart's contributions which, notwithstanding their important differences exhibit significant coincidences, are paradigmatic of what is nowadays held to be legal positivism. For reasons that will later become apparent, I shall use Hart's conception. Now we can ask: To what extent can "principles^1" and/or "principles^2" be accomodated within an approach which views the law of a community as a set of rules or norms identified by means of an accepted rule of recognition (or a presupposed Grundnorm) which specifies the criteria the former must satisfy in order to qualify as part of the system?

The remainder of this article is an attempt to answer this question.

6. THE "MODEL OF RULES"

Legal positivism in general, and Hart's work in particular, has lately been the target of a severe criticism. It has been argued that Hart's approach is blind to the central role principles play in the daily business of courts. Such is, roughly summarized, the position taken by Professor Ronald M. Dworkin in the lucid and stimulating essay referred to in section 1. Dworkin's view is that Hart resorts to an unsatisfactory model for the understanding of the law. According to the former the latter's conceptual framework—the "model of rules"—may be characterized as follows:

(a) The law of a community is a set of rules. A rule is a pattern of the type of "speed on highways shall not exceed 50 miles an hour" or "a valid will requires the signatures of three witnesses." Rules are applicable in an all-or-nothing fashion. This means that in the circumstances contemplated in a rule is given, either that this is a valid rule of the system and therefore it determines the result completely, or that it is not, in which case it contributes nothing to the decision of the case. If a rule prescribes that legal consequence C shall follow when conditions x, y and z are given, then it cannot be that these obtain and x ought not to be applied. A rule may have exceptions, in which case it is not accurately stated if it does not list them all. Theoretically, at least, it is always possible to do so. The more exceptions we list, the more accurate the statement of the rule will be. A rule is either applied or not applied. If two rules conflict with one another one of them cannot be valid.

(b) To say that a rule is valid means that it satisfies the criteria set out in a supreme rule or "rule of recognition" which is accepted by the community. These criteria do not refer to the content of the particular rules but to their origin or pedigree, that is to say, to the manner in which they have been adopted. In Hart's words the rule of recognition "will specify some feature or features, possession of which by a suggested rule is taken as a conclusive affirmative indication that it is a rule" of the system.

In this regard, Dworkin suggests that the sole criterion that the rule of recognition may incorporate is that of enactment by competent authorities through a deliberate act of law making. If this is so it follows that there are no other norms apart from statutes and those contained in precedents. Once others are admitted, e.g. general or judicial customs, the theory of the rule of recognition has to be dropped altogether.

(c) Nothing can be a part of the legal system that is not a rule thus characterized and identifiable. More specifically, the law of a community does not contain or include principles such as that which prescribes the application of more stringent standards of responsibility to car manufacturers than to manufacturers of less dangerous items or that which provides that no one shall be permitted to benefit from his own wrong. Principles are logically different from rules.

(d) When judges must deal with a case which does not come clearly under a rule, the law does not supply them with any guidance. In such cases, judges must exercise their own discretion, not being guided by any legal standards. If the decision thus rendered imposes an obligation or confers a right the situation cannot be described in terms of the enforcement or acknowledgment of a pre-existing obligation or right.

Such is very succintly the "model of rules" which Dworkin expounds and criticizes. Before going to examine what he suggests to replace it, let us pause to consider what is the essence of his criticism. The concept of rule which positivists have adopted and the method they employ for the identification of the units of law conceal from them the rich world of principles. These are basically what we have labelled "principles[2]": social, economic, political objectives or goals, and demands of positive justice, equity and morality. Principles are not to be equated with rules, and cannot be identified in the simple fashion suggested above. They only qualify as law if "law" is defined in a manner unacceptable to positivists <u>á la</u> Hart. To admit that principles are part of every legal order is much the same as to decree the death of positivism and of its "model of rules." Only when we have admitted that principles are part of the law wherein they play a crucial role, only then shall we be able to do justice to the complexity of legal phenomena and to grasp in all its import the task performed by judges, who are always guided by principles, and only exercise discretion in the most trivial sense of the word. Positivism, with its "model of rules," offers but an impoverished image of the legal order and a distorted view of the judicial activity. This is a consequence of approaching legal phenomena from such a narrow perspective.

Let us now consider what we are offered to replace the "model of rules."

7. AN "ANTI-POSITIVIST" MODEL

In lieu of the so-called "model of rules," Dworkin advocates a conception with the following characteristics:

(a) The law of a community is an aggregate of rules and principles. Unlike the former, the latter may be described in this way:

(i) Principles do not require a specific behavior but set "goals to be reached," "generally an improvement in some economic, political or social feature of the community" (policies)--like the principle of the greater responsibility of car manufacturers--, or --like the principle that no one should benefit from his own wrong--they embody a "requirement of justice or fairness, or of some other dimension of morality" (principles in the strict sense of the word).

(ii) Principles are not applicable in an "all-or-nothing" fashion. They do not set out either conditions that make their application necessary or consequences that follow automatically when the stipulated conditions are given. Rather they state reasons for taking a decision, but do not necessitate a particular decision. Other principles may be applicable which incline in the opposite direction. When we say that a given principle is a principle of our law, we only mean that officials must take it into account as a consideration inclining in one direction or another. And this is the reason why counter-instances of a principle are not treated as exceptions to it since, not even in theory, could they be subject to an exhaustive enumeration.

(iii) Principles have a dimension of "weight" or "importance." To solve the conflict between two principles their relative weight or importance has to be taken into account in the context of the case in hand. A principle which has been displaced by another carrying greater weight survives intact although it has not prevailed in this occasion.

(iv) The test of the origin or pedigree is not suitable for the identification of principles. The rule of recognition is only useful to identify rules which, like statutes or precedents, are the product of a deliberate act of law making. The concept of validity, which operates with regard to rules created in this way, does not operate with regard to principles. It is a concept which operates in much the same way as rules do, in an "all-or-nothing" fashion, a manner which is incompatible with the dimension of "weight" or "importance" which principles have. Principles are not identifiable as legal principles on the basis of such a simple (or simplistic) verification as the one advocated by the theory of the rule of recognition, but on the basis of a complex argumentation which requires an adequate grasp of a wide variety of standards, beliefs and attitudes.

(b) The aggregate of rules and principles which make up the law excludes judicial discretion. Judges are always guided by principles. It is true that they do not dictate results in the way rules do but, in all and every case, even the most difficult, principles provide sufficient and ample guidance. Consequently judges are never compelled to base their decisions on extra-legal standards.

(c) It is not only rules that confer rights or impose obligations. This same function is performed by the combined operation of a set of principles. Perhaps we could say that there is a legal obligation to do something when the reasons on which the existence of that obligation is based outweigh the reasons which argue in the other direction.

This "anti-positivist" model, the central idea of which is that policies and demands of positive justice and morality are part of the law, does away with the very useful distinction between formal and material sources of law. Law becomes a continuum of social factors and ingredients. There is neither place nor role for the rule of recognition. Policies and demands of positive justice and morality are identifiable as parts of the law on account of their actual or probable impact on the decision of a case. The law of a community is a heterogeneous aggregate of standards and criteria of various origins and textures. Any attempt to give further precision to the word "law" would require that artificial distinctions be introduced in a continuum which does not tolerate them.

We still have to see whether this conception of law is the only possible alternative to the "model of rules" which it purports to replace. Is it true that we _have to_ choose between that rigid and austere conceptual framework and this overwhelming amorphous mass? I shall try to prove that this is not the case, and at the same time I shall also try to answer some of the questions that were left unanswered in the foregoing discussion.

8. A MODEL OF SPECIFIC RULES, STANDARDS AND PRINCIPLES

I think that anyone wanting to vindicate the spirit of Hart's conception without doing violence to its letter could reject both the "model of rules" and the "anti-positivist" model on the basis of considerations running along the following lines.

(i) Rules. The law of a community is a set of rules. The word "rule" does not apply exclusively to patterns like "speed on highways shall not exceed 50 miles an hour" or "for a will to be valid it must bear the signature of three witnesses." These are *specific* rules and specific rules are merely one of the various types of rules which make up a legal order. Apart from them there are standards of a very general nature which limit in a peculiar fashion the administrative bodies charged with their application, and there are, besides, variable standards which, like that of "due care," do not require a specific conduct from their addressees. These standards are rules in a broad sense of "rule." (Cf. Hart, *The Concept of Law*, pp. 127-30 and 249-50). There is nothing in the meaning of "rule" that precludes a pattern of the type of "no one shall be permitted to benefit from his own wrong" from being considered a rule of a legal order. Whether there is or is not a rule of a given legal order depends on other considerations.

(ii) *Rules and exceptions*. All rules, even specific ones, have an open texture. That is why they are subject to exceptions which "are not exhaustibly specifiable in advance" (Hart, *op. cit*. p. 135). "A rule that ends with the word 'unless...' is still a rule" (Hart, *op. cit*. p. 136). "(W)e should not cherish, even as an ideal, the conception of a rule so detailed that the question whether it applies or not to a particular case was always settled in advance, and never involved, at the point of actual application, a fresh choice between open alternatives..." (Hart, *op. cit*. p. 125).

(iii) *Rules and principles*. It follows that there is not a "logical difference" between legal rules and patterns of the type of "no one shall be permitted to benefit from his own transgression." It is not true that rules are always applicable on an "all-or-nothing" fashion. Nor is it true that rules permit, at least in theory, the exhaustive enumeration of all their exceptions. For this to be so, one should be able to foresee all the possible attending circumstances, which is plainly impossible. On the other hand, problems of conflicting rules are not always solved by denying the validity of one of them. Nor infrequently a decision--which may well assume the form of a compromise--must be based on the relative "weight" of each rule in the context of the specific case at hand (Hart, op. cit. pp. 125-26). The dimension of weight is not the exclusive property of patterns such as the one which says that "no one shall be permitted to take advantage of his own wrong."

(iv) Open texture of the rule of recognition. It is true that the rules of the system can be identified by refering to an accepted rule of recognition which disjunctively states the criteria that they must satisfy. As all legal rules, the rule of recognition has a "fringe of vagueness," an open texture. (Hart, op. cit. pp. 120, 94-95, 144-45; "Scandinavian realism," Cambridge Law Journal, Nov. 1959, pp. 239-240).

On this account, it is possible that we should have reasonable doubts as to whether a particular rule is part of the law of a community. The question "is this a rule of the system?" does not always admit of a categorical answer.

Even though we state the rule of recognition with the greatest possible accuracy it will never be able to determine how many times a given pattern of conduct has to be observed in order to qualify already as a general legal custom. Nor will it determine the number of times judges have to base their decisions on a general rule for it to qualify already as a judicial custom (a rule of la jurisprudence). Nor will it determine how many times a statute must be disregarded in order to count no longer as part of the system (in those legal orders whose rule of recognition assigns an abrogatory character to desuetudo).

The fringe of vagueness which surrounds the rule of recognition does not, of course, preclude the existence of clear central cases with regard to which the criteria it incorporates make it possible to assert categorically that a given rule is part of the system. It is in those cases where the presence of the characteristic or characteristics set out in the rule of recognition "is taken as a conclusive affirmative indication" that the rule in question is to count as a rule of the system. But from the existence of central cases it does not follow that all cases are equally clear. The situation is perfectly compatible with the existence of doubtful cases with regard to which the rule of recognition does not provide a definite answer one way or the other.

This indeterminacy or vagueness of the rule of recognition may affect the identification of specific rules, variable standards and any other type of rules whose membership to the sytem we want to ascertain.

(v) *Rule of recognition and principles. "Principles of positive law."* The rule of recognition, with its core of settled meaning and its surrounding area of doubtful cases, constitutes the ultimate criterion offered by the system for the identification of its particular rules, be they specific rules, standards or rules of other types. Nothing in the rule of recognition as such precludes our considering as part of the system patterns such as that which says that no one shall be permitted to benefit from his own wrong, as long as they meet the requirements specified in that ultimate rule.

Speaking in more general terms, there is nothing to prevent us from recognizing as parts of the system those second-order patterns addressed primarily to judges, which serve to adjust, widen, restrict, etc. other rules and exhibit a certain measure of topic-neutrality ("principles^{1}"). Nor is there anything that prevents those aims, objectives and policies and those demands of positive justice and morality which I have called "principles^{2}" from being parts of the law, as long as they are susceptible of being formulated as "principles^{1}" and operate as such.

When they meet the conditions set out by the rule of recognition, both "principles^{1}" and "principles^{2}" (when reducible to the former) are legal principles, in the sense that they are parts of the system on the same ground as any other rules belonging to it. In order to avoid misconceptions, I shall henceforth call those principles identifiable by reference to the rule of recognition "principles of positive law."

(vi) *The criteria of the rule of recognition and "principles^{1}."* It is possible that the special features of the second-order patterns which I have labelled "principles^{1}" render inapplicable, with regard to them, some of the criteria which may be provided by the rule of recognition.

It can be argued that the single precedent which, in some common law jurisdictions, suffices to establish a rule as belonging to the system, is not enough to incorporate a "principle^1." The single precedent, linked as it is to the specific facts of a given case, is usually construed as establishing a narrow rule controlling only that type of case. Such a narrow rule does not therefore show the topic-neutrality or relative indifference to content which is characteristic of "principles^1." Under that construction the case of *Riggs vs. Palmer*, viewed as a single precedent, would only permit us to infer the specific rule that he who kills another in order to inherit from him shall not receive the inheritance.

It seems doubtful that general custom, as distinguished from judicial custom (*la jurisprudence*), could become a source of "principles^1" since, as we have seen, the latter are primarily addressed to judges. Judicial custom (*la jurisprudence*) seems to be the most appropriate form of "positivization" for "principles^1." In other words, among the criteria provided by the rule of recognition, that of the *usus fori* seems to be the best suited to their special nature. Yet, it is worth mentioning that nothing in their nature prevents "principles^1" from being incorporated into the law of a community through legislative sanction.

On the other hand, we may conceive of a system in which the rule of recognition provides criteria which are only valid for the identification of "principles^1" but not for other rules. Those criteria could conceivably be, for instance, legal tradition and the weight of *la doctrine* (opinion of legal writers).

(vii) *Legal principles which are not part of the law*. Principles which do not satisfy the criteria of the rule of recognition are left outside the law, be they "principles^1" or "principles^2." They are not "principles of positive law." They may be called "legal" insofar as they refer to the law but not as parts of it. Their occasional use by judges may confer on them, at best, the status of candidates for admission to the system. They will be considered as parts of it once that use reaches generality, consistence and normative character strong enough to warrant the assertion that the patterns followed are already judicial custom. It is its _normative_ status and not its mere actual or probable impact on the decision of a case, that counts to determine whether a given principle is a part of the legal system.

It is true that only through an arbitrary decision can we determine with any precision how many times and by how many courts a principle must be applied, for it to be considered part of _la jurisprudence_ of the community to which the judges concerned belong. This indeterminacy does not preclude, however, the existence of clear cases of "principles of positive law" and, moreover, it does not affect principles exclusively. We have seen that it may also affect rules of other types.

(viii) _Permissive sources and "principles of positive law."_ Not all demands formulated in a rule are conceived of in terms of obligation, their being or not being so conceived of depends on the degree of social pressure brought to bear to exact conformity and on the type of reaction to which deviations give rise. (Hart, _The Concept of Law_, pp. 83-85 and 213). Thus, there is a rule of etiquette that says that we must send a telegram when somebody we know gets married, but it would be at least an exaggeration to say that we have the duty to do so. This, which applies to rules in general, also holds good for legal rules, and the more so to those addressed to judges. Not all canons and standards applicable to the judicial task are conceived of in terms of duties. There are rules which judges must follow to render a _good_ decision, but neither the social pressure brought to bear upon them to secure conformity is insistent enough, nor the critical reaction and disapproval deviations face are strong enough for us to conceive of their requirements as duties.

It is precisely on this account that we can distinguish between mandatory and permissive sources of law. "In systems where a statute is a formal or legal source of law, a court in deciding a case is bound to attend to a relevant statute though no doubt it is left considerable freedom in interpreting the meaning of the statutory language... But sometimes much more than freedom of interpretation is left to the judge. Where he considers that no statute or other formal source of law determines the case before him, he may base his decision on e.g. a text of the Digest, or the writings of a French jurist... The legal system does not _require_ him to use these sources, but it is accepted as perfectly proper that he should do so. They are therefore more than merely historical or causal influences since such writings are recognized as 'good reasons' for decisions. Perhaps we might speak of such sources as 'permissive' legal sources to distinguish them both from 'mandatory' legal or formal sources such as statute and from historical or material sources" (Hart, _op. cit._, pp. 246-47).

In the absence of mandatory sources the permissive ones provide a guide or orientation which serves to justify a concrete decision, but the fact of not conforming to them does not entail the automatic disqualification of the decision which disregards their "requirements."

The demands of _some_ legal principles seem to be more easily conceived of as the "requirements" stemming from permissive sources. This does not affect the indisputable fact that in every legal system there is a large central nucleus of specific rules, standards and principles, observation of which is mandatory.

(ix) _Judicial discretion_. On those matters for which the legal system--with its specific rules, standards and principles--fails to provide adequate guidance, judges must base their decisions on reasonable criteria which lie outside the law. The case is left to their discretion. In this context "discretion" does not mean or suggest "whim" or "fancy" but rather "prudence," "good judgment." In other words, they must ground their decisions on reasonable criteria which are not part of the law.

It does not seem plausible to claim, as Dworkin does, that the existence of principles which, admittedly, are often in conflict with other principles of similar "weight" suffices to eliminate the area of freedom the law leaves to judges. Unless we postulate the existence of principles of higher rank which determine when a certain principle should be given precedence over another, and then go on in this fashion _ad infinitum_.

(x) _The model of specific rules, standards and principles and the definition of "law."_ It may seem that in rejecting the rigid concept of rule advocated by the "model of rules," in acknowledging the marginal indeterminacy of the rule of recognition, and in accepting the distinction between mandatory and permissive sources and their respective roles, we have gone too far in blurring the concept of law and have ruled out the possibility of defining it.

But we should not worry over this. Faced with the persistent question "what is law?", perhaps it may well be that the most sensible course is to admit from the outset that "nothing concise enough to be recognized as a definition could provide a satisfactory answer to it. The underlying issues are too different from each other and too fundamental to be capable of this sort of resolution" (Hart, op. cit., p. 16; cf. also p. 234).

9. CONCLUSION

I have attempted to ascertain whether patterns or criteria called "legal principles" are part of the law as it is viewed from the standpoint of legal positivism. In furtherance of this aim I have surveyed the various meanings of that expression to find which ones, if any, are compatible with that approach and at the same time worth our interest. As a paradigm of legal positivism I turned to its most recent version, namely that expounded by Hart in The Concept of Law. The other contemporary version, which is Kelsen's was tacitly discarded because it does not permit us, as Hart's does, to take the steps that are necessary to accomodate principles within the legal order and to assign them the rank they deserve.

It may be argued that Hart's version, precisely because it permits those steps to be taken, and to the extent that it does, is not a genuine form of legal positivism but something else. In other words, that a defence of Hart's approach only goes to prove that, in order to accomodate principles, legal positivism has to be "embellished" or dressed up beyond recognition.

Whatever the weight attached to this charge, the fact remains that Professor Dworkin's anti-positivist attack levels its artillery against a so called "model of rules" which differs substantially from Hart's theory which it purports to refute. Although the titles of the latter to be called positivist may be questionable, it surely remains intact after the attack for the simple reason that this has been misdirected.

NOTES

* This article presents certain ideas to be elaborated in a book now in preparation. I wish to acknowledge the generous assistance of the Guggenheim Foundation for the grant of a fellowship which enabled me to pursue these studies at Oxford during the academic year 1968-69. Published in Buenos Aires, by G. R. Carrió, 1971.

1. Essays in Legal Philosophy, ed. R. S. Summers, (London: Blackwell, 1968), pp. 25-60; first appeared in 35 University of Chicago Law Review 14(1967) under the title "The model of rules."
2. The patterns I list by way of example have been taken from Argentine law reports. I do not know whether similar criteria are followed in common law jurisdictions. Moreover, it may well be the case that differences in legal terminology obscure, in this context, the substancial identity of criteria.
3. Cf. Lon L. Fuller, The Morality of Law, ch. II (New Haven: Yale Univ. Press, 1964).

Gregorio Klimovsky

LOGIC AND THE HYPOTHETICAL-DEDUCTIVE METHOD*

The main purpose of this paper is to examine the function of logic within scientific systems, and to indicate possible analogies between the epistemological nature of scientific hypotheses and that of the rules of logic.

An old tradition distinguishes two main types of scientific disciplines: "formal" and "factual." Among the former we find traces of the idea of demonstrative science advanced by Aristotle in his *Posterior Analytics*; whereas in the latter the empiricist point of view takes precedence, making of science a summary of observations obtained through perception. Naturally, the limit which separates one from the other mostly depends on the philosophical position which is adopted. A rationalist would not accept the existence of factual sciences and, moreover, he would consider the expression "factual science" to be a *contradictio in adjecto* which would obviously clash with the very definition of science; from such point of view, all sciences--from mathematics and physics to ethics, for example--should be adapted to the demonstrative structure. On the other hand, a radical empiricist would observe that formal sciences are just generalizations from facts which sensible or practical experience offers us. The characteristic common to both extremes is the way of thinking according to which mathematics lies on one side and the natural sciences on the other; sometimes, doubts are expressed about whether physics should be divided or not into two disciplines: mathematical physics, as a "formal" science, and experimental or natural physics, as a "natural" one.

Nowadays, this distinction does not arouse any enthusiasm, but if we bring the matter up it is just because a certain conception of logic is related to it. According to such a conception, the use of logical procedures in science--especially those that belong to the deductive type--, is essential to formal disciplines but not to empirical ones. Furthermore, logic itself, when viewed as a

J. J. E. Gracia et al. (eds.), Philosophical Analysis in Latin America, 73–92.

scientific discipline, would be classified as formal. Another version, which fits better with what is considered at present to be an empirical science, has as an essential component the use of logical procedures among factual disciplines, but without changing the intrinsic formal character of logic. We hold, on the contrary, that it is perfectly coherent to believe in the importance of logic for empirical science without acknowledging essential differences between the types of knowledge that logic and factual sciences provide. Throughout this paper we shall discuss some reasons which may lead some to agree with us.

It is almost an obvious fact, nowadays, that deductive procedures have the same importance within the disciplines traditionally considered formal as it does in the factual ones. There was a time, however, when it was necessary to be quite insistent about this idea.[1] The reason why the concepts of logic and formal sciences were so deeply related lies in the fundamental role that the idea of "demonstration" plays in formal disciplines. In a way, if it is considered that both types of disciplines strive for truth, in the formal ones this truth would be bound to the notion of "theorem," that is, of "demonstrated proposition." In a radical empiricist conception, which makes the propositions of natural science mere summaries or generalizations of singular truths, there exists nothing similar to demonstration. But this conception is now an old-fashioned one. At present, it is acknowledged that scientific factual disciplines have their adequate expression in "hypothetic-deductive systems." The testing of hypotheses--through which the validity or acceptability of scientific theories is controlled (this very possibility is taken, sometimes, as a defining characteristic of scientific hypotheses)--is carried out by deducing observational consequences. Without deduction there would be no way of refuting theories or corroborating hypotheses. Therefore, the rationalistic points of view would not be wrong after all, since we would find similar deductive structures both in mathematics and in the natural sciences. But, since some epistemologists assert that there is a difference between them, this is the right moment to ask about it. Let us examine this point in some detail.

A deductive organization of a scientific discipline, both in the traditional way of thinking and in the contemporary one, would consist of a body of propositions[2] out of which some are accepted as the starting point of the deductive structure (the principles) and the others are obtained as logical consequences of deductions or chains of deductions starting from such principles (the theorems, or derived propositions). These propositions would refer to certain objects or entities whose study is the very purpose of the discipline. So, if the difference between formal and factual sciences is admitted, it will be necessary to seek a solution in order to set them within the structure already described. Three possible criteria can be suggested in order to establish the difference--each of them being based on the following aspects of the problem:

1) The nature of the objects to be studied;
2) The semantic function of the propositions:
3) The epistemological "status" of the principles.

Let us examine what each of these alternative choices involves. We shall begin with (1). In order to establish a difference between formal and factual sciences, it might be admitted that there are two types of entities: "formal" and "factual." This is quite tempting, and if we accepted it, we would place ourselves within an epistemological tradition which comes from Plato. For reasons which we shall examine in (2) this compromise is not considered, at present, necessary to the philosophy of mathematics and, on the other hand, it is not viewed with pleasure by empiricist and nominalist thinkers. However, it is not in itself a contradictory conception and we see that nothing crucial to it prevents acceptance of the modern theories of sets as theories about formal objects, for example.[3] The question of whether the terms of the theories of sets stand for entities, "really" or "virtually,"[4] brings about a similar discussion regarding whether the theoretical terms of a high level scientific hypothesis are "really" designative or only possess an instrumental value. In a way, one could consider the symbols of sets as a particular kind of theoretical terms. It is true, as Nagel[5] observes, that the discrepancy can have no scientific value if the fact of accepting one thing or the other does not affect the capacity, demonstrability or contrastability of the theory. But from a philosophical point of view, the difference is important, and it is still more important within science itself, if it is admitted that not only predictions but also explanations are important in science and that a scientific proposition doesn't have the same explanatory power, if its terms are understood in an instrumental way[6] or in a realistic manner. In that sense, to interpret the terms of mathematics as referring to formal entities could give its statements an explicative value which does not appear in the conception which views mathematics as a mere instrumental artifice auxiliary to natural science.

We do not intend to discuss whether formal entities exist or not. It is worth noticing, however, that the difference between the two types of entities would be really interesting only in the case that this difference should affect the epistemological "status" of the principles admitted by the respective types of scientific disciplines. If the reasons which guarantee the truth of the principles are the same, both for the theories which deal with formal objects and for those which refer to concrete objects, it would not be worth distinguishing methodologically between them (and all our discussion would be useless). It would be like distinguishing chemistry from medicine. Therefore, the possibility of using aspect (1) would be significative only if the choice proposed in (3) is important. As (3) will be analyzed shortly, we shall postpone the argument for a while and consider (2).

The possibility of finding a difference of semantic character between the propositions belonging to formal and to factual sciences is one of the most original discoveries of contemporary epistemology and its philosophical consequences are as remarkable as extensive. We are referring to axiomatic methods.[7] In a way, the "official" attitude adopted at present concerning the eventual existence of a difference between formal and factual sciences is based on the following: the propositions of factual sciences are genuine propositions, that is, they describe facts, events, or states of affairs, whereas the propositions of formal sciences are such in a merely syntactic sense, ("syntactic" is used here in Morris'[8] and Carnap's sense),[9] that is to say, they only refer to the properties of linguistic expressions which involve neither users nor meanings. The deductive systems of mathematics are mere calculi, that is, rules to pass from combinations of signs to other combinations of signs. If they are understood in this way, these systems are called "axiomatic systems"; the "principles" turn into "axioms" (this word does not keep its primitive sense of primary, evident and apodictic truths but only means a "starting point in the game of obtaining linguistic expressions through transformation of other expressions"), the logical consequences of the principles in "theorems" (this word only means "a result obtained by transforming expressions into other expressions starting from axioms"). Even though propositions are considered to be merely syntatic and a mere succession of symbols, it is important, in order to understand the use of this method, that its structure should belong to that structure of the genuine propositions used in the ordinary language of factual science. In other words, its grammatical structure must be such that if its terms are given a designation, a proposition with a semantic sense will be obtained. Let us see why.

The propositions of an axiomatic system can't be considered to be true or false. "True" and "false" are semantic notions which imply a correspondence or inadequacy with the facts.[10] The only notion which corresponds to the axiomatic system is "demonstrable": a proposition is demonstrable, or it is a theorem, when it can be deduced from axioms. The rules of deduction are given by logic. This might be surprising since it is supposed that logic deals with deductions between real propositions and not with the transformation of syntactic groups of symbols. However, this doesn't sound strange, if we bear in mind that we have made it a condition that their grammatical form should be correct and should correspond to the same rules of formation which allow building factual propositions. The rules of logic are "formal" in the sense that they guarantee deduction by taking into account the syntax of the expressions[11] and not the referents of terms (this is equivalent to saying that the rules of logic do not take into account the "content"). Therefore, it is possible to make deductions in the axiomatic systems. The only requirement is to associate with each term of the system, not a reference, since we are working with expressions syntactically considered, but a semantic or grammatical category at least indicating the type of referent the term may have if we want to provide a referent for it. For, if we did not do it, we could not speak of a correct grammatical form, because in order to decide whether an expression is correct according to its form, without knowing the referents of its terms, it is necessary to know which category of things may be named by those terms and also if the categories are linked in "the correct way." It is worth mentioning a possible extension of the idea of "axiomatic system": that of syntactic system;[12] in these systems any expression can be used instead of propositions syntactically correct, and instead of rules of deduction, "syntactic rules of inference" could be used, that is, arbitrary procedures for transforming expressions into expressions. Although the notion of a syntactic system is mostly used in contemporary logic, the usual notion in mathematics is that of an axiomatic system. It is time to ask what axiomatic systems are for and to determine the advantages they may have with respect to syntactic systems in general.

If axiomatic systems are studied according to their intrinsic syntactic interest, they do not differ much from games such as chess, for example, where there are initial positions, movements, and privileged positions; here we would have axioms, rules of inference and theorems. The fact is that it is also possible to understand an axiomatic system semantically by giving a description to its terms (but observing their categories). When this is done, the propositions are transformed into genuine statements which can be true or false. If the axioms come out true, in which case the theorems also do, for having been obtained through deductions, we say that the interpretation is "adequate" or that we have a "model."[13] The main interest of the axiomatic method is that one and the same system admits of different models, so that its investigation is a way of considering simultaneously deductive aspects of different factual fields (each of these fields corresponding to each of the possible adequate interpretations). The advantages of axiomatic systems, as opposed to syntactic ones, is that in order to recognize that an interpretation is adequate, we need only to realize that the axioms have been transformed into true propositions. For the requirements of observing the categories and grammatical forms and of using correct logical deduction are enough to guarantee that the products of the transformation of the axioms--that is to say, the theorems--, will also be true. In syntactic systems, it is also necessary to prove that the "rules of syntactic inference" preserve truth (which, if they are arbitrary, it may not be correct or it may be very difficult to establish).[14] All this shows the central role logic plays in the axiomatic method; logic helps us obtain theorems and to guarantee that they are true in all the interpretations in which the axioms are also true.

Due to the fact that different fields of investigation regularly possess analogous structural properties, the idea of using common axiomatic systems--so that such fields are models common to those systems--, is advantageous and explains why the method is spread out. It is interesting to notice that the Aristotelian notion of "demonstrative science" and the conception of "formal science" come together in the notion of "axiomatic system." But when it is required that the scientific language be meaningful and semantically informative, it is necessary to surpass this methodology by paying attention to interpretations and models, that is to say, to factual science. But in a model, the axioms become principles again, and the theorems become derived propositions. This means that the very moment the scientific language comes to the significative, the difference between factual and formal fades away again. Otherwise, if formal disciplines exist (in the sense associated with the axiomatic method) they would not be sciences to be legitimately expressed in communicative or informative language; otherwise, genuine disciplines which describe particular or general facts would not be formal.

Coming back to the problem concerning the relation between logic and science it is not appropriate to point out that "if we understand logic as having to establish correct rules of deduction, it is possible to wonder whether logic, as a discipline in charge of such a task is formal or not." This question seems to be rather obvious, since it could be argued that most people know that logic, or at least some part of it, is "formal." But this is a confusion between two senses of "formal": when it refers to the categories and not to the references of the terms which figure among the premises and the conclusions to which the rules allude, and when "formal" is used in the sense of an axiomatic system. Although it is possible to build axiomatic systems for logic, it is worth noticing that logic itself is a collection of significative statements about which there are admissible transformations of authentic propositions that preserve truth. In this sense we can say, paradoxically, that formal logic is not a formal discipline, i.e., that it isn't an axiomatic system. Arguments somewhat similar to those which we have just made, would also show that, contrary to what many formalistic epistemologists adduce, mathematics is not restricted to the axiomatic method and the development of the corresponding axiomatic systems, but that there is also a non-formal mathematics (Russell adduces something similar against Peano's axiomatic in order to prove that the ordinary sense of arithmetic--when we use numbers to count sets in everyday life--is not the one we would find in a formal axiomatic system).[15]

Now, those who insist on conceiving logic and mathematics as formal disciplines in the traditional sense of a "formal science," will draw some interesting conclusions from the last paragraph. They can adduce that the distinction between axiomatic systems and models (or, instead of models, factual disciplines in general) do not completely represent the distinction between "formal" and "factual," since there would be a logic and mathematics which, in that sense, would be factual despite the fact that, according to the tradition, we would have to consider them as having a formal character which neither physics nor biology possess.[16] If this is the case, it will be necessary to look for a non-semantic difference. One possibility would be to go back to the distinction between formal and non-formal objects, as in (1). Logic would be in charge of formal objects such as propositions and deductions; and mathematics in charge of numbers and sets. But we have already noticed that this approach will not be very promising, if the difference does not show itself in the epistemological status of the principles of logic and mathematics. For if the guarantee of truth we have, when we investigate deductions or sets, is not different from the guarantee we have when we study pendulums or genes, the difference between formal and non-formal disciplines would be of subjects and nothing else. As we have already said, the difference between mathematics and physics would be as great as the difference between physics and biology. But things do not seem to be like this; those who believe that the duality between formal and factual is worth keeping, think that in our case the difference of subjects involves a difference in the epistemological status of the principles of both types of sciences. It is clear, however, that after discussing (1) and (2) we are forced to go to (3).

Those who try to establish a dividing line between formal and empirical science believe that there is a crucial difference between the epistemological "status" of the principles of both sciences. The truth of the principles of the formal sciences would be necessary, whereas that of the principles of the factual sciences would be contingent. We must bear in mind that we have admitted that the structure of both kinds of sciences is deductive, so now we have to discuss the principles. Let us begin by stating that contemporary epistemology accepts the distinction between the two kinds of sciences provided that only one type of intrinsic necessity is admitted for scientific statements, and this type is the one which corresponds to the logical laws or to mere tautologies (including "definitional" and "analytical" truths). If this is the case, formal science could be characterized as that science which investigates deductive systems whose principles are logical truths; factual science would be, then, the science which studies deductive systems whose principles are contingent statements. We realize that this distinction is epistemological and not merely logical, when we consider that the type of method, through which we get to know the first type of principles is true, is different from the method through which we know the validity of the second type (singular factual facts are involved in the justification of the second type but not in that of the first type). Taking this into account, and remembering that deductive factual systems can be obtained as models of syntactic systems, we again face the difference between logical interpretations and descriptive or factual ones put forward by Carnap;[17] formal sciences would be logical interpretations of syntactical systems whereas factual sciences would be descriptive interpretations of such systems. This shows that formal sciences are not trivial, since logic and mathematics are useful in spite of the non-factual character of the truth of their principles, or perhaps due to such character. Once we have admitted the difference, we can examine some of its consequences.

In the case of factual sciences, it is worth noticing that observation and experimentation provide empirical methods to justify singular propositions whose terms designate elements of the empirical basis, that is to say, elements belonging to the sphere of entities whose knowledge is considered to be direct knowledge for the discipline in question. Unfortunately, the principles of the empirical sciences are not propositions of that kind; they are not singular but general or, at any rate, they make reference to non-observable objects which are not in the empirical basis.[18] In such a case there is no direct method to verify or refute such principles. Contrary to an old and inaccurate tradition concerning the epistemological nature of the principles of factual science, they cannot be completely verified and they are, in a way, even less certain[19] than the singular propositions we have already discussed, which, from now on, we shall call "observational propositions." For that reason, the truth value of the principles is not in general conclusively known. Hence, the "status" of the principles is generally that of hypotheses, namely, the "status" of propositions whose truth value is unknown, but which are supposed to be true (the supposition will be put to test because of its implications, which will indicate its aptitude to describe what is real). The deductive systems in question turn into hypothetic-deductive systems in which the principles are fundamental hypotheses of the system, and in which there are two kinds of theorems: a) derived-hypotheses: that is to say, general or theoretical propositions which are deduced from the fundamental hypotheses (their truth or falsehood is, therefore, also unknown, but we are forced to accept them as true, provided that we keep supposing that the fundamental hypotheses from which they are deductive consequences are true) and b) observational consequences: they are observational propositions which are deduced from the principles. The observational consequences must also be admitted as true (this is because, as we have said, the consequences of the principles--the principles being supposed to be true--must also be admitted as true; "admitted," in this case means "supposed"), but contrary to the derived hypotheses, they are susceptible of conclusive verification or refutation by means of a direct access to the empirical bases. If the observations coincide, that is to say, if the observational consequences which have been contrasted up to now, are true, then the fundamental hypotheses and the whole system are said to be "corroborated." Contrariwise, if any observational consequence is false, we hold that some of the fundamental hypotheses (even though we can't say which) are false and the whole system is thus "refuted." This brings about important consequences: 1) the hypothetic-deductive systems are not accepted once and for all; 2) it is not conclusively known whether their hypotheses are true or not; 3) when they are defective, we can conclusively know that they have false hypotheses; 4) they can be checked through experience by means of the observational consequences of their fundamental hypotheses; 5) they do not admit an inductive justification, that is to say, there is not a procedure to verify the fundamental hypotheses, starting from observations (although it is possible to go from generalizations to particular cases by means of deductive means, the opposite case is not possible. The logical deduction guarantees that the truth of the principles imposes the truth of the consequences, but not the other way around, because logical correction does not hinder the case in which the premises are false and the conclusion is true. As we know, there is no deductive or logically warranted procedure called "induction," in spite of what the tradition states).[20] Besides, instead of speaking of an empirical science, we should speak of various hypothetical-deductive systems which are admissible (there are some people who prefer to speak of "theories"). In general, there are various non-equivalent and even incompatible systems within the same discipline. The empirical test eliminates some of them, but many always remain. Hypothetic-deductive systems are not used in isolation; a system of physics presupposes geometry, a system of biology presupposes chemistry. Whenever a system presupposes another, it is considered that the fundamental hypotheses of one are presupposed hypotheses of the other. When a theory is subject to the test of experience and the result is negative, we must take into account that the "guilty hypothesis" may be present among the hypotheses which were presupposed.

This methodology is attractive because it deprives factual science of dogmatism and it is like a course of consecutive approximations to the knowledge of what is real (by means of the refutation of imperfect theories and their substitution for more refined ones). Choosing the principles involves not only induction, but also an intellectual creation which does not end in speculation, for it can be tested by experience. But, again, logic plays an important role in all this, since without it there would be no linking between principles (fundamental hypotheses) and observational consequences. This is the right time to ask again about the knowledge logic provides and to compare it with the hypothetic-deductive knowledge which we get from factual sciences. Are these types of knowledge different? In this case factual sciences would possess a non-factual ingredient, the rules of logic which link principles and theorems, whereas formal sciences would offer principles as logical in nature as that of the rules of deduction used to obtain their theorems. Formal sciences would offer, in this way, a kind of purity and homogeneity which factual sciences would not possess. Besides, they would be certain, since it is supposed that knowledge of the logical rules and laws have an absolute degree of certainty, which we do not find in the knowledge of empirical hypotheses.[21]

From where is knowledge of the truth of logical laws and rules obtained? To answer that logic itself can be organized as a deductive system so that such rules and laws are obtained as theorems, is not conclusive, since the system would start, at any rate, from principles and would use special rules of deduction, and it could be possible to question the reasons on the basis of which such rules and principles are admitted. We come up to the moment when we must ask how we know the truth of the first logical principles and of the first logical rules. As the deductive-hypothetic method cannot be applied here—at least this is what is ordinarily conceived—, and is not possible to speak of logical justifications without reasoning in a circle, it seems that we should resort to a third procedure, neither logical nor factual. There are several procedures through which this stage of knowledge may be grounded; we will not detail them, since we think that the history of formal sciences forces us to discard this way of thinking. In fact, the logical antinomies, especially in the field of higher logics and set theories, have clearly proved that the so called "certain" procedures to validate logical principles and rules are a myth.[22] Systems of formal logic with "evident" principles have been refuted because of their very own theorems (this would be the most infamous way in which a system can succumb), not so much for contradicting the facts but for internal inconsistency. Nowadays it is usual to have several systems of logic which have alternative and non-equivalent laws and rules coexisting in the field of contemporary formal sciences. Two-valued, many-valued, intuitionistic or probabilistic logics in elementary logic, theories of types, Zermelian or Quinean systems in theory of sets or higher logic, show great variety. But this plurality would be reflected in the existence of various arithmetics, since, according to the adopted logic, the principles and rules allowed to establish the theorems would not be the same. This would also occur in the case of hypothetical-deductive systems: even if we started from the same fundamental hypotheses, we would get to different derived hypotheses, for the rules of deduction are not always the same, and, what is even worse, we would not get to the same observational consequences. Therefore, we would have as a result that the corroboration or refutation of a theory may vary according to the underlying logic. This means that our knowledge of empirical or factual reality will accord with our logical knowledge, and this apparently goes against the kind of independence which should exist between the logical and the factual.[23] It is really interesting to think that if we obtained a false observational consequence, starting from fundamental hypotheses, there would be three types of "suspects" to consider. We could think that some of the fundamental hypotheses are false. As we have already stated, there might also be a false presupposed hypothesis. But, and this is the novelty, we could also think that all the hypotheses in question are true, but that we have used an incorrect rule of logic which has led us from true principles to false observational consequences. Therefore, the internal inconsistency would not be the only procedure through which we could discard a system of logic; it could be possible that we might be led to discard it because of its inadequacy to help us to correct predictions starting from scientific principles. It might be annoying to think this way for those who consider the "testing" of observational consequences as a method which is used to check the principles of scientific theories. From this we gather that one not only tests the principles but also the logic which has been used.

There is no doubt that a logical rule or law, or just a simple logical truth, can have a hypothetical character. Therefore, its truth or falsehood is not known by itself (the opposed aprioristic belief has been disproved by the history of logic and for the fate of apparently evident and "verified" systems), but it is presupposed in order to build up logical systems and, above all, to be able to have the hypothetical-deductive systems at work. In order to complete this idea it is necessary to ask if, in spite of the fact that all the deductive systems of science belong to the same hypothetical-deductive type, there would be a difference which would allow us to distinguish the logic of the empirical systems, the formal systems from the non-formal ones.

In the first place, we must bear in mind that we always have the possibility of keeping the word "formal" for axiomatic and syntactical systems, leaving the word "factual" for all those deductive systems that use hypotheses. By the way, we must notice that it will be necessary to use syntactic systems and not merely axiomatic ones in order to formalize the systems of logic, since the transformational rules of expressions cannot be unequivocally pre-determined, but have as much arbitrariness as the hypotheses of logic one wants to choose. Otherwise, we could say that, if we establish a syntactic system so that a possible interpretation of its propositions is logical, knowing whether we are or not in the presence of a model is a question of hypotheses. On giving a certain interpretation one can suppose that the axioms are valid and that the rules of inference are truth-preserving.[24]

But if we insist on saying that the distinction between "formal" and "factual" applies to hypothetical-deductive systems, then it is not clear how we can do it. Either we can resort to the difference of subject, which we have already doubted as non-significative, or there is no difference between the epistemological "status" of the principles, or both. The difference we could use is that of examining the hierarchical relation between the two systems. It can be stated that a hypothetical-deductive system is formally higher than another one, when the latter presupposes hypotheses of the former in order to develop itself, but not vice versa. In this sense, it can be suggested that logical systems are formally higher than those of physics, because they presuppose logic, but not vice versa.[25] To this criterion we can add another: it may occur that a hypothetical-deductive system does not possess observational consequences (i.e., empirical content), if so, its theorems would be derived hypotheses. If this is the case, and there is some hypothetical-deductive system with empirical content (that is, with observational consequences) which presupposes the hypotheses of a first system, we can say that this system is simply formal. This criterion sounds attractive, but--as well as the other one--, it involves certain difficulties and binds us to take some precautions. It is true that logic would be a system (or set of alternative systems) presupposed by every system. But there are mathematical systems which presuppose other systems, or which obviously presuppose logic, and one would not consider them to be less formal for that reason. It is not clear either that the systems of logic do not have observational consequences (or empirical content); it is easy to see how we can extract observational consequences from logical laws. It is also true that we are tempted to say that the only observational consequences which can appear in this way are tautological and what is really interesting is to extract non-tautological consequences. But this is a mistake. The tautological character of logical principles is merely hypothetical and because of this reason the character of the consequences is also hypothetical. In order to measure the force of this hypothetical aspect there is nothing else to do but to test the observational consequences. If they happen to be true, the hypotheses will remain and we will have corroborated (though not verified) their tautological character, as well as the character of the observational consequences.

As a result of this discussion we can conclude that the difference between "factual" and "formal" sciences does not seem to be genuine (or, as we have seen in (1), it is hardly relevant from an epistemological point of view). If science is expressed through systems of propositions semantically significative, these systems would be hypothetical-deductive, including those of logic and mathematics. In all these systems, the propositions and the rules of deduction would have a hypothetical character, in the sense of supposing them to be true or to preserve the truth, respectively. There are also observational propositions through which an entire system of rules and hypotheses is tested. In case of refutation, either a rule or a hypothesis fails and the problem is, then, to identify it. The "logical laws" which make up a system of logic and the "logical rules" do not escape the "testing." This explains the multiplicity of non-equivalent or incompatible systems that we have at present. If it is accepted that syntactic and axiomatic systems constitute a science, then the only relevant difference between formal and empirical science is that which exists between these systems and the hypothetical-deductive ones. The latter will form the semantic and informative part of science. The syntactic systems are important in virtue of their models. If we transform their expressions into genuine propositions we change them into hypothetical-deductive systems. The formal systems of logic, that is to say, the syntactic systems of logic, are important because they transform themselves into hypothetical-deductive systems of logic.[26]

NOTES

* Published by the Instituto de Lógica y de Filosofía de las Ciencias de la Universidad Nacional de La Plata, SIF.

1. The insistence on the hypothetical-deductive character of factual theories is relatively recent and it is mainly due to Duhem, Poincaré and Popper. A classic argumentation in such a character can be found in Popper (13) and (14).
2. In this paper, the word "proposition" refers to what we are now accustomed to calling "sentence"; it does not refer to an abstract or mental entity.
3. Several contemporary logicians and mathematicians share this way of thinking (Gödel is one of them) and adduce that the existence of several set theories points out the defects of all of them with respect to the exact determination of the properties of sets. According to what we argue in this paper, and if it is true that set theories can be regarded as hypothetical-deductive systems referring to the properties of sets, this would not be incompatible with the existence of alternative systems for the theory; this is the case with respect to physics, where there are alternative theories for the same topic.
4. A set is "real" if the symbol which denotes it is really designative; when it is merely an incomplete symbol--and it is eliminable by means of contextual definitions--we may say it is virtual, (naturally, this is an incorrect way of speaking), though it is in vogue nowadays; signs, not entities, should be classified this way, since, in the opposite case, we have the paradox of hypostasizing virtual sets as existent, that is, those which, precisely, should not exist. See Quine (12), page 218.
5. See Nagel (11), pp. 141 and ff.
6. An instrumentalist position would be, for instance, that

of Skolem, as a consequence of the paradox which bears his name. See Fraenkel-Bar-Hillel (8), p. 107. With regard to Hilbert's position, he seems to oscillate between two points of view, admitting on one hand, the ideal nature of the conjunctistics terms, but holding at the same time that the true designative part of formal mathematics is made only by the natural numbers.

7. The axiomatic method is, in Hilbert's words, the "official" method of contemporary mathematics. In fact, we must recognize that the genetic and constructive methods find their expression in the conjunctistic building up of mathematics, which would constitute the other characteristic aspect of contemporary mathematics. But the theory of sets must base itself axiomatically, and this in the end seems to point out that the first method is the most important one.
8. See Morris (10), p. 13.
9. See Carnap (3), p. 16.
10. Since it is a decision about the use of words, there is nothing which prevents us from using "truth" in a merely syntactic sense, as a synonym of "syntactically demonstrable." Thus, Carnap, for example, uses in (4), p. 168, the denomination "C-truth." Epistemologically speaking, this does not seem to be useful, it brings about unnecessary confusions (since the word "demonstrable" already fulfills the functions of that notion) and it does not constitute a true elucidation of the Aristotelian concept of "truth." For "elucidation" see Carnap (5), p. 3.
11. If we consider "form of expression" to be the result of substituting in it its designative terms for variables; a logical law is a form of expression such that if its variables are substituted for designative terms, a true proposition is obtained. A logical rule is a sequence of <u>formulae</u> such that if its variables are substituted for designative terms and if all the propositions which are obtained but the last one are also true, then the last one is also true. But in order to guarantee that the substitutions do not lead to blunders, it is necessary that the variable range coincide with a semantic category so that no relations appear when we speak of properties, or no properties appear when individuals are mentioned. The logical and grammatical rules point out how to join the semantic categories to avoid blunders. Hence, if it is desired that in

axiomatic systems the passage from the axioms to the theorems be done in such a way that the maintenance of the truth in all possible interpretations is guaranteed, then it is necessary to indicate the category of the signs, since in a way these signs act as variables such that each interpretation assigns a value as if it were transformed into designative constant. The indication of the semantic categories in the signs is the only "semantic" trace which we find in axiomatic systems, apart from the logical, non-designative vocabulary (connectives, operators, etc.) which necessarily must be present in order to allow logical mechanisms to function. In syntactic systems these semantic components fade away completely.

12. See Carnap (6); and Carnap (4), p. 15.
13. "Model" is a word which has several and even disparate meanings (for example, some epistemologists tend to call model what we call "axiomatic system"). We use it here in its "mathematical-logical" sense, which seems to be transforming itself into the usual meaning. See Suppes (16), p. 163.
14. The rules of inference of a syntactic system are arbitrary. When an interpretation is chosen it won't be possible to know at once if they are transformed into correct procedures of deductions, since they are not associated to categories or forms. (See note 11). Therefore, it will be necessary to demonstrate a "meta-theorem," and that may be a very difficult task.
15. See Russell (15), pp. 8 and ff.
16. That is to say, if we begin by distinguishing axiomatic systems from authentically deductive ones, we should have a division of the latter into formal and factual ones. As all the deductive systems are semantic systems, the difference does not seem to come from semantics.
17. See Carnap (3), p. 22.
18. This means that they would use what epistemologists call at present "theoretical terms" which we have mentioned before. The entities which these forms designate are called "theoretical entities." See Nagel (11), pp. 106 and ff.
19. Popper in (13), pp. 93 and ff. considers observational propositions to be as insecure as hypotheses. This is the right moment to make clear that our observational statements can be atomic or molecular propositions.

Some authors--Nagel, for example, in (11), p. 31--seems to refer only to atomic ones. Thus, Popper would be right in saying that observational statements cannot be deduced from scientific laws unless other statements have been used in the premises. But observational statements can be tested using truth tables if the atomic ones have been contrasted, for which reason we see nothing that prevents using them. Besides, by virtue of the theorem of deduction, (see Mendelson (9), p. 61), if there are observational statements in the premises, they can be taken as antecedents of a molecular conditional which is the conclusion (see Carnap (7)).

20. Popper's exposition on the subject in (13) is already a classic. However, we must notice that nothing prevents the existence of hypothetical-deductive systems for the notion of probability or for inductive inferences, and that these inferences are used as underlying systems in science. In this case, the situation would not differ much from the one we are about to describe, consequently Popper's demonstration will not be so conclusive.
21. But this way of thinking would be inaccurate in the event that what we finally state is true.
22. See Beth (2), part VI, and the systems presented in Fraenkel-Bar-Hillel (8). See Beth (1), p. 117.
23. In other words, that a proposition may be logically deduced from another is not the ground of its truth, although it is the means to get to know it in most cases. The truth of an empirical statement lies in its accordance with the facts. Intrinsically, this would mean to do without logic.
24. Actually the models of axiomatic systems would always be hypothetical-deductive, for which reason it would be more exact to say that they are hypothetical models (that is, that we are not sure that they are models, we just suppose it).
25. This way of thinking would coincide with Gonseth's idea about "logic as physics of any object whatever."
26. All this has the consequence for logic, or rather, for logical systems, that (1) and (5) apply to hypothetical-deductive systems. That is to say, that a logical system is not accepted once and for all, we do not know if its principles are true, it can be false, and therefore this can be known in a conclusive way, it can be controlled by experience and it does not admit

inductive justification. (It is worth adding that it is not aprioristic either.)

BIBLIOGRAPHY

(1) E. W. Beth. Les Fondaments Logiques des Mathematiques. Paris and Louvain: Gauthier-Villars and Nauwelaerts, 1950.
(2) E. W. Beth. The Foundations of Mathematics. Amsterdam: North Holland, 1959.
(3) R. Carnap. Foundations of Logic and Mathematics. Chicago: University of Chicago Press, 1939.
(4) R. Carnap. Introduction to Semantics. Cambridge, Mass.: Harvard University Press, 1942.
(5) R. Carnap. Logic Foundations of Probability. London: Routledge and Kegan Paul, 1937.
(6) R. Carnap. The Logical Syntax of Language. London: Routledge and Kegan Paul, 1937.
(7) R. Carnap. "The Methodological Character of Theoretical Concepts." Ed. Feigl and Scriven. In Minnesota Studies in the Philosophy of Science, vol. 1. St. Paul: University of Minnesota Press.
(8) A. A. Frankel and Bar-Hillel. Foundations of Set Theory. Amsterdam: North Holland, 1958.
(9) E. Mendelson. Introduction to Mathematical Logic. Van Norstrand C., 1964.
(10) C. W. Morris. Foundations of the Theory of Signs. Chicago: University of Chicago Press, 1938.
(11) E. Nagel. The Structure of Science. New York: Harcourt, Brace and World, 1961.
(12) W. Quine. O sentido da nova lógica. São Paulo: Martins, 1944.
(13) K. Popper. The Logic of Scientific Discovery. London: Hutchinson, 1959.

Carlos Santiago Nino

THE LIMITS OF THE ENFORCEMENT OF MORALITY THROUGH THE CRIMINAL LAW*

What are the limits of the state power to punish people for the performance of some acts? This is an old and complex question. One of its facets is related to the issue of whether or not the mere immorality of an act provides a *prima facie* reason for resorting to the threat and imposition of punishment against the person who intends to perform or actually performs that act. The competing positions on this particular issue have extensive ramifications which reach out into several aspects of the adjudication of criminal responsibility and presuppose radically different views of society. The purposes of this article are, first, to identify and to distinguish the social conceptions underlying the opposite answers to the aforementioned question, and, secondly, to show some of the implications that each of those answers has for a system of criminal law.

1. The Perfectionist and Liberal Views on the Legal Enforcement of Morals

The enforcement of morals through the law in general, and particularly through the criminal law, has been the theme of a deservedly famous debate which took place in English-speaking countries some years ago. The main protagonists of the debate were Lord Devlin and Professor H. L. A. Hart.[1]

Illuminating as this discussion was, it did not go, however, deep enough into the articulation of the competing views about the limits of legal interference with individual conduct, although this articulation is needed in order to infer specific guidelines for, among other issues, criminal legislation and adjudication.

At the beginning of his essay *Law, Liberty and Morality*[2] Hart makes two important distinctions which reveal

J. J. E. Gracia et al. (eds.), Philosophical Analysis in Latin America, 93–113.

the possibility of extending the scope of the controversy beyond the limits set out by its protagonists. The first distinction points to the fact that the controversy is a moral one about whether the law should enforce morality. Hart makes it clear that the point of view from which the polemic is developed is the point of view of a critical morality, that is to say, the moral principles that one holds valid independently of their social currency. The second distinction is that the modern polemic deals not with the problem of _which_ morality may be enforced by the law (that was, according to Hart, the subject of older controversies), but with the problem of whether or not the fact that a morality has gained acceptance in a certain community has any significance sufficient to justify its enforcement. In sum, the subject of the controversy is whether or not our critical morality supports the contention that the law should enforce the positive morality of society, whatever its content.

Hart advances convincing arguments against two variants of that contention: an extreme one, defended by Stephen, which sees the enforcement of positive morality as something valuable in itself, and a more moderate one, supported by Lord Devlin, which considers that the enforcement of positive morality is a necessary means of preserving the social structure. So far, so good. But Hart consciously leaves aside, as he anticipates in those preliminary remarks, the consideration of a less vulnerable position which might reject the moral positivism of Stephen and Lord Devlin, but, nevertheless, agree with them in asserting that the law should prohibit many other immoral acts apart from those that harm others. The importance of taking account of this position, whether _démodé_ or not, is not only its possible inherent appeal but also the fact that it might cast doubt upon the intelligibility and consistency of some formulations of the contrary outlook, which holds that only harmful acts may be punished.

Consider this possible series of arguments:[3]

(a) One must concede that the mere existence of some moral consensus is not by itself a good reason for its legal enforcement. Thus, the claim that the law should enforce 'morality' must be interpreted as referring to a "valid critical morality."

(b) The claim thus interpreted is analytic. It states that, for a legal system to be justifiable according to any critical morality, it must enforce the principles of that critical morality. Nobody would deliberately deny this tautology. So, it is absurd to maintain that the law should not enforce or should be neutral toward moral standards. Surely utilitarians do not maintain this; they accept implicitly that the law should enforce the content of what they consider to be a valid critical morality. For they maintain that an act is immoral when it is harmful to other people and that the fact that an act is harmful (i.e. immoral) is a relevant, though not necessarily a conclusive, reason for making it a crime.

(c) If there is a disagreement between a utilitarian and another over the legal prohibition of acts such as incest or homosexuality, what is it but a clash between different moral systems? It is a discussion not about whether or not the law should enforce some moral system, but about which moral system should be enforced. What is not possible is to admit simultaneously that acts such as incest or homosexuality are indeed immoral and deny that their immorality is relevant to the question of whether they should be crimes. So, the contention that only harmful acts should be punished can only be reasonably upheld by a supporter of a utilitarian morality.

(d) Perhaps the issue is clouded by the nature of the examples given. It is improbable that people who, although not utilitarians, object to the punishability of acts such as homosexual intercourse or incest, would really think that those acts contravene a valid critical morality; they would probably think, instead, that those acts are deviations from mere social customs or manifestations of a pathological condition. Whether this is so or not, the matter could be greatly clarified if we were to adduce other cases whose immorality is less controversial. Consider, for instance, the possibility of there being harmless acts of racism. A consistent utilitarian would surely stick to his principles and maintain that even those acts are not immoral and, thus, should not be prohibited by law; but many people would resist the conclusion that harmless racist acts are not immoral; if so, they should be prepared to accept that laws should, in principle, prohibit them.

If this line of argument is sound, it will make some formulations of a well-known liberal view about the permissible content of criminal laws seem unintelligible. For this view is not necessarily tied to utilitarian morality but, nonetheless, it proclaims the value of excluding harmless acts from the sphere of legal interference. It is true that probably many liberals do not consider that mere unorthodox sexual practices are immoral, but surely most non-utilitarian liberals would maintain that actions motivated by racial attitudes are immoral whether harmful or not. However, many of those liberals would not only accept but strenuously emphasize that those immoral harmless actions should not be prohibited by the law; they would make a virtue out of proclaiming that they are anxious to protect the right to perform morally heinous but harmless actions. How can this position be justified or even consistently and intelligibly formulated? Apart from the problem that the moral justifiability of a legal system seems to require it to reflect the prescriptions of the chosen critical morality, there is also the problem that the justification of laws against harmful acts seems to lead, ultimately, to the claim that to inflict harm is immoral, which is, *ex hypothesi*, also true of some acts that are not harmful.

The first step toward the clarification of what can be taken as the paradigmatic liberal viewpoint about the relationship between law and morals, is to realize that at least one of the contentions contained in the foregoing argument is unwarranted. This is the assertion that the statement that the moral justifiability of a legal system requires it to enforce valid moral standards, is an analytic one. This is not so. In order to see this, it is enough to consider the case of an anarchic ideology according to which any sort of deprivation of goods, such as life, freedom, etc., is in any circumstance immoral. So, it condemns as immoral not only murder, theft, etc., but also legal punishment and the law which authorizes it. This would be a logically consistent morality. A legal system is objectionable from the point of view of a critical morality if it violates some of its principles (in this example, if it authorizes punishment), but not if it omits remedies, sanctions, etc. for the violation of those principles, unless the morality in question included a principle that the rest of its principles should be enforced in certain specified ways. Obviously, laws and penalties, on the one hand, and the acts to which those penalties are attached, on the other, are separate items that could be evaluated independently by a moral system; the morality or the moral necessity of the punishment of an act cannot be derived from the immorality of the act itself, except by the interposition of a moral principle to that effect. So, the clash between different moral systems about this matter is not necessarily the result of differences over what acts are considered immoral, but could be due to differences over the principles that should govern the relationship between immorality and crime.

Therefore, to maintain that certain acts are immoral but that the legal system should not interfere with them is a logically consistent position. But, of course, the fact that a position is logically conceivable says very little about its plausibility, and the plausibility of this position is far from obvious. Why is it asserted that the immorality of an act is not, in itself, a prima facie reason for making it punishable? Is it not, in principle, valuable to prevent the occurrence of recognized immoral acts? Is it not even that when the liberal is pressed to justify why harmful acts are to be prevented he will in the end state that to cause harm is immoral?

Well, the plausibility of this position is dependent on its formulating a persuasive general explanation which enables us to distinguish among different sorts of immoral acts which ones should be punished. This explanation must be compatible with a non-utilitarian morality, since it is a non-utilitarian liberal position that is in question; as we have seen, a utilitarian needs only call attention to the content of his morality, since he may accept that the immorality of any act (i.e. its harmful character) is a relevant, though not conclusive, reason for justifying the law in interfering with it. Besides, I think that if that explanation gave some credence to the idea that, in some sense of 'immorality,' the immorality of actions has some relevance for their legal prohibition, it would be more plausible; I agree with some authors[4] that to hold, without qualifications, that the law should be neutral toward moral standards that we accept as valid is unreasonable.

There is one particular type of liberal doctrine which has the above-mentioned features. That is to say, it is not committed to a utilitarian morality and, consequently, is compatible with the view that some harmless acts are immoral (although whether or not this is exemplified by deviant sexual behavior is another matter). Even when it maintains that harmless acts are not to be legally interfered with, it need not give blanket support to the idea that the law should be neutral toward a valid moral system; furthermore, it is independent of the antithesis between moral positivism and moral idealism.

This type of liberal doctrine, which undoubtedly underlies many of the discussions on this subject, is the well-known one that is currently opposed to the philosophy variously called 'perfectionism,' 'Platonism,' 'totalitarianism' (in the original sense and not in the current pejorative one), etc. It is not easy to formulate a precise and sound account of those assumptions of the liberal, anti-perfectionist, view which have a direct bearing upon the issue of the relationship between law and morals. However, I am inclined to describe tentatively those assumptions (and, by contract, those of its rival) along these quite crude lines.

Almost all moral systems, including many utilitarian ones, allow for two different kinds of moral judgments. One kind evaluates actions according to whether or not they contravene moral rules which prohibit bringing about certain states of affairs; they decide whether an action is right or wrong on the exclusive basis of its properties and consequences and with independence of any evaluation of the personality of its agent. The other evaluates actions taking into account their bearing upon the quality of the agent's personality. Judgments of this second sort decide whether the agent's actions (which may or may not cause harm to others) are self-degrading, that is to say, whether the performance of those actions allows the attribution to the agent of a defect of moral character. Obviously, this kind of judgment takes into account certain ideals of human excellence. It could well be the case that a morally wrong action is not taken as depraving the agent just because his ulterior purpose satisfies an underlying ideal of human excellence. Conversely, we can praise a man for a supererogatory act that realizes to a high degree a personal ideal in spite of or precisely because of the fact that the action was not prescribed by moral standards. So, moral judgments of this dimension are not restricted to the verification that some act violates certain moral standards; they determine whether those acts degrade the agent so much that he falls short of some minimal ideal of human excellence. Evidently, it is in relation to this second kind of moral judgment that the inquiry about the agent's intentions, motives and psychological traits is essential; the extent to which a morally wrong act degrades the agent in relation to certain ideal of a good person is dependent on whether or not that action was intentional, what were the motives for performing it, and what connection it had with his psychological constitution.

A utilitarian morality does not necessarily exclude moral judgments about the bearing of certain actions upon the quality of the agent's personality. Its peculiarity would be that those judgments would have to be limited to actions that involve harms to other people; but a utilitarian could conceivably concentrate the focus on the extent to which a harmful action involves a degradation of its agent. In contrast, when a moral system condemns actions not harmful to other people (as well as prohibiting harmful acts) it provides instances in which <u>only</u> moral judgments of the second sort are relevant. The whole point of standards which proscribe harmless actions is to prevent the degradation of the agent in the performance of those actions and the failure to achieve ideals of human excellence. An obvious example of those standards is that which condemns unorthodox sexual practices because of particular ideals, sexual chastity, or owing to the belief that the indulgence in sexual intercourse is to be allowed only for the purpose of reproduction; others are the standards which condemn harmless racialism, occultism, consensual submission to conditions of slavery, begging, etc.

This account is obviously quite crude and further refinement is needed before it will be at all philosophically fecund. However, I think that a non-utilitarian liberal might make use of it to strengthen his position. He could say that behind the rejection of harmless immoral acts, there is the claim that legal systems must not be concerned with the evaluation of people's personalities in accordance with ideals of personal goodness. He may state also that this does not imply that moral standards should not be legally enforced; they can be, insofar as the enforcement does not take into account the bearing of supposedly morally wrong actions upon the moral fiber of their agents. This would, of course, have the consequence that moral disapproval of actions whose only result is self-degradation should not be enforced, but this responds to a more general contention than the mere claim that only harmful actions should be legally prohibited. The contention that the moral evaluation of people's personalities should not be legally relevant is also applicable to cases of harmful actions (and, so, it is an issue even when a utilitarian morality is adopted); it implies that the prohibition of acts harmful to others must not be qualified by considerations of their effect on the moral fiber of the agents. This is quite important because there may well be many people, including utilitarians, who accept that only harmful acts should be punished but concede that the effect of the act on the agent's personality is relevant to the implementation and distribution of that punishment. The rejection of this contention is based on the fact that the same reason that militates against punishing harmless acts, militates against making the punishability of harmful acts dependent, in some way or another, on judgments about the moral degradation of the agent. And the reason is that the law should not embrace ideals of human excellence and discriminate among people in virtue of their moral worth and the quality of their _modus vivendi_; the law should treat the saint and the morally depraved equally, judging them in accordance with the face value of their actions. This, as we shall see later on, has a considerable bearing upon the treatment of subjective attitudes.

The justification of this anti-perfectionist doctrine lies beyond my concern.[5] It is undoubtedly quite complex and everybody is familiar with most of the arguments advanced. The justification may be connected with the adoption of a certain social ideal establishing the legitimate functions of social organization and the contention that these functions can be complied with insofar as the members of the community do not interfere with each other's life plans, whatever be their quality and whatever be the moral integrity of the people who subscribe to them. So, the social ideal defended is one that, unlike perfectionist conceptions of a good society, is not interlocked with the realization of personal ideals. The outlook can also enhance the value of moral experiment and the positive attractiveness of a pluralistic society whose members pursue different ideals of human excellence. The conception may rest on the tenet that personal ideals are not properly imposed by coercive means but by way of discussion and persuasion. It could, but need not, correspond to a relativistic view about the validity of ideals of personal goodness.

Instead of discussing the soundness of these justifications, I want to show the consequences that this liberal conception and the conflicting perfectionist outlook have upon the framing and, particularly, upon the application of criminal laws. I want to point also to the affiliation of current ideas about the matter with each conception.

2. Perfectionist and Liberal Assumptions in Criminal Legislation and Adjudication

Some of the consequences, and corresponding difficulties, that the adoption of one or the other of these opposite ideologies involve, arise at the stage of legislation, others, at the stage of adjudication (even when they may well be the object of legislative provisions). I shall make a very brief review of some well known implications of the anti-perfectionist and perfectionist outlooks upon the framing of penal laws and, then, I shall concentrate my attention on consequences that are generally confronted in adjudication.

Obviously, a perfectionist conception has a clear view about the scope of criminal laws. Any immoral act can in principle be converted into a criminal offence (or made the object of other kinds of legal interference), whether or not the act represents a harm to other people's interests or rights. The law must be an instrument for inducing people to embrace valuable ways of life and to realize ideals of human excellence. So, the perfectionist can avoid discussion of the social effects or possible harmfulness of the acts intended to be made into crimes and urge the punishability (depending on the personal ideals adopted) of acts such as deviant sexual behavior, consumption of drugs or alcohol, begging and vagabondage, occultist practices, racist attitudes, contempt for patriotic symbols, the evasion of compulsory education, etc. (to mention only the most plausible ones). This position has some independence of that concerning the justification of punishment. Even when punishment is justified in relation to its deterrent effects, this justification can be combined with the idea of discouraging people from pursuing a degrading life.[6] However, this conception has a natural tendency to assign a more transcendental mission to punishment: either pure retribution or the symbolic reaffirmation of the moral values at stake.

The rejection of moral self-degradation as a reason for punishing actions accounts for the liberal concern to determine whether the actions proposed as candidates for punishment involve the invasion of other people's interests or rights. This confronts liberals with several well known difficulties.

One of those difficulties, which I shall mention only in passing, is put forward by the contention that self-degrading actions performed in public or which become public involve necessarily harms that the law must seek to prevent. Therefore, the distinction between acts that merely contravene personal ideals and acts that harm other people is sometimes replaced by the distinction between 'private' and 'public' acts.[7] One of the 'harms' most often alleged as resulting from public self-degrading actions is the repugnance which is aroused in many people who observe or come to know of the performance of those actions. In the face of this allegation, the question is the extent to which attitudes of intolerance intervene in the generation of those unpleasant feelings and, if so, whether a liberal legal system must protect some people against harms that tolerant persons do not suffer (perhaps the only personal virtue that liberalism encourages is tolerance). But, of course, other sort of harm, which presents different problems, can also be said to result from public self-degrading actions.

This difficulty is connected with another which is the object of extensive controversy, the justifiability of paternalistic criminal laws such as those declaring punishable attempted suicide, consensual homicide, the evasion of compulsory education, drug addiction, not wearing seat belts in cars, etc. While perfectionism may advance straightforward justifications for these legal interferences, liberalism confronts the obvious obstacle that the notion of harm is dependent on the wants of the person supposedly affected. The current way of surmounting this obstacle is by saying that those laws are not intended to prevail over the overall wants of the person concerned, but to protect their future opportunities of choice against present choices that would preclude them. The idea is to preserve the capacity of men as choosing beings independently of the quality of their choices. But this suppression of certain choices in order to maintain the opportunities to undertake others is far from being clearly justified. Obviously a quantitative criterion is preposterous: for some decisions could be the supreme exercise of a man's capacity to choose (such as the decision as to whether or not to commit suicide), so that its suppression cannot be outbalanced by the possibility of making many other trifling decisions. Besides, it is obvious that almost all choices that men make preclude them from undertaking an indefinite number of others in the future. Therefore, the contention must be supplemented by some distinction of the choices that can be suppressed, at least in relation to their character as real choices if not to their content. Authors like Gerald Dworkin[8] have gone some way toward this, by indicating several circumstances in which decisions would not express the real will of individuals. Obviously, this involves the risk, as is pointed out by Berlin[9] and Dworkin himself,[10] of confusing what an individual would want, if he were in full possession of his capacity to make decisions and were supplied with all the relevant information, with what he <u>should</u> want according to some ideal of human excellence. Nevertheless, some decisions which a person makes in situations of compulsion, when emotionally disturbed, immature or not fully informed, and which one can predict, on the basis of some psychological generalisations, that he will probably regret later, should undoubtedly be discounted in the consideration of what constitute harms against which men ought to be protected. There is even some basis provided by certain theories of personal identity, for treating the self of an individual, who is under the disabilities just mentioned, as different from that at some other stage in his life, in which those disabilities are not present, and for protecting him against harmful acts performed by his previous self as though against a third party.[11]

Another difficulty that could appear at the stage of framing criminal laws concerns the justifiability of excluding the possibility that a bare mental state could be a crime. A perfectionist need not necessarily favor legal interference with people's mere inner states of mind; most probably he would reject criminal laws of that sort on varying grounds, for instance, the difficulty of producing satisfactory evidence. However, it is not completely inconceivable that he might advocate that people should be punished for their bare beliefs and intentions, since the mere fact of entertaining them could involve a moral depravation, and punishment may serve as a public condemnation of it. Obviously, this is extremely repugnant to a liberal frame of mind. However, the task of justifying, from a liberal point of view, the non-punishability of mere mental states is not as easy as it seems _prima facie_ to be.

The rationale traditionally advanced is that it is necessary to build a fence to protect freedom of thought and freedom of speech, so that there is some scope for developing and testing ideas and reasoning without anything to contend with other than rational argument and persuasion. Someone might, however, object that this applies only to some mental processes and their expressions, the holding of beliefs and opinions, which cannot be entertained or abandoned at will and have not, by themselves, effects in the external world. He might say that this does not apply to intentions, which can be suppressed when the fear of punishment intervenes and which involve practical dispositions with potential effects in the external world. He might add that, insofar as we are interested in preventing harms to other people, it is absurd to discriminate between different stages of the processes which produce those harms, erecting a barrier against legal interference with their mental stages. It is foolish to wait for a minute bodily movement before interfering with the production of harm, since the prospect of harm could entirely be based on the intention of the individual and the external activity in question may add nothing to the confirmation of that prospect. There is nothing sacred, he might claim, in the "mental" in opposition to the "physical," and it is preposterous to take people's bodily movements as if they represented a sort of green light that allows us to act without violating their right to a free inner life. Perhaps the critic would propose the following flight of fancy in order to refute what he sees as the myth of the inherent sacredness of mental processes: if we lived in a world in which men had the power of telekinesis, would we have any doubt that it was legitimate to interfere with some mental processes in order to prevent outward harmful effects and to deter other people from forming similar intentions? Of course, our world is not like that. We cannot bring about the death of our enemies through merely intending it. But intentions are still important steps toward causing harm and people should be punished for their intending it, not because that intention makes the person blameworthy, but in order to prevent its materialization and to deter others from entertaining it.

What can be said against these criticisms? I think that they succeed in throwing doubts on traditional appeals to the inviolability of mental life based on claims of freedom of thought. Nevertheless, there are well known arguments in support of the non-punishability of intentions. The most obvious one is that advanced by Blackstone,[12] notwithstanding that his treatment of subjective attitudes is not exactly anti-perfectionist; it rests on the difficulties of getting reliable evidence about the existence of bare intentions. There is much to be discussed on this issue but, nonetheless, there is no doubt that the punishability of mere intentions would introduce many uncertainties and greatly increase the possibility of punishing innocent people. H. Morris[13] points out, besides, the absurdity of laws formulated so as to prohibit mere intentions. If we are interested in preventing subsequent acts, it is sufficient to prohibit those acts, since the person persuaded by that prohibition cannot have the intention to act but decide, at the same time, not to act. If the prohibition of the act is effective, its effect on potential criminals is to prevent them, not only from carrying out their intention but also from persisting in it or ever forming it in the first place. This is true, but, nevertheless, there is the possibility of combining prohibitions of certain _actions_ with authorizations to punish those who merely intend to act (as in the case of attempts where it would be similarly absurd to _prohibit_ the mere attempt). The decisive argument is one also suggested by Morris[14] (although he gives it a transcendence connected with assumptions of Natural Law to which I cannot subscribe): Some intentions are firmer than others; the firmest of them logically imply taking steps to accomplish the intention in question. This degree of intention constitutes an attempt, and its punishability is justified, at least under certain conditions, by the considerable danger that harms may result from it. Less firm intentions are not so dangerous mainly because of the likelihood of a change of mind, and the evils that would ensue from punishing such intentions generally outweigh the advantages of suppressing the low risks involved in them. Among those evils it is necessary to count not only the possibility of punishing innocent people because of the unreliability of the evidence, but also the partial lack of inducement to refrain from materializing intentions already formed. For, once having the intention, the perspective of punishment would not operate as a motive against carrying it out; punishment would only deter at a subliminal level, preventing the formation of intentions. (This result could be partially avoided by fixing different degrees of punishment for mere intention and for attempt, but the perspective of a conviction as soon as an intention is formed would, in many cases, make a further dose of punishment not discouraging enough.)

These issues are mainly related to the framing of criminal laws. But the liberal and perfectionist outlooks also have different implications and raise different problems when applied to adjudication.

The most important of these implications are to be found in the treatment of subjective attitudes. I have said elsewhere[15] that the requirement that the person to be punished must have <u>consented</u> to undertake a liability to suffer that punishment (the principle of assumption of punishment), satisfies demands of fairness in the distribution of burdens imposed by the need for social protection. This, I think, is the only relevance that a liberal view may concede to the subjective attitudes of the agent. However, a perfectionist conception would assign to the requirement of certain mental states a much deeper significance. Insofar as the legal system is designed to promote personal ideals and the moral evaluation of personality is considered relevant for ascribing legal responsibility, the motives and intentions of offenders must be central items in the determination of whether the rationale for punishing people applies in a particular case. Under this conception, to punish somebody in disregard of his state of mind is not only unfair; more than that, it implies perverting the true function of criminal law which is to denounce people with vicious dispositions, and provide occasions for their moral regeneration and the edification of others. It might be admitted that it is all to the good that punishment diminishes harm to society, but it would be stressed that that function can, in any case, quite well be satisfied if particular punishments are meted out only to those who have caused harms as a result of their wicked dispositions and not in pursuance of good motives. The harms that we must fear are those which are caused by depraved people; the rest are either justifiable or unavoidable. Thus, the necessary antecedent of punishment is a blameworthy or guilty mind; the punishable act must be the expression of a vicious personality. From this central requirement, the subjective attitudes that are needed for ascribing criminal responsibility can be inferred. Obviously an anti-perfectionist conception of the law disassociates itself from this. Intention and other mental states are not required because it is necessary to evaluate the agent's personality as expressed in the act and to determine whether that act reflects a vicious mind. If my consensual justification of punishment is adopted, to assert that a defendant has acquiesced in the eventuality of being punished does not involve holding him blameworthy any more than to assert that a contractor has consented to some consequences of the contract involves considering him reprehensible. The considerations that lead us to reject moral blameworthiness as a sufficient and proper reason for punishing somebody also lead us to reject it as a necessary condition for punishment. To require such a condition, even if only in addition to other conditions, would imply that the law should treat people differently according to their moral worth and, consequently, that legal adjudication should embrace ideals of personal goodness.

These competing conceptions of the relevance of subjective attitudes for criminal responsibility lead their supporters, obviously, to require different mental conditions. For a perfectionist conception the motives that an agent has for engaging in a crime should be highly relevant, since the general plan of action, of which the act in question was a part, is decisive for judging the moral blameworthiness of the agent. In contrast, the requirement of mental states within a consensual justification of punishment is independent of the motives of the agent; whatever are the final purposes or goals of the offender (with the exception of some specific ones which merit some special treatment),[16] the fact that his act meets the conditions of the principle of assumption of punishment preempts the possible unfairness of such a measure when it is justified by the need for social protection. On the other hand, this latter anti-perfectionist rationale for requiring subjective attitudes preeminently requires knowledge that the conduct is punishable, which the perfectionist would only exceptionally require in cases in which the only moral significance of the action is that it is legally prohibited; in most cases the moral blameworthiness of the agent can be judged independently of his knowledge that the action is an offence. Needless to say, the differences between subjective attitudes (such as those between intention and knowledge), which are highly relevant to moral blame, may be immaterial to an anti-perfectionist rationale for requiring mental attitudes.

The perfectionist conception also has definite implications for the definition of situations where the offence would be justified. It can not only justify what it may consider to be only "technical crimes" (i.e. crimes performed not to fulfill some wicked purpose but perhaps to realize some ideal of human excellence), but can also favor a definition of traditional justifications, such as self-protection, necessity, exercise of a right, etc., so as to require a valid motive. Thus, for a perfectionist view it is, for instance, out of the question that a man who by chance repels his aggressor without intending to act in self-defence (since he did not realize that he was being attacked and his own act was, in fact, meant as an attack) should be exonerated. He has the same guilty intention that the man who actually is the aggressor. A similar solution will be applied to the case of a man who performs what is, in general, an offence without knowing that he had a special right or license to act as he did and believing that the act was unlawful. So, the distinction between justification and excuse loses much of its relevance, since the legitimacy of an action is partly dependent on the motives and knowledge of the agent. Although it is obvious that a liberal strategy for adjudication will not characterize justifications so as to make them dependent on the evaluation of the agent's reasons for acting, it is impossible to determine exactly how it will define them, and, indeed, what relevance it will concede to the agent's knowledge and intentions, until a rationale of justifications has been presented.[17]

Of course, the opposition between a perfectionist and a liberal outlook about the relationship between criminal responsibility and moral blameworthiness also has relevant consequences for the selection and gradation of the penalty to be applied in particular cases. The perfectionist conception maintains that there is a _prima facie_ case for adjusting the penalty which an offender must suffer in accordance with the degree of his blameworthiness, which would be determined by his motives and other features of his crime. This is obviously so when perfectionism is coupled with a restributive or denunciatory theory of punishment. But it is also the case even when it is coupled with a theory which makes the rehabilitation of actual offenders or the deterrence of potential offenders the goal of punishment, since the assessment of the offender's moral qualities will indicate how much and what sort of punishment is needed to induce the defendant to adopt the endorsed models of life and to discourage other potential lapses from human excellence. Besides, perfectionism contains no _a priori_ reason for limiting the data that are relevant for evaluating the personality of the agent to the circumstances of the act for which he may now be punished. The agent's "criminal record" and general behavior may be decisive. A liberal conception of the criminal law rejects the foregoing criteria for selecting and gradating penalties, though it is not at all clear what it substitutes for them.

3. _Concluding Remarks_

While one may envisage virtually all the implications of the perfectionist conception of the criminal law, it is not so easy to adumbrate the implications of the liberal conception without further discussion. My opinion is that a full development of a criminal law system requires additional principles, some of which I have defended elsewhere.[18]

One of the most important issues in relation to which the liberal program is in need of much more elaboration is the definition of self-regarding actions. As we have seen, the punishability of allegedly self-degrading conducts--like the consumption or possession of drugs--might be defended on the basis of the argument that they generally involve harms or risks (other than that constituted by the injured feelings of the intolerant) to third parties, like, e.g., the harms which ensue from the imitation of bad example, the failure of the agent to contribute to his family's support or to the social wealth, the burden that he might import for public health services, etc.

In fact, even Mill acknowledged that it is difficult to find a self-degrading conduct which would not result in some harm to third parties:

> The distinction here pointed out between the part of the person's conduct which concerns only himself, and that wich concerns others, many persons will refuse to admit. How (it may be asked) can any part of the conduct of a member of society be a matter of indifference to other members? No person is an entirely isolated being; it is impossible for a person to do anything seriously or permanently hurtful to himself without mischief reaching at least to his near connexions, and often far beyond them... I fully admit that the mischief which a person does to himself may seriously affect both their sympathies and their interests, those nearly connected with him, and in a minor degree society at large...[19]

This acknowledgement seems to make ineffectual the liberal distintion between two dimensions of morality and, consequently, of two kinds of immoral conducts to the effect of legal interference with them: it seems that any actions which seriously contravene ideals of human worth tend also to violate standards of intersubjective morality, and are thus proper objects of legal interference. But I think that an adequate articulation of the liberal conception will save that distinction on the basis of a more precise characterization of the notion of harm to third parties. Pending that characterization, I dare suggest the following very general guidelines to be taken into account in it:

(i) When a harm to a person other than the agent is alleged, one must separate those cases in which the harm is caused by an act of the agent other than that which involves self-degradation, or by the fact that this act is done in special circumstances. For instance, the spread of drug addiction thorugh imitation or incitements is not caused by the mere consumption of the drug by a certain person, but by the fact of doing it in public or by the different act of offering the drug to others. These acts may be punished without doing violence to the liberal tenet.

(ii) One must also probe into the <u>causal relation</u> between the act and the alleged harm in order to determine whether the latter is really ascribable to the agent of the self-degrading conduct. For instance, if the harm has as its immediate cause a fully voluntary act of the very "victim," who is a responsible person,—like in many cases of imitative drug-consumption,--this is normally considered as a circumstance which "breaks" the causal chain leading to the former act,[20] in this case that of the "corrupting" agent. Besides, the ascription of causal effects to omissions--like the failure to contribute to the social welfare--requires a justified duty to act.[21]

(iii) Even when the harm is in fact causally attributable to the self-degrading conduct itself--and not to another act or surrounding circumstances--the distinction implicit in the liberal view can still be preserved and be operative if we reformulate so as to demarcate alternative results of a balance between, on the one hand, the importance that the conduct in question may have for the plan of life chosen by the agent, and, on the other, the extent and degree of interference that that conduct has for other people's own life plans. According to this reformulation, a self-regarding conduct would be an action such that if it were prevented, the central life-plan of the agent would be radically impaired to a considerably greater extent than any peripheral disturbance that its performance may provoke in other people's life-plans. This balance does not collapse into a utilitarian calculation insofar as we refuse to give full weight to the aggregative sum of other people's nuisances <u>vis à vis</u> the radical impairment involved in hypothetical restraint of the agent (thus, while, e.g. the homosexual tendencies of a man, may be restrained insofar as they involve abusing young boys, no amount of nuisance of even very many--like the offensive sight of tranvestites or special costs for the public health service, etc.--may justify the preclusion of such a central item of somebody's way of life).

As I have said, this subject requires further elaboration, but, nevertheless, the kernel of the liberal conception of the criminal law is clear enough: it is not even a _prima facie_ reason for punishing an action the fact that it taints the moral character of the very agent because it involves a failure to satisfy some model of human excellence or the choice of a repugnant life-plan. Only the circumstance that the act at stake impedes the choice or materialization of plans of life of people other than the agent grounds the power to punish him within the framework of a liberal society.

NOTES

* This article is a modified version of a chapter of my thesis accepted by the University of Oxford for the Doctor of Philosophy degree. I express my gratitude to my supervisors D. J. M. Finnis and Professor A.M. Honore and to my examiners Professor H. L. A. Hart and P. F. Skegg for their illuminating comments and objections.

1. See Lord Devlin, _The Enforcement of Morals_ (London, 1965); H. L. A. Hart, _Law, Liberty and Morality_ (Oxford, 1963) and _The Morality of the Criminal Law_ (Jerusalem, 1964). Other contributions to the modern debate include, among many others, B. Mitchell, _Law, Morality and Religion in a Secular Society_ (Oxford, 1967); R. Dworkin, "Lord Devlin and the Enforcement of Morals," _Yale Law Journal_, vol. 75, p. 986; A. R. Louch, "Sins and Crimes," _Philosophy_ 43 (1968), 163 ff.
2. _Op. cit._, pp. 17 and 22 _et. seq_.
3. For the formulation of some of these arguments I have taken into account the position of Louch in _op. cit_.
4. For instance, B. Gert, _The Moral Rules_ (New York, 1973), p. 139.
5. I deal with it in two forthcoming articles: "Liberty, Equality and Causality" and "The Derivation of Two Principles of Liberalism."

6. For instance, Plato ascribes to Protagoras, in a famous passage, the idea that punishment should not be retributive but should 'look at the future' (an idea that Socrates does not contest in the dialogue), while maintaining, at the same time, that the object of punishment is to teach virtue. Protagoras, trans. by B. Jewett, Dialogues, Vol. I (Oxford, 1953), p. 324.
7. Article 19 of the Argentinian Constitution of 1853, which summarizes the liberal thesis in a beautiful way, focuses, however, the distinction between acts which are against "private morals" with the distinction between "private" and "public" acts, when saying: "The private actions of men which by no means offend the public morals and order and do not harm others, are reserved to the sole judgment of God and lie beyond the authority of magistrates."
8. In "Paternalism," in R. Wasserstrom (ed.), The Morality and the Law (Belmont, California, 1971).
9. In "Two Concepts of Liberty," in A. Quinton (ed.), Political Philosophy, (Oxford, 1967), p. 150.
10. Op. cit., p. 119.
11. See D. Parfit, "Later Selves and Moral Principles," in A. Montefiore (ed.), Philosophy and Personal Relations (London, 1973).
12. Commentaries, Bk. IV, Ch. II, cited by H. Morris in "Punishments for Thoughts," in Skummers (ed.), Essays in Legal Philosophy (Oxford, 1971), p. 110.
13. "Punishment for Thoughts," p. 95.
14. Op. cit., pp. 102 et. seq.
15. In "The Consensual Theory of Punishment," Philosophy and Public Affairs, forthcoming.
16. See my Los límites de la responsabilidad penal (Buenos Aires, 1980), ch. V.
17. I deal with the subject in Los límites de la responsabilidad penal (Buenos Aires, 1980) and La legítima defensa (Buenos Aires, 1982).
18. See mainly Los límites de la responsabilidad penal and "A Consensual Theory of Punishment," Philosophy and Public Affairs.
19. "On Liberty," in R. Wollheim (ed.), Three Essays (Oxford, 1975), pp. 98-99.
20. See, H. L. A. Hart and A. M. Honore, Causation in the Law (Oxford, 1959), pp. 38 et. seq.
21. I defend this view in "Da lo mismo omitir que actuar?" La Ley 1979-C-801

Raûl Orayen

ON THE INCONSISTENCY OF MEINONG'S ONTOLOGY*

My aim in this paper is to give an analysis of the logical difficulties which are present in some of the main theses in Meinong's ontology. These theses have been criticized by logicians since 1905, the same year Russell published "On Denoting" in Mind, posing serious objections to Meinong's doctrines and giving an alternative theory, by setting up his well-known theory of descriptions. Nevertheless, a new study of this question is justified, as we shall see later, because Russell's arguments are not as conclusive as one might think them to be at first glance. One of his arguments is based on a misinterpretation of the theory he attacks, and the rest of them lose some of their strength due to Meinong's replies to his objections.

This paper has two parts. The first will be devoted to the consideration of the Meinongian theses relevant to this discussion. In the second, I shall analyze the classical Russellian criticisms, showing their limitations; and finally, I shall propose another critical view that will result in a new proof of the inconsistency of Meinong's ontology. In fact, my criticisms are not just or especially aimed against Meinong. Their scope is more general and they are offered as the basis for the rejection of certain ontological and semantic conceptions which Meinong himself made explicit and developed extensively, but of which he is not the only exponent.

1. MEINONG'S THEORY OF OBJECTS[1]

1.1 Definite descriptions and objects. Kinds of objects.

Meinong's theory of objects is an ontological theory, not a semantic one. However, it does embody several semantic theses. Some of them can be summarized in the following statement:

J. J. E. Gracia et al. (eds.), Philosophical Analysis in Latin America, 115–139.

T^o: Every meaningful linguistic expression, which can be the grammatical subject of a meaningful sentence, denotes (designates, refers to) an object.

This thesis is very general. To test its validity, we will take into consideration a less general one, which is logically entailed by it and which concerns exclusively those expressions which, since Russell, are known as "definite descriptions." Definite descriptions are expressions like "the present president of France," "the round-square," "the man who squared the circle," etc. The distinctive characteristic of these expressions is that each one of them contains a monadic predicate _F_, such that the expression denotes a given individual, if and only if that individual is the only one of which the predicate _F_ is true.[2] E.g., "The present president of France" contains the monadic predicate "present president of France" and denotes an individual if and only if that individual is the only one which has the property of being the present president of France. In English, the unicity requirement is usually indicated by the presence of the definite article.

Obviously any definite description is a meaningful linguistic expression which can be the grammatical subject of a meaningful sentence. Hence, from T^o we can deduce the assertion:

T^1: Every definite description denotes (designates, refers to) an object.

Considering that T^o logically implies T^1, any rejection of the latter will also be a rejection of the former. For this reason, we shall concentrate on T^1, analyzing only cases in which definite descriptions occur; these are very common in Meinong.[3]

It can be easily shown that T^1 takes us far away from our intuitions. We would usually consider that there is an important difference between "the present president of France" and "the present King of France," namely that the first expression denotes an object but the second one does not, because there is no present King of France. Nevertheless, according to T^1 _both_ expressions denote objects. How is one to explain, then, the profound difference between them? According to Meinong, the difference does not consist in the fact that one of them denotes while the other does not, but in the fact that they denote objects of very different kinds: the first denotes an existent object and the second a non-existent one. Such considerations take us into the Meinongian classification of objects.

The first Meinongian category of objects includes those that exist. According to Meinong, the notion of existence is a basic and undefinable one, but sufficiently clear to be understood. Physical entities like tables and chairs and psychical phenomena, like being mad, are existent objects. A characteristic shared by all existent objects, and only by them, is temporality, i.e., they are always linked to specific temporal determinations. The most commonly used descriptions in daily life (the present president of France, etc.) refer to these kind of objects.

Existence is a kind of being but not the only one. According to Meinong, there are objects which are part of the world and have some kind of being, but it would be absurd to claim that they exist. Such is the case of facts, numbers, "pure" characteristics and ideal relations, e.g., the likeness between two paintings.[4] These objects do not exist but subsist. Subsistence is a kind of being, broader than existence, since according to Meinong existent objects do subsist while there are subsistent objects that do not exist.[5] Descriptions such as "the likeness between painting A and painting B" and "The existence of The Antipodes" denote subsistent objects. The second description denotes a fact: the fact that The Antipodes do exist. But even if The Antipodes do exist, their existence (because it is a fact) only subsists.[6]

Russell appears to think that for Meinong every object has either existence or subsistence. But it is not so. The two Meinongian categories exhaust the realm of being, not the realm of objects. There are objects which neither exist nor subsist, nor have any other kind of being. Posing the question in puzzling terms, Meinong says: "There are objects of which it is true that there are not such objects."[7] Among these, we find the present King of France, the round square and the man who squared the circle. These objects, owing to factual or necessary reasons, do not have any being at all, but they "are there" in some obscure and difficult sense.

Meinongian ontology is completed with these objects. The realm of objects is twofold; it contains objects which have being and objects with no being at all. The former can be existent or just subsistent.

At this point it would be convenient to consider a possible interpretation of Meinong's thought before going on to the next problem.

It could be thought that, when Meinong asserts "there are objects such as the round square, and the man who squared the circle," he just tries to express in a somewhat confused way that "the round-square" and "the man who squared the circle" are meaningful expressions or express ideas. In this case, he would accept them as objects only as a _facon de parler_. The quoted assertion "would only commit him ontologically" to linguistic expressions and maybe also to meanings or ideas. If this interpretation is correct, the criticisms of those who have considered him as "the greatest entities multiplier" in the history of philosophy,[8] would be unfair, and our objections in the second part of this paper, would not be valid. But certainly, Meinong's formulations are completely incompatible with this interpretation. In fact, Meinong made explicitly clear that words "express" ideas and are not used to refer to them but to transcendent objects which differ from those ideas and are the denotations of the words in question. Meinong says that if someone uses the word "sun," it could be inferred with some probability that he has a certain idea in his mind; but the speaker wants to refer to a transcendent object, not to that idea. This object is the denotation (_bedeutung_)[9] of the word. So, when Meinong says that there are objects such as the round-square and the man who squared the circle, he does not merely say something about words or ideas, but about their denotations. He accepts that there are denotations for "the man who squared the circle," "the round-square," etc. and our formulation of thesis T^1 should be understood with this in mind. In some sense, therefore, for Meinong there are entities lacking being such as those previously mentioned. Even more: Meinong supports an extremely realistic conception of these objects, and thinks that they "can be there" even when they are not apprehended by any mind whatsoever.[10] The objects lacking being are neither linguistic fictions nor mere conceptual entities. Hence, in what follows we shall take as absolutely correct the interpretation which shows Meinong as an authentic "entities multiplier."

Once this point has been made clear, we can set forth some problems about these strange entities. Could an object lacking all being have nevertheless some properties? Or perhaps, because of its lack of being, it lacks every property? And if it has properties, which ones does it have? Let us consider now the Meinongian answer to these questions.

1.2 The Principle of Independence of Being-so from Being

The answer to these problems is given by a principle which Meinong took from Mally, the so called Principle of Independence of Being-so (*Sosein*) from Being (*Sein*). According to this Principle, an object can have properties even if it lacks any being.[11] It is possible for an object not to be, but to be in a certain way. For instance, the round-square is not; nevertheless the round-square is round and also square. Similarly the man who squared the circle, squared the circle and so forth. However, those beingless objects could be shown to be indeterminate in many respects. E.g., the golden mountain has no determinate height. There is no answer to the question as to whether it is higher or lower than Everest. Hence, what are the properties that a beingless object could have? Meinong gives another formulation of this Principle that makes this point clear: "The principle of the independence of *Sosein* from *Sein* tells us that what is not in any way external to the Object, but constitutes its proper essence subsists in its *Sosein*--the *Sosein* attaching to the Object whether the Object has being or not."[12] In other words, whether the object has being or not, it does have those qualities or properties which constitute its own essence. For these reasons, the golden mountain, no matter how indeterminate it may be with respect to its height, has the properties of "mountainness" and of "goldness."

It is very important for us to comment on two theses present in these Meinongian doctrines, which follow from the previous considerations.

Consequence I. Meinong claims that every subject-predicate assertion is true when its grammatical subject is a definite description and its predicate refers to a property which is "alluded to" in the given description.[13] We have seen some instances of this: the round-square is round, the man who squared the circle squared the circle. It is obvious that Meinong accepts these assertions because he thinks that the properties that a definite description "alludes to," are part of the essence of the object the description denotes. Therefore, whether the object lacks being or not, it has the properties alluded to in the description that denotes it according to the Principle of Independence of Being-so from Being (henceforth PIBB).

Consequence II: Meinong thinks that the being or the existence of an object is alien to its nature. So he adheres to the Kantian thesis according to which existence is not an authentic predicate of things. The nature of an object is given by a being-so determination series, according to which the object is such and such. But by virtue of PIBB, existence is not part of the object's nature; it cannot be counted among the being-so determinations[14] because the being-so of an object is independent from its being.

Both consequences of PIBB will be relevant to our critical analysis, developed in the second part of this paper.

1.3 Meinong's Reasons

Before going on to criticize the assertion T^1, we shall be briefly concerned with both the influences that affected Meinong at the time he presented his thesis, and with the arguments he tried to use to justify it. In other words, we shall be concerned with what some philosophers of science would call "the context of discovery" and the "context of justification" of this doctrine.

Meinong took his first steps in philosophy under Brentano's direction, and accepted very soon his theory on the intentionality of psychic phenomena. According to this well-known conception of Brentano, the peculiar characteristic of mental phenomena consists in their being directed somehow toward objects. Each mental state possesses something that serves as its object, something toward which it is "directed," and that object is conceived as something transcendent to the mental phenomenom itself. To be "directed" to objects is a characteristic feature of the psychic; it means that all psychical phenomena and only they have this feature. The adherence to this doctrine was, no doubt, one of the reasons that led Meinong to accept T^1 and to postulate objects lacking any being. Undoubtedly the expression "the round-square" is correlated with a psychic phenomenon. The expression is not a meaningless noise for us: we associate it with a thought which we could call "the round square idea." Assuming Brentano's doctrine, we must conclude that this idea "is directed" to a transcendent object different from the idea itself. It seems natural to think that the object is precisely the round square. So, there is such an object as the round square. But as this object has no being, the application of Brentano's theory to this particular case leads to the postulation of beingless objects. Naturally, similar considerations could be made in the case of any definite description, whether or not something denoted by it exists. Every description is associated with an idea, which seems "to be directed" toward a denotation of the description. Assuming Brentano's thesis, we reach the conception summarized in T^1.

The thesis of the intentionality of psychic phenomena is a controversial one. It has been suggested that, for instance, even though "intentionality" can be given only in psychical phenomena, there could be psychical phenomena without such a feature. Hence, "intentionality" would be a sufficient but not a necessary condition of the psychical. We will not discuss this point here. We shall only indicate that, even though some ontological ideas of Meinong could have had their origin in Brentano's doctrine, this could not serve as a proof of them, among other reasons, because it is a controvertible theory.

Meinong based his main arguments to prove that there are non-existent objects on the analysis of true assertions about non-existence. According to Meinong, when true assertions are made about non-existence, there will exist facts that correspond to them. The existence of true assertions of the form "x does not exist" leads him to the conclusion that there are facts of non-existence. But, according to the Meinongian doctrine of facts,[15] every fact is a higher level object "built on" other objects. Every fact is "about" objects. But, about which objects is the fact represented by "the round-square does not exist"? It seems that the round-square is the only object involved herein. Therefore, this fact is "about" an object that has no being: the round-square. So, there are beingless objects. "The round-square" denotes an object, but a beingless one. Assuming this argument, it is easy to prove the truth of T^1. This thesis asserts that every description denotes, and those cases of descriptions that denote existent objects are really obvious. Now, if a description "D" does not denote any existent object, the statement "D does not exist" will be true and will be correlated to a fact. Such a fact is "about" some object. But the denotation of "D" seems to be the only candidate, and therefore, it is concluded that this description denotes an object but a non-existent one.

Curiously enough, even though Russell was a staunch enemy of these doctrines from 1905 onwards, he had arrived at similar conclusions in his Principles of Mathematics before Meinong: his reasons are analogous to Meinong's. Russell thinks that assertions of the form "A is not" should be either always false or meaningless, "because if A were nothing, then one could not say that it is not."[16] In other words, if in a meaningful sentence "A" occurs as the grammatical subject, then "A" denotes an object, otherwise no such meaningful assertion could have been made. But Russell differs from Meinong in concluding that every object has being, even though he does not believe that every being exists. The round-square would not only be an object but would also have being.

It can be noted that Meinong and Russell based their conclusions on an analysis of facts and assertions of non-existence. As is well known, later semantic analyses given by Russell and others have shown that it is possible to construe facts and assertions of non-existence without having to accept assertion T^1 or to appeal to non-existent objects. Nevertheless, the possibility of alternative analyses does not show that T^1 is false. In what follows, we shall be concerned with other arguments which tend to prove its falsehood.

2. INCONSISTENCY OF THE THEORY OF OBJECTS[17]

2.1 Russell's Criticisms

In "On Denoting" Russell posed three classic objections to Meinong's theory of objects. We shall discuss each one in turn.

First criticism. Let us suppose that object A is different from object B. For Meinong, this fact can be expressed by:

"The difference between A and B subsists"

Russell pointed out that in such a case if A is not different from B, such a fact can be expressed by the sentence:

"The difference between A and B does not subsist"

In consequence, we would have found an object which neither exists nor subsists. It seems that Russell thought he had found by this means an incoherence in Meinong's theory because he believed, as we said before, that for Meinong every object either exists or subsists. But since this is not so, "to find" an object which neither exists nor subsists does not represent any internal difficulty to Meinong's theory.

Second criticism. Meinong accepts that the present King of France, the round-square, etc., are non-existent objects. But Russell points out an interesting inconvenience. Let us take a description such as "the existent round-square." We said before (see 1.2 Consequence I) that Meinong accepts as true those statements whose subject is a definite description and whose predicate refers to a property "alluded to" in the given description. In that case, it seems that he should accept the truth of the sentence "The existent round-square exists." This result is paradoxical: Meinong denies the existence of the round-square, but would have to accept the existence of the existent round-square.

Nonetheless, Meinong could cope with this difficulty. In the preceding considerations, the term "existent" was treated as if it were a predicate and was included inside a description. In that case, existence is treated as a property and can be part of the nature of an object. But all this is expressly rejected in Meinong's theory, due to the validity of PIBB (see 1.2, Consequence II). Meinong had reasons to reject the truth of "The existent round-square exists." These reasons are the same he uses to criticize the ontological proof of God's existence: existence cannot be treated as a property.

What happens then, with the description "The existent round-square"? Meinong, by virtue of these considerations, could have considered it meaningless. Instead of doing that, he interpreted it as meaningful, but he supposed that the term "existent" used in this way did not have a meaning closely related with the verb "to exist." Hence, he accepted that in a certain sense it could be true that the existent round-square is an existent object; but that was not the same as saying that such an object exists. This last point is quite obscure. Maybe it would have been more consistent with his theory to take the alternative course of denying meaning to the description.

Meinong's answer is questionable. But it is interesting to note that Russell, at the time he was developing his theory of descriptions, held precisely that existence is not an authentic predicate of objects and also held this in order to reject the ontological proof of God's existence. It follows that, apparently, Russell should have accepted Meinong's reply by virtue of his own logical ideas about existence.

Third criticism. The most serious objection Russell posed to the theory of objects consists in the accusation that the doctrine leads to contradictory assertions. His argument can be formulated this way:[18]

Let us take the sentence "The square which is not square is square" and "The square which is not square is not square." By virtue of consequence I of PIBB (see 1.2), both assertions are accepted in the theory of objects. Nevertheless, they are mutually contradictory. Symbolizing with the letter "x" the expression "the square which is not square" both can be reformulated in the following form, which makes evident the contradiction:

I) x is square

II) x is not square

Russell argues then that Meinong's theory of objects requires imposing a limitation to the validity of logical principles. The principle of non-contradiction loses its unlimited validity.

Meinong did not completely solve this difficulty and recognized that his theory required imposing a limitation to the principle of non-contradiction. However, he argued that the limitation was not important, because the principle alluded to fails to apply only in case of contradictory, impossible objects. According to him, logical principles are applied without restrictions in the realm of existent and possible objects, but it would be unreasonable to regard them as valid in the realm of impossible objects.[19] When the problem is so stated, it cannot be denied that the difficulty seems to lose some of its strength. Mathematical intuitionism has made us see--according to a certain philosophical conception--the possibility that some logical principles are not valid within certain special ontologies. The principle of excluded middle would not be valid in the realm of mathematical objects. In the same way, it could be thought that the principle of non-contradiction is valid for all existent or possible objects, but is not valid in the realm of impossible objects, such as the square which is not square. This is briefly Meinong's answer.

Nonetheless, if we want to preserve at all costs the unlimited validity of the principle of non-contradiction, it seems that we could achieve this goal by modifying the theory of objects in such a way that contradictory objects were no longer postulated. As a result, the theory would have, instead of T^1, the weaker following thesis, which is a logical consequence of T^1:

> T^1': Every non-contradictory definite description[20] denotes (or designates, or refers to) an object.

Meinong did not weaken his theory this way. But nevertheless we shall analyze the validity of thesis T^{1}' because of its intrinsic interest and because it will throw more light upon the theory of objects. It is clear that if Meinong's thoughts are correct, T^{1}' leads no longer to a contradiction. He thought, indeed, that the violations of the principle of non-contradiction could only appear in connection with impossible objects (denoted by contradictory descriptions). On the other hand, it is obvious that Russell's last objection cannot be applied to T^{1}', since this objection arose from examples with contradictory descriptions.

In what follows I shall show, that thesis T^{1}' also leads to contradictions. This will allow us to prove that in Meinong's theory of objects, contradictions about possible objects, even existent ones, can be pointed out, without having to allude to impossible objects. Therefore, Meinong was wrong in thinking that his theory respects logical principles both in existent and possible realms.

2.2 A new proof of inconsistency

Meinong's theory of objects contains thesis T^{1}, which logically implies thesis T^{1}'. Therefore, if we prove that T^{1}' leads to contradictions in which no impossible object is involved, this will also hold for T^{1} and for the theory of objects itself. But here we are not interested in analyzing *just* the validity of Meinong's ontology, but of any theory holding thesis T^{1}'. That is why we shall derive directly a contradiction from T^{1}' without taking as premisses other ontological assumptions of Meinong. In order to do that, using T^{1}' we shall first prove the following:

LEMMA 1

If "F" and "G" are predicates such that:

a) the propositional function "x is F" is consistent; and

b) "x is F" logically implies "x is G"

then, the statement "The F is G" is true.

This Lemma intuitively states the truth of every statement whose grammatical subject is a non-contradictory definite description, and whose predicate refers to a property already "alluded to" in the given description. This lemma can be considered a mere application of Consequence I of PIBB (see 1.2) to the case of non-contradictory descriptions. Nonetheless, here we shall prove it from T^{1}, without using Consequence I, because we want to show that T^{1}' leads to serious difficulties on its own (and not only in conjunction with other ontological assumptions).

The proof is trivial. Suppose that two predicates "F" and "G" fulfill (a) and (b). In that case, by virtue of (a), the expression "The F" is a non-contradictory definite description (see note 20). Applying T^{1}' we can prove that "The F" denotes an object. Let us call "a" this object. By virtue of the current meaning of the statement "The F is G," the following equivalence holds:

(E) "The F is G" is true if and only if *a* is G.

But, by the simple analysis of the meaning of the expression "The F," we know that this description denotes an object only if that object is F. In effect, given the common rules of use, a definite description denotes an object only if the object fulfills certain requirements among them, that of having the property expressed by the predicate with which the description was constructed. But "The F" denotes the object *a*. Then *a* is F.[21] By virtue of (b), it is also true that *a* is G. Therefore, using the equivalence (E), we can conclude that "The F is G" is true.

The Rule of Substitutivity of Identicals

In our next demonstration we shall apply a logical rule based on the so called "Law of Leibniz" or "Identical Indiscernibility Principle." According to this principle, an individual *x* is identical to an individual *y* if and only if both have exactly the same properties.

This principle allows us to support the rule known as the "rule of substitutivity of identicals" and which we will simply call "RS." According to RS, if it is known that the individual *a* is identical to the individual *b*, and also that a has a certain property, we can conclude that b has such a property too. The scheme is:

a is identical with b

a is F

---

b is F

Inconsistency Proof of T^{1}'

Let us take two linguistic expressions "b" and "F" such that:

I. "b" is a non-contradictory definite description.

II. "F" is a predicate.

III. The statement "b is F" is not self-contradictory.

IV. The statement "b is not F" is not self-contradictory.

By virtue of these conditions, then, the expressions "The F identical to b" and "the non-F identical to b" are two non-contradictory descriptions. Therefore, we can apply Lemma 1 to them and accept that every statement whose subject is one of these two descriptions and whose predicate refers to a property "alluded to" by the subject-description will be accepted as true. Then, the following statement should be accepted:

1) The F identical to b is identical to b.

2) The F identical to b is F.

By applying RS, is deduced:

3) b is F.

But by virtue of Lemma 1, these two statements should be accepted too:

1') The non-F identical to b is identical to b.

2') The non-F identical to b is non-F.

By applying RS, we infer:

3') b is non-F

Combining results (3) and (3') we get:

4) b is F and b is non-F.

We have obtained a contradiction from T^{1}'.

Examples and Conclusions

The proof will be better understood if we analyze an example formulated in ordinary language. In order to do that we shall replace "the present president of France" for "b" and "mortal" for "F." In order that the language be more natural we shall abbreviate "non-mortal" by "immortal" and we shall add the predicate "man," as follows:

1) The mortal man identical to the present president of France is identical to the present president of France.

2) The mortal man identical to the present president of France is mortal.

Then,

3) The present president of France is mortal.

1') The immortal man, identical to the present president of France is identical to the present president of France.

2') The immortal man identical to the present president of France, is immortal.

then,

3') The present president of France is immortal,

and, from (3) and (3') it follows:

4) The present president of France is mortal and immortal.

This proves that $T^{1'}$ is inconsistent. The same is valid for T^{0} and T^{1} which logically imply $T^{1'}$. It has been proved also that <u>Meinong's theory of objects violates the principle of non-contradiction even in the case of existent and possible objects</u>. Let us remember that in order to formulate the proof inside Meinong's theory, it is not necessary to demonstrate Lemma 1, because (1), (2), (1') and (2') follow directly from <u>Consequence I</u> of PIBB.

2.3 <u>Observation on the range and legitimacy of the preceding proof</u>

The preceding proof shows that Meinong's ontology is not inconsistent <u>just because</u> of its postulation of contradictory objects. The theory is contradictory even if we eliminate the postulation of impossible objects and remains unaltered in other respects. This is interesting because other authors have thought exactly the opposite. Anthony Kenny, for example, has recently suggested that the contradictions emerge in the theory of objects just because of the admission of impossible objects. Therefore, he considers that there is no inconsistency in the postulation of beingless objects, as far as there are no impossible entities admitted among them.[22] The considerations of the preceding section show the opposite. For Kenny, the most serious objection one can pose to "Meinongian pure objects" is the one indicated by Quine in "On What There is" (see note 17).[23] My proof has in fact some relationship with Quine's objection, because both are focused on the identity relation. But Quine's objection is not a proof of inconsistency. For this reason the difficulty we have pointed out in the preceding section seems to be more serious.

Kenny seems to believe that if the theory of objects is reformed, eliminating from it the postulation of impossible objects, one obtains a consistent doctrine. Wonfilio Trejo goes farther: he thinks that Meinong's ontology, <u>such as it is</u>, does not violate the principle of non-contradiction. His reasons are that the contradictions appear only with respect to impossible objects whose existence is not asserted in the theory.[24] His observations constitute an interesting, although questionable, defense of the consistency of Meinong's ontology. Nonetheless, this defense can no longer be attempted in the case of our demonstration, since in it, contradictory statements about an existent entity (see the example), without mentioning impossible objects, are derived from assertions accepted by Meinong.

To conclude, we shall analyze a possible reply to our proof. In the course of our argumentation, we used twice RS. RS is employed in reasonings containing two premises: the first one is an identity statement in which two singular terms occur, "a" and "b," linked together by the predicate "is identical to." The second premise is a statement in which "a" occurs. RS allows us to obtain as a conclusion the statement that arises when "b" is substituted for "a" in the second premise. Nevertheless, the inference procedure so described "fails" in some instances and allows us to pass from true premises to a false conclusion. Let us consider two examples from Quine:[25]

Example A

1) Cicero is identical to Tulio.

2) "Cicero" contains seven letters,

then,

3) "Tulio" contains seven letters.

Example B

1') Giorgione is identical to Barbarelli.

2') Giorgione was called so because of his stature,

then,

3') Barbarelli was called so because of his stature.

Premises from examples A and B are true, but their conclusions are false. How could this happen if they are apparently examples of RS application? Analyzing the examples, it can be seen that the proper names that occur in (2) (2') are used there in a special way. In (2) the proper name is not used to refer to its denotation, but when quotation marks occur it is used to refer to itself. Indeed, (2) is about a linguistic expression, the expression "Cicero." Case (2') is more complex; there the proper name is used to refer to its denotation, but refers to itself too. In fact, (2') is about the expression "Giorgione" but also about its denotation, and this will be easily noted if we observe that it is equivalent to the assertion: Giorgione was called "Giorgione" because of his stature.

Therefore, in (2) the proper name is not used <u>to talk about its denotation</u> and in (2') the proper name is used not <u>only to talk about its denotation</u>. In both cases, it is said that the proper name has not a <u>purely referential use</u>. When a definite description is used in this way, we will also say that its use is not purely referential. When a singular term (proper name or description) does not have a purely referential use in a context, we will say that the context is referentially opaque.

We can now explain why in A and B falsehoods were inferred from truths. If two singular terms "a" and "b," have the same denotation (i.e., a=b) and "a" occurs in a true statement <u>p</u>, but "a" has not a purely referential use in <u>p</u>, substitution of "b" by "a" in <u>p</u> does not necessarily result in a true statement. This is so because the original statement and the one obtained by substitution do not merely talk about the denotation of "a" and "b." Therefore, there is no guarantee of their <u>saying the same about the same object</u>, though "a" and "b" denote the same object.

Summary: RS is not applicable to referentially opaque contexts; these are just the contexts in which it can fail. In consequence, to ascertain whether our use of RS was legitimate, we must ask ourselves whether any of the statements (1), (2), (1') or (2'), used in the exemplification of our inconsistency proof, constitutes a referentially opaque context. If the statements are understood <u>in their usual meaning</u>, the answer is clearly negative. Moreover, Meinong's explanations make it clear (see end of I) that he accepts statements such as (1), (2), (1'), (2') using the singular terms in a purely referential form. Therefore, the use of RS in our proof of the inconsistency of the Theory of Objects is a legitimate one. Naturally, it is also legitimate to use RS in order to prove the inconsistency of T^{1}' because from this thesis are inferred statements (1), (2), (1'), (2') with their normal meaning. Nevertheless, we must clarify a point. In this paper we have criticized assertion T^{1}', understanding it in a sense very close to the customary one.[26] In particular, we have supposed that the definite descriptions we considered, were those of ordinary language. But obviously someone could use the linguistic clothing of T^{1}' to mean something different without incurring the ontological postulations we have criticized here. For instance, someone could assert T^{1}' merely intending to say that every definite description has a meaning. He could also assert T^{1}' with respect to definite decriptions in an artificial language, in which empty descriptions are provided with an arbitrary denotation (for instance, the number 0). On the latter case T^{1}' would be true and the contradictions we have demonstrated would not arise from it. Even Lemma 1 could not be derived, because in such a language it is not always fulfilled that denotation of "The F" has property "F." But naturally, T^{1}' taken under one of these "anomalous senses," is far from expressing the semantic-ontological conceptions we have criticized here.

To sum up: what we were interested in discussing was the validity of certain semantic-ontological theses about the "ontological commitments" of ordinary language. Within this language, the fact that referentially opaque contexts are contained in statements similar to those used in our proof of 2.2 cannot be plausibly interpreted; thus we can legitimately use our proof to evaluate those theses. If a Meinongian ontological theory is to escape our criticism, it has either to defend the opacity of certain predicative contexts of colloquial language--in which case it becomes not very plausible, or its theses should be formulated with respect to a language different from the ordinary one--but in that case it would not express the doctrines we were concerned with in this paper.

1982 POSTSCRIPT

At the time I wrote the preceding article, I was not <u>only</u> interested in criticizing Meinong's theory; I sought as well to examine the validity of a <u>certain kind</u> of semantic-ontological theory, paradigmatically exemplified by Meinong's conceptions. Therefore, my analysis was very extensive and I attempted to reach the following general conclusion: any theory of the kind we have analyzed (i.e., those which involve T^{1}', or even stronger theses such as T^{o} or T^{1} in connection with ordinary language) admits a proof of inconsistency, similar to the one developed in 2.2, unless it provides good reasons in favor of the opacity of certain predicative contexts. After the publication of my paper, several important articles and books have appeared devoted to the re-examination of Meinong's theory, or to developing some theories I call "Neo-Meinongian," defending some basic intuitions of that author from a modern conceptual framework, and sometimes formalized with the aid of some conceptual resources of present-day logic.[27] The study of this recent bibliography has convinced me that the general conclusion I reached in my essay is not totally valid just as it was formulated. For reasons I shall expound elsewhere I consider especially important Castañeda's investigations on these issues. Castañeda has developed a Neo-Meinongian theory which contains T^{1}, and T^{1}' as a particular case, in connection with ordinary language. But even when the RS rule and the indiscernibility of identicals are within the theory, without assuming the opacity of contexts like those used in 2.2 (see, in n. 27, references to Castañeda and Rapaport papers, in which the proof and the way to avoid it are analyzed for some Neo-Meinongian theories), the theory would be able to reply, in a simple and elegant way, to every objection posed in part 2 of my paper, including my own proof of inconsistency. The way Castañeda's theory meets these difficulties is interesting, because his answers neither give ad hoc solutions (they are rather corollaries of general principles of the theory, particularly from a certain key distinction among different kinds of predication), nor imply obscure assumptions or controvertible restrictions to the principles of logic (as happens in the answers of Meinong that I was concerned with). The logical difficulties spelled out in my essay do not appear in Castañeda's theory (neither in other similar ones), this fact shows that the general conclusion I attempt to reach must be formulated more weakly. The proof of inconsistency, proposed in section 2.2, does not necessarily apply to every theory holding T^{1}' without adequately defending the opacity of certain contexts: there are refined logical resources that allow us to hold T^{1}' but prevent the proof of inconsistency, without adducing the context opacity. I do still maintain that the proof applies to Meinong's ontology, where there are not the theoretical resources that recent theories have, to meet the difficulty; due to this reason and to the increasing interest in Meinong recently, I think that it is a good time for the publication of this paper. However, the problem remains as to how precisely to determine the field of Meinongian or Neo-Meinongian theories affected by my proof of inconsistency, since my general thesis on this subject was disconfirmed.

I would like to point out that, although Castañeda's theory is resistant to the kind of objections posed in part 2 of my paper, this fact does not discharge it from logical difficulties. In a recent work, I pointed out a curious difficulty in Castañeda's theory, which concerns the operation of counting objects. Regarding this question, the theory has anti-intuitive consequences that disagree with our basic Arithmetical intuitions. I showed in my paper that the first solution Castañeda attempted in order to solve the difficulty was defective. (Castañeda whole-heartedly recognized the adequacy of my objection to his old answer and an alternative solution was immediately given by him. My article was presented at the III Simposio Internacional de Filosofía de la Universidad Nacional Autónoma de México. On that occasion, Castañeda outlined his new solution to this problem. My paper entitled "Objetos inexistentes," will be published together with Castañeda's reply in the Proceedings of the mentioned Symposium). My latter paper on Castañeda's answer is part of an extensive investigation, which I am presently developing in order to evaluate the recent Neo-Meinongian theories, but I have no conclusive answers to this question to date.

NOTES

* This paper was written in July 1971 but was not published until October 1972 in Cuadernos de Filosofía (Facultad de Filosofía y Letras, Universidad de Buenos Aires), no. 14 (1970), 327-344. After its publication, Professor Héctor-Neri Castañeda wrote several papers developing an interesting and sophisticated "neo-Meinongian" theory. These papers, some letters and several discussions with the

author in February 1979 gave me enough encouragement to proceed with the investigation of these problems. As a consequence of this, I have changed my mind in several respects and I am preparing other papers on neo-Meinongian approaches. Nevertheless, I think this is a good opportunity both to add a 1982 Postscript to show in which respects some of the theses developed in the present article have changed and to include some news about literature which has appeared during the last years on the subjects dealt with here. I would like to acknowledge my profound indebtedness to Professor Castañeda for his help and for the considerable enrichment that my understanding has acquired on the ontological questions I discuss here thanks to him. Finally, I would also like to acknowledge Lourdes Valdivia and Raymundo Morado, researchers in the IIF, UNAM, Mexico, for the translation of this paper into English.

1. Meinong expounds the main ideas of his theory of objects in his paper "The Theory of Objects," included in R. M. Chisholm, (ed.), Realism and the Background of Phenomenology (Glencoe, Illinois: The Free Press, 1960). A valuable exposition of Meinong's doctrines can also be found in J. N. Findlay's Meinong's Theory of Objects and Values (Oxford University Press, 2nd edition, 1963). Other aspects of Meinong's ontology are studied in Gustav Bergmann, Realism: A Critique of Brentano and Meinong (The University of Wisconsin Press, 1967).
2. This formulation is not an entirely adequate definition of the concept of definite description. An analysis of the problem and a definition of the concept can be found in section I of my article "Tres dificultades en la teoria de las descripciones de Bertrand Russell," published in Crítica, Revista Hispanoamericana de Filosofía 7 (1975), 69-104.
3. It is not completely certain that Meinong accepted an affirmation as general as T^o. On the other hand, there is no doubt that Meinong did accept a thesis like T^1. This is another reason why we shall focus our criticisms on T^1.
4. For Meinong not every relation is ideal. Some are actual and have existence. With respect to characteristics, Meinong's position is more complex. The "pure" characteristics, divorced from the particulars in which they are exemplified, subsist

intemporally. But, on the other hand, they exist "in their examples." With respect to these points, cf. Findlay, op. cit., ch. V.

5. For Meinong, "existent" and "subsistent" do not have analogous meanings to "real" and "ideal." The golden mountain, for example, is a real object, but it neither exists nor subsists. According to Meinong's use of these concepts, a real object is any object whose nature does not exclude it from existence: if it does not exist it is because of factual reasons. Ideal objects are those which can subsist but do not exist. Cf. Findlay, op. cit., pp. 115-6.
6. Cf. Meinong, op. cit., p. 80.
7. Ibid., p. 83.
8. Ryle expressed this opinion about Meinong. Cf. Oxford Magazine, 26, October 1933 (quoted in Findlay, op. cit., p. xiv).
9. Cf. Findlay, op. cit., p. 61.
10. Ibid., p. 45.
11. Cf. Meinong, op. cit., p. 82.
12. Ibid., p. 86.
13. This consequence can be more precisely formulated this way: it is accepted that, if "x is F" logically implies "x is G," then the affirmation "the F is G" is true.
14. Cf. Meinong, Ueber Moeglichkeit und Wahrscheinlichkeit (Leipzig: Barth, 1915), pp. 278 and 282, Gesammelte Abhandlungen, vol. II (Leipzig, Barth), pp. 493-4 (Cited by Findlay, op. cit., pp. 50, 105).
15. This is a part of his doctrine of the "objectives." Cf. Findlay, op. cit., cap. III.
16. Cf. Russell, Principles of Mathematics (New York: The Norton Library, 1964), p. 449.
17. The objections of Russell on which we shall comment here were expounded in his article "On Denoting"; the article is included in Feigl and Sellars, (eds.) Readings in Philosophical Analysis (New York; Appleton Century Crofts, 1949). A very clear exposition of the discussion between Russell and Meinong is found in chapter III of Thomas M. Simpson's Formas lógicas, realidad y significado (Buenos Aires: Eudeba, 1964). Quine has presented objections of another kind to ontological theories similar to those of Meinong, although these objections do not represent a proof of inconsistency. On this point, cf. Quine, "On What There Is," included in Quine, From a Logical Point of View

(Cambridge: Harvard University Press, 1953).

18. I shall follow Simpson's presentation (*op. cit.*, pp. 98-9) which avoids a certain difficulty in Russell's presentation.
19. Cf. Findlay, *op. cit.*, p. 104.
20. We say that a definite description is not contradictory when it is not constructed out of incompatible predicates. More exactly, a description of the form "The F" is not contradictory if the propositional function "x is F" is consistent.
21. In this step, the use of T^{1}' is indispensable. Without this assumption we would not be able to suppose that "The F" denotes an object and, therefore, we could not conclude that it denotes an object with property F. Without T^{1}' we could only conclude that, if "The F" denotes an object, such an object has the property F.
22. Cf. Anthony Kenny, "Descartes' Ontological Argument," included in Joseph Margolis (ed.), *Fact and Existence* (Oxford: Blackwell, 1969), p. 33.
23. Cf. Anthony Kenny, *op. cit.*, p. 34.
24. Wonfilio Trejo, "Russell: Descripción y existencia," *Crítica, Revista Hispanoamericana de Filosofía* 2 (1968), especially pp. 89 and 93.
25. The examples have been taken from "Reference and Modality," included in Quine's *From a Logical Point of View* (see note 17). The following considerations on referentially opaque contexts represent a compact synthesis of Quine's ideas expounded therein.
26. The word "object" is perhaps the only one not used under a common meaning in our formulation of T^{1}'. In fact, to give a "Meinongian" sense to thesis T^{1}' it is not correct to understand "object" in the same sense as "existent object" and it could be thought that both expressions usually mean the same thing. On the other hand, the other expressions of T^{1}' are understood with their usual meanings. The verb "to denote," for example, refers to the familiar relation between linguistic expressions and the entities named by them.
27. See Reinhardt Grossmann, *Meinong* (London: Routledge and Kegan Paul, 1974; Héctor-Neri Castañeda, "Thinking and the Structure of the World," *Philosophia* 4 (1974), 3-40, "Individuation and Non-Identity; A New Look," *American Philosophical Quarterly* 12 (1975), 131-40, "Identity and Sameness," *Philosophia* (1975), 121-50, "Philosophical Method and the Theory of Predication and Identity," *Nous*

12, (1978), 206-7, note 9. See also W. Rapaport's doctoral dissertation, Intentionality and the Structure of Existence, Indiana University, 1976, and Terence Parsons, Nonexistent Objects (Yale Univ. Press, 1980), Richard Routley, Exploring Meinong's Jungle and Beyond (Camberra: Australian National University, 1980), Karel Lambert's Meinong and the Principle of Independence (forthcoming, Cambridge University Press).

Eduardo A. Rabossi

MEANING, FORCE AND EXPLICIT PERFORMATIVES*

I. Let me start with some general and rather obvious remarks about Austin's doctrine of explicit performative sentences.

An explicit performative sentence is a sentence, say _p_, satisfying the following two conditions, (1) _p_ is of the form 'I X...,' where X is a verb in the first person singular, present, indicative, active, and (2), normally, the verb X is the name of the doing (the act, the action) one would be performing in uttering _p_ (cf. HTD, 32). Accordingly, when uttered in appropriate circumstances,

(1) I (hereby) order you to go,
(2) I (hereby) warn you that aunt Matilda's cat is dangerous, and
(3) I (hereby) promise to pay you back one hundred pounds,

are explicit performative sentences.

The study of this sort of sentences is somehow basic to Austin's approach to language: a place _prima facie_ well-deserved. According to Austin, explicit performative sentences masquerade, mostly, as statements of fact, i.e., as having a descriptive or constative function. Neither grammarians nor philosophers have managed to see through this 'disguise.' But a perspicuous look at their functioning reveals a different and, indeed, peculiar role. True, sentences like (1)-(3) are indicative sentences. But the _grammatical form_ does not imply that they report, describe or inform about an actual state of affairs. Their being indicative merely means that they are coined in the English indicative grammatical form. Austin's conclusion is well-known: to utter an explicit performative sentence, say 'I X...,' is not to describe my doing X nor to state that I am doing X; it is to X. In Austin's own words:

J. J. E. Gracia et al. (eds.), Philosophical Analysis in Latin America, 141–163.

> ...it seems clear that to utter the sentence (in, of course, the appropriate circumstances) is not to describe my doing of what I should in so uttering to be doing or to state that I am doing it: it is to do it. None of the utterances cited is either true or false. (HTD, 6).

That explicit performative sentences are neither true nor false is taken as obvious and in no need of argument.

Three points call for attention here. First, the family of relevant cases, in which saying something is doing something (in an interesting sense of 'doing something'), is wider and richer than the class of explicit performative sentences. For example, consider such utterances as,

(4) Guilty (uttered by the spokesman of a jury),
(5) Out (uttered by an umpire in a baseball game).

Obviously, (4) and (5) differ from explicit performative sentences on mere grammatical grounds. But, more attractively, they also differ in the way they presuppose conventions and in the nature of the conventions thus presupposed. Take the case of 'Guilty.' It is clear that uttering 'Guilty' is giving a verdict insofar as this linguistic feat is part of a system of non-linguistic rules, conventions or practices which constitute saying 'Guilty' as the act of giving a verdict. In other words, it is by convention--in fact, by a set of non-linguistic conventions, e.g., those concerning criminal procedure--that to say 'Guilty' in appropriate circumstances constitutes giving a verdict: the utterance plays a precise role in a convention-constituted procedure. These sorts of cases, although related to, are, nevertheless, different from, explicit performative sentences. (Cf. Warnock [1971], 70-74).

Moreover, there are cases of prima facie explicit performative sentences, that do not fit exactly into the conditions so far introduced. Take, for instance,

(7) Passengers are warned to cross the track by the bridge only (HTD, 57).
(8) Warning. No smoking in this area;

and

(9) Notice is hereby given that trespassers will be prosecuted (HTD, 57).

It is clear that (7) is in the third person plural, that in (8) and (9) the verb is 'impersonal,' and that in (9) the verb is in the passive voice. We might deal with this sort of cases in an Austinian way, that is, as showing that person and voice are not essential to explicit performative sentences. But we prefer to adopt here a different approach, we will treat them as somehow typical cases which may be considered together, with due allowances, with those exemplified in (1)-(3).

In short, there are cases in which _saying_ something is doing something, that are _different_, in some respects, from explicit performative sentences. However, this fact does not prevent the expression 'explicit performative sentence' from picking out a class of utterances that is adequately neat and able to be contrasted with the fairly large class of non-performative ones.

Second, it is difficult to deny the originality of Austin's claim that descriptive performative sentences are peculiar in not being descriptions of my doing or statements that I am doing 'what I should be said in so uttering to be doing.' But it is important to notice that this claim is associated to an analytic strategy that is far from being original. In fact, Austin's claim may be interpreted as a new application of the time-honored distinction between _grammatical form_ and _logical form_. Austin's starting point is a class of utterances in which the grammatical form--the indicative mood, furnished with some kind of philosophical import--wrongly suggests a descriptive or constative function. But grammar--as philosophers have learned so often--is not a reliable guide, as far as philosophy is concerned. Sentences like (1)-(3) have a peculiar function of their own. And consequently, they display a typical _logical form_ (in a broad sense of this obscure and fruitful expression). It is not easy to discern such form, but a possible guess is,

(9) P (...),

where 'P' is a variable ranging over verbal phrases with a first person personal pronoun followed by a performative verb (first person singular, present, indicative, active), and '(...)' is a blank to be filled in by some sort of content corresponding to what counts--in a sense to be discussed next--as an _implicit_ or _primary_ performative sentence. Such proposals as 'P-radicals' (M. Furberg), 'neustics' (R. Hare), 'illocutionary force indicating devices' (J. Searle) and 'performative prefixes' (J. Cohen, some linguists), point to a parsing of explicit performative sentences which is similar to that suggested in (9). There are some hints in Austin that favor this interpretation. Discussing whether utterances with performative features are meant to be true or false, Austin points out,

> ...we could distinguish the performative opening part (I state that) which makes clear how the utterance is to be taken, that it is a statement (as distinct from a prediction, and c.), from the bit in that clause which is required to be true or false. However, there are many uses which... we are not able to split into two parts in this way, even though the utterance seems to have some sort of explicit performative in it, thus 'I liken x to y,' 'I analyze x as y.'" (HTD, 90).

In spite of exceptions, this sort of parsing seems to be presupposed in Austin's analysis of explicit _performative verbs_, that is, verbs which make explicit the illocutionary force of an utterance or, alternatively, what illocutionary act is performed in issuing the utterance. (Correspondingly, the logical form of a non-explicit performative sentence will be 'P*(...).' The star marks the absence of an (explicit) performative prefix. Recently, some linguists have argued that every indicative, imperative and interrogative sentence has a performative prefix in deep structure. (This is a most attractive thesis that deserves to be mentioned in this context).

This way of interpreting Austin and/or the proposal that follows from it, may be challenged--and, in fact has been challenged, on different grounds. In what follows we will accept the view that Austin's doctrine of explicit performative sentences amounts to the proposal of a peculiar _logical form_ for such sentences, and will take (9) as rendering such form. We will have something to say about this suggestion later on.

Third, explicit performative sentences are approached by Austin from a point of view that may be properly called 'dynamic.' Not only explicit performatives are historically a latter development from mere primary utterances (HTD, 71), but they are distinguished, in concrete speech-situations, from _implicit_ and _primary_ (or _primitive_) performatives. It is not easy to disentangle the logic of this distinction, but a possible 'reading' may be this. Suppose a person S issues in a conversation the utterance,

(10) Go.

In doing so, S may achieve practically the same as he achieves by the utterance of (1). No doubt, if someone were to ask what S did (in uttering (10)), a proper answer would be that S ordered someone to go (cf. HTD, 32). In these sorts of cases we will say that (10) is an _implicit_ performative, and (1) the corresponding _explicit_ performative sentence.

But as far as the mere utterance of (10) is concerned, it is always left uncertain whether S is ordering someone to go or, say, advising someone to go. Or there may be nothing in the context of the utterance of (10) by which we can decide whether it is a performative at all (cf. HTD, 33). In other words, given the utterance of (10), or of a similar expression, we may have to decide about its possible force and, in case of an affirmative answer, about the specific force meant by the speaker (i.e., whether S meant (10) as an order, an advice, or what not). In this sort of cases we will say that (10) is a _primary_ performative (or a _primitive_ performative), and (1) the corresponding _explicit_ performative sentence.

Austin's way of drawing the distinction between _implicit_ and _primary_ performatives may suggest, wrongly, that he is willing to distinguish between (a) successful non-explicit performatives, and (b) non-explicit performatives likely to be successful. But in fact, he is suggesting a more important distinction, namely, the distinction between the smooth functioning of a non-explicit performative (that is, _implicit_ performatives), on the one hand, and the potentiality of forces inherent to any non-explicit performative (_primary_ performatives), on the other (granted, of course, that the utterance is a performative at all). And this latter sort of performative are intimately related to Austin's way of introducing the distinction between _meaning_ and _force_. It is important to notice that Austin speaks for the first time of such distinction on occasion of contrasting _implicit_ and _explicit_ performatives (HTD, 32-33). Later, after dwelling on the expression 'explicit performative' (as opposed to 'primary performative'), he concludes,

> Language as such and in its primitive stages is not precise, and it is also not, in our sense, explicit: precision in language makes it clearer what is being said--_its meaning_: explicitness, in our sense, makes clearer the _force_ of the utterances, or 'how...the utterance is to be taken.' (HTD, 73).

Austin insists that 'making explicit' is not the same as describing or stating (in a philosopher's sense) what I am doing, but making plain "how the action is to be taken or understood, what the action is." (HTD, 70). Finally, once the locutionary/illocutionary distinction has been introduced (remember that meaning and force are associated to locutionary and illocutionary acts, respectively; and that Austin suggests that when we have an explicit performative we also have an illocutionary act), Austin remarks,

> ...the whole apparatus of 'explicit performatives'... serves to obviate disagreements as to the description of illocutionary acts. It is much harder in fact to obviate disagreements as to the description of 'locutionary acts.'" (HTD, 114, footnote 1).

In short, the notion of 'explicit performative sentence' is not a _static_ notion produced out of mere taxonomic delight. More importantly, it is a 'dynamic' instrument, both practically and theoretically, that plays a basic role in a doctrine of language such as Austin's.

II. The notion of 'explicit performative sentence' has been a favorite target for criticism. Enemies and partisans of

Austin's views about language have found that explicit performatives are an excellent platform for launching attacks on such crucial notions as performative/constative, locutionary/illocutionary, meaning/force, illocutionary/perlocutionary, conventional/nonconventional, and consequently, on the concepts involved. A good survey of such discussions will be equivalent to a revision of Austin's doctrine and to an assessment of its developments and refinements.

In what follows I intend to pursue a modest task. I will pick out two _types_ of criticism concerning explicit performative sentences, and will try to elucidate their content and to assess their critical import. The first _type_ of criticism ('CR I,' in what follows) challenges Austin's way of drawing the meaning/force and the locutionary/illocutionary distinctions, and points out difficulties arising when this conceptual apparatus is applied to explicit performative sentences. The second _type_ of criticism ('CR II,' in what follows) dwells on Austin's point about the non-statemental character of explicit performative sentences, and draws conclusions about the 'verification' of explicit performative sentences, about the meaning of performative verbs and, even, about the logical form of such sentences. Both _types_ of criticism are variously stated by philosophers. There is also a variety of views concerning their critical weight. This lack of agreement is, of course, a revealing fact. But it cannot obscure the real import of CR I and CR II.

III. Consider the primary performative, uttered by S,

(11) The eleven-thirty train is late.

Suppose someone asks what did S mean by (11) (or, what is the meaning of (11), or what is the meaning of what S said when uttering (11), etc.). According to Austin, a correct answer will consist in a description of the locutionary act performed by S. Such answer will clear up, if necessary, syntactic or semantic ambiguities, in order to describe the intended interpretation of the 'linguistic' meaning of (11). And it will also clear up, if necessary, the reference of the demonstrative elements of (11). However, this answer is far from being the 'whole' answer, as far as the original question is concerned. And it is not the whole answer because we do not know yet whether S meant (11) as a warning, an advice, a report, etc. In other words, we do not know yet what sort of illocutionary act was performed by S in uttering (11), what force--as different from 'linguistic' and 'referential' meaning--S intended (11) to have. So, according to Austin, a proper answer will include also a description of the illocutionary act performed thereof.

Consider now the explicit performative sentence, uttered by S,

> (12) I (hereby) warn you that the eleven-thirty train is late.

What would be a proper answer to the question about the meaning of (12) (or about what S meant by (12), etc.)? Notice that in this case the distinction between locutionary and illocutionary acts—and, consequently, the distinction between meaning and force--is far from being as neat as in (11). Once we know the 'linguistic' and the 'referential' meaning of (12), it is difficult to identify something additional to that knowledge--something about the way in which what is said is intended to be taken--that should count as relevant additional knowledge about what S meant by (12). In general terms, force is made explicit by means of the formula 'I X...,' which is a linguistic item. To grasp the intended meaning of an instance of such formula we only need to rely on our knowledge of the 'linguistic' and 'referential' meaning of its linguistic elements. But, if this is the case, the basic distinctions between locutionary and illocutionary acts and between meaning and force seem to collapse, or, at least, seem to be in need of revision. Surprisingly enough, the difficulty is detected in the highly privileged case of explicit performative sentences.

This is a possible way--and a fairly neutral one--of introducing CR I.

It is important to notice that there are variants of CR I. J. Cohen ([1965]), for instance, argues (among some other things) that: (a) in explicit performative sentences the personal pronouns and the verb of the performative prefix, 'I X...,' must have a reference and a sense, respectively; (b) if the locutionary act is defined in terms of the phonetic and phatic acts, the locutionary act of uttering, say, (11) must be different from the act of uttering (12); (c) if expressions like 'I protest' have a meaning as well as a force when uttered alone, they hardly would lose their meaning when subordinate clauses are added to them; and (d) if an utterance like (11) gives a warning that is rendered explicit by (12), and if the warning is part of the meaning of (12), it is unreasonable to suppose that the warning is also part of the meaning of (11).

Some of these arguments are misguided. Take (c). It is difficult to imagine a situation in which the expressions 'I protest,' 'I warn,' 'I order,' etc., would be "uttered alone," unless they were used as short formulations of sentences like 'I protest the decision of the umpire,' 'I warn you that the cat is dangerous,' 'I order you to go,' etc. But this is not the sort of cases Cohen is in need of to make his point valid. Actually, the only way in which we might give a content to the claim that 'I protest,' etc., may be "uttered alone" is by way of a convention-constituted procedure in which, somehow, the issuing of 'I protest' constitutes the act of, say, expressing a protest (Imagine regulations concerning debates, for instance). But this peculiar use of 'I protest' excludes the addition of subordinate clauses. In short, Cohen's assumption about relevant cases in which 'I X' is "uttered alone," is void: either 'I X' is uttered for short of 'I X...' or 'I X' is not, strictly speaking, an explicit performative sentence. Take now (d). Even granting that there is a sense according to which warning is part of the meaning of (12), it is incorrect to suggest that, in some obscure way, the warning is also a part of the meaning of (11). According to Austin's doctrine, a primary performative is an expression _meant with_ a certain force. Such force _may_ be rendered explicit by the use of the corresponding explicit performative prefix. But making explicit the force of a primary performative does not consist in displaying _something_ hidden or implicit in its _meaning_. There is nothing in the meaning of the explicit performative to be displayed in that way. (In fact, the meaning of the primary performative is restricted to the linguistic items that would be included in a description of the locutionary act performed). So, Cohen's argument (d) seems to be groundless.

Arguments (a) and (b) merit a different assessment. Although far from being conclusive they are the sort of arguments that compel defenders of Austin's doctrine to be explicit about some important topics. Are we to say that in spite of grammatical appearances, 'I' and 'you' do not have a referring role in (12)? How about the meaning of 'warn'? Which is, exactly, the content of the locutionary act performed in uttering (12)? These are questions that spring out of Cohen's arguments (a) and (b). But, is it necessary to accept _a priori_ that these questions are _relevant_ questions? I think that the proper move for an Austinian philosopher is to object to the assumption that they are relevant. And the best way to show, briefly, their irrelevance, is by reminding the critic of some basic tenets of Austin's proposal.

Austin's doctrine is, basically, an attempt to systematize the _forces_ of linguistic utterances. The importance of Austin's views about explicit performative sentences lies, precisely, in his suggestion about the peculiar nature of performative prefixes as force-indicating devices, 'I warn you that,' in (12), allows the hearer to understand how the locution 'The eleven-thirty train is late' is to be taken. It is a linguistic device that helps the hearer to get a correct uptake of the locution. But performative prefixes are more than mere markers or signals to help linguistic transactions. Warnings--as well as promises, advices, predictions, reports, bettings, etc.--are _doings_ that call for a special _pendant_ at the level of _sayings_. When used in the first person, present, indicative, active, the issuing of such verbs as 'warn' _constitutes_ the action that that verb _names_ or _describes_ when used in the infinitive, or in other persons and times. This is, of course, a peculiar linguistic function: there are _sayings_ that are _doings_ (in an interesting sense of 'doings'). This function is not posited out of transcendental arguments. It comes out of a fair description of how language works. True, 'I,' 'you,' 'warn' may be viewed as _grammatical_ items performing a _grammatical_ function. But when (12) is analyzed from a _philosophical_ point of view, the 'anomalous' case of explicit performative sentences--'anomalous' insofar they are indicative sentences--is put into focus. And there is no point, then, in asking the sort of questions suggested by Cohen's arguments (a) and (b). As there might be no point in asking

Russell about the reference and sense of the grammatical subject of 'The present king of France is bald,' once the logical form of the sentence has been displayed. In fact, Russell's theory of descriptions and Austin's doctrine of explicit performative sentences are philosophical attempts to exhibit logical structures lying behind grammatical forms I am not suggesting, of course, that this parallel may be carried further than this. But I think that even restricted to this minimal point, it is a revealing parallel that helps to see a basic outcome of Austin's doctrine (cf. Schiffer [1972]), 104-105). In short, Cohen's arguments (a) and (b) give rise, interestingly enough, to these sorts of comments. They are not real challenges to Austin's position, which must be assessed on grounds different from those suggested by Cohen. Actually, CR II is closer to a relevant criticism in this respect.

An alternative variant of CR I is found in J. Searle's important discussion of Austin's doctrine (cf. Searle [1968]). Searle points out difficulties concerning explicit performative sentences, but his conclusions are far from being as drastic and negative as Cohen's. Searle does not think, for instance, that the locutionary/illocutionary and the meaning/force distinctions are bogus ones. Both distinctions play a crucial role in his refinements on Austin's doctrine. Searle's qualms about explicit performative sentences concern the way in which the existence of such sentences prevents--so he thinks--a "general" functioning of the distinction between locutionary and illocutionary acts ("general" in the sense of "marking off two mutually exclusive classes of acts"). According to Searle, meaning determines (at least one) illocutionary force of the utterance of the explicit performative sentence. The utterance of the sentence with certain meaning has certain particular force. But if so, the description of the act as a (happily) performed locutionary act, is a description of the illocutionary act involved; "they are one and the same act." Searle argues that when meaning uniquely determines a particular force, there are not two different acts (namely, a locutionary and an illocutionary act) but two different labels for the same act (a certain illocutionary act). In short, in the case of explicit performative sentences it is impossible to abstract meaning from force. (Such an attempt would be like "abstracting unmarried men from bachelors.") And this is equivalent to asserting that _some_ locutionary acts are illocutionary acts. Conclusion: the locutionary/illocutionary distinction is not completely general.

Searle's argument is far from being acceptable because it relies heavily on a doubtful rendering of Austin's views about the _description_ of linguistic utterances. This is an important point, and I shall dwell on it for a moment.

When S utters (11) his uttering may be described as the performance of a certain locutionary act, i.e., the act of _saying_ 'The eleven-thirty train is late,' and it also may be described as the performance of a certain illocutionary act, i.e., the act of warning that the eleven-thirty train is late. In order to describe S's uttering as 'the locutionary act of saying...,' we must take into account the _grammatical structure_, the _lexical components_ and the '_linguistic' and 'demonstrative_' meaning attached to them. The description of S's uttering as 'the illocutionary act of X-ing' presupposes taking into account _S's intention_, as expressed, somehow, in tone of voice, emphasis, etc., the _circumstances of the utterance_, and the _conditions for the successful performance of the illocutionary act involved_. (In the case of warning, some such conditions are (a) that what the addressee is warned about is (for him) important, (b) that the effects (consequences) of what the addressee is warned about are (for him) undesirable, (c) that the addressee is in a position to avoid (correct, change) what he is warned about, (d) that the person who issues the warning is in a position to state, if challenged, his reasons for issuing the warning, and the consequences that would follow if the addressee pays no attention to what he is warned about, etc.). The features which we must take into account in order to describe an utterance as 'the locutionary act such-and-such' or 'the illocutionary act such-and-such,' I dub 'conditions for the description of the utterance.' It is clear that the conditions for the description of the utterance as a locutionary act are of a _different_ sort (of a different nature) than the conditions for the description of an utterance as an illocutionary act.

Well then, how many acts does S perform when uttering (11)? Clearly, two acts, namely, the locutionary act of _saying_ 'The eleven-thirty train is late' and the illocutionary act of warning that the eleven-thirty train is late. These two acts are not independent from one another, but intimately related in a peculiar way: in performing one (the locutionary act) S performs the other (the illocutionary act). Though peculiar, this is not so extraordinary as it may appear. (Suppose we see soldiers across the fields. We may describe what we see as 'Army Z crossing the border' or as 'Army Z starting hostilities.' And we may also say that in performing one act (crossing the border) Army Z was performing the other (starting hostilities). Notice how different are the conditions for the description of both acts. There are, of course, many examples of this sort).

Suppose S utters (12), and we are asked to describe the act(s) performed by S in so doing. If we keep in mind the thesis that Austin's doctrine of explicit performative sentences amounts to displaying their logical form, we may say that on the occasion of uttering (12) S performed the locutionary act of saying 'The eleven-thirty train is late' (remember the suggestion that in explicit performative sentences the locutionary act is restricted to the primary performative involved; the performative prefix being a linguistic device for showing how the utterance is to be taken, i.e., for exhibiting the force of the utterance), and the illocutionary act of warning that the eleven-thirty train is late. Prima facie, there is no difference in the descriptions of (11) and (12). But an important difference exists at the level of the conditions for description. The conditions for the description of the illocutionary acts involved in the utterance of explicit performative sentences are restricted to what we called 'conditions for the successful performance of the act.' The conditions concerning the intention of the speaker (as expressed in tone of voice, emphasis, etc.) and the conditions concerning the circumstances of the utterance, are dropped out because unnecessary:

> ...the performative rules out equivocation and keeps the performance fixed, relatively. (HTD, 76).

This partial dropping out of some of the conditions for the description of the illocutionary act explains, somehow, Searle's proneness to think that in the case of explicit performative sentences the description of the locutionary act and the description of the illocutionary act are the same, and to suggest as an explanatory principle: "force is part of the meaning." It is true that the conditions for the description of illocutionary acts involved in explicit performative sentences are the rather 'formal' ones concerning the 'successful performance of the act.' This does not mean that there are no conditions at all; nor does it amount to a confused way of introducing the distinction between <u>trying</u> and <u>succeeding</u> in the performance of an illocutionary act. Actually, the very possibility of drawing this distinction presupposes the existence of such conditions for the successful performance of the act. In other words, the description of an utterance as 'trying to perform such-and-such illocutionary act' or as 'a successful performance of such-and-such illocutionary act' is possible because there are conditions for the description of such utterance as an illocutionary act of a certain <u>sort</u>. It is easy to overlook this conceptual priority. And I think Searle has overlooked it.

Notice that it follows from Searle's analysis that the expression,

> (13) <u>In</u> saying 'I (hereby) warn you that the eleven-thirty train is late', S was warning,

is void of any informative content. I am arguing that (13) is not thus void. Consider, for instance, the following utterances of the '<u>in</u> formula,' that resemble (13) only in its linguistic dressing,

> (14) <u>In</u> saying 'I (hereby) warn you that the lack of a daily target practice will affect your marksmanship,' S was warning. (The addressee is a convict, a former marksman; an Olympic champion).

> (15) <u>In</u> saying 'I (hereby) warn you about your wife's views on abortion,' S was warning. (S is not able to state what her views on abortion are, nor how they differ from her husband's).

> (16) <u>In</u> saying 'I (hereby) warn you that the ice is thin near the coast of Greenland,' S was warning. (S addresses the sentence to a fellow executive of the car industry, while discussing

business in Paraguay).

In none of these cases is the '_in_-formula' true. The very point of warning seems to be absent (in several ways) from (14)-(16). But what prevents (13) from a similar fate? In fact, nothing in particular. It might well be the case that (13) was a false assertion about the force of S's utterance. And this is important to realize: (13) _may_ go wrong. If this is the case, (13) is not _void_. Now, a Searlean philosopher might argue at this point that cases like (14)-(16) are ruled out of consideration by a restriction concerning the _seriousness_ of the utterances of explicit performative sentences ((14)--(16) are taken as including non-serious utterances). But this is an _ad hoc_ restriction lacking theoretical interest. Leaving aside the vague content of 'a serious utterance of,' as used by Searle, the restriction about 'seriousness' does not hinder the possibility of cases like (14)-(16): it only rules them out of consideration. And this, I suggest, is a most doubtful strategy. The Searlean philosopher might argue, alternatively, that what is really at stake here is the distinction between _successful_ and _unsuccessful_ performances of the act, and that this is a much less interesting distinction than the locutionary/illocutionary one. But, as I have suggested before, the drawing of the distinction between _successful_ and _unsuccessful_ utterances presupposes the existence of the 'conditions for the successful performance of the illocutionary act (involved).' And these conditions are essential for the description of an act as an illocutionary act (of a certain sort). It follows from this that the description of an act as an illocutionary act can never be the same as the description of an act as an locutionary one.

In short, Searle's arguments are far from being conclusive. It is doubtful that in the utterance of explicit performative sentences, "the description of the act as a locutionary act _is_ a description of the illocutionary act involved." I have been trying to show that the conditions for the descriptions of locutionary and illocutionary acts are different and, consequently, do not conflate even in the extreme case of explicit performatives. Further, it is not true that in such cases the locutionary and the illocutionary acts "are one and the same act." If the conditions for the description of locutionary and illocutionary acts are different, they cannot identify "one and the same act." Finally, there seems to be something queer in Searle's point about force being part of meaning (a point suggested already by Cohen). The point is made when stating CR I. But this is objectionable: one of the things CR I is suppposed to prove is, precisely, that force is part of meaning.

The foregoing discussion of these two poignant variants of CR I is far from exhausting their basic claims. It suggests, however, some fairly acceptable points about the way in which such claims might be tackled. This is a most interesting outcome: there are adequate grounds for doubting that CR I makes definite objections to Austin's doctrine of explicit performative sentences; that is, to Austin's contentions about the peculiar role of the performative prefix, and the way in which saying 'I X...' is X-ing. The question now is whether CR II raises stronger claims against Austin's approach.

IV. CR II challenges a basic feature attributed by Austin to explicit performative sentences, viz., that such sentences are neither true nor false. Austin takes this point as obvious and in no need of argument, an attitude that is the natural outcome of his views about the way in which explicit performative sentences behave. In uttering, say, (12), S does not state that he is warning, nor does he describe his act of warning; in uttering (12), S makes explicit his warning and, in saying 'I warn you that...,' S performs (in appropriate circumstances) the act of warning. This way of viewing the function of explicit performative sentences is basic to the claim that they are 'masqueraders,' that is, sentences that wrongly suggest, through their grammatical form, a statemental character. And this point leads--as suggested in Sections I and III--to the thesis that Austin's doctrine is concerned with the logical form of explicit performative sentences.

The variants of CR II are rather large in number, and couched in fairly elaborate terms. However, in spite of notorious differences defenders of CR II standardly argue that (a) explicit performative sentences are statements and, hence, true or false utterances, and (b) their alleged peculiarities may be explained away by recourse to features commonly shared by sentences other than canonical performatives. Thus, we are told that explicit performative sentences are sentences verifiable by their use (E. J. Lemmon [1963]), or statements of intention _and_ assertions about the speaker's action (P. S. Ardall [1968]), or plain statements (expressions that may be assessed as true or false, or as a lie) (J. Heal [1974]), or sentences describing the act in the grammatical _perfective_ aspect (instead of the grammatical _imperfective_ aspect) (D. Wiggins [1971]), or constatives (in _logical_ form) not uttered, as explicit performative sentences, with their full conventional force (S. Schiffer [1972]), or perfectly _ordinary_ first person present indicative sentences (G. Warnock [1970]), etc.

In what follows I shall restrict myself to a brief discussion of Warnock's sensitive and, unsophisticated statement of CR II. I will not claim that my critical remarks, if valid against Warnock, would apply _mutatis mutandi_ to any variant whatever of CR II. But, since Warnock's moves are very typical of CR II, I will claim that my criticism suggests a way of dealing with the basic mechanism presupposed by the variants of CR II.

Warnock starts by drawing a sharp distinction between 'performative utterances' (remember examples (1) and (2) of Section I), and 'explicit performative sentences.' Concerning 'performative utterances' Warnock points out that

> ...there is a class of conventional acts which can be, or normally are, or even necessarily are, done _by_ the utterance of certain conventionally prescribed words... the utterance is 'operative' in a special way... There is however, no special verbal form that such utterances have to take. ([1970], 88).

Explicit performative sentence differ from performative utterances in two major respects: (a) explicit performatives are all of a certain verbal form, and (b) it is not necessarily, nor even often, by _convention_ that to issue the _explicit_ performance sentence "is to do the thing." Warnock insists rightly I think, that 'performative utterances' are a sub-class of conventional _doings_.

> ...which have in common that, in virtue of non-linguistic conventions, to issue them (happily) _counts_ as _doing_ this or that. ([1970], 74).

Now, the idea of _saying_ something which, by _convention_, counts as _doing_ something, differs from the idea of _making explicit_ what one is doing with the help of a performative verb. Warnock thinks that Austin was unclear about the differences that exist between both ideas, and points out, further, that Austin took explicit performative sentences as 'masqueraders' because he followed the "model" of 'performative utterances.' But explicit performative sentences--Warnock asserts--are not 'masqueraders.'

> ...they are to be construed exactly as their form of 'grammar' suggests that they should be,... conventions do not (necessarily anyway) come in at all here... in saying, for example, 'I advise you to resign,' I do indeed make it explicitly clear that I am offering you advice, but... I do so just by saying, truly or falsely, that I do. ([1970], 81).

In short, explicit performative sentences are to be construed as perfectly _ordinary_ first person present indicatives.

Warnock's main contentions may be summarized as follows: 1. Austin presupposes the general principle that _to say that_ one does something is not, and cannot be, _to do_ it. The principle applies in those cases where what one does, and whether one does it, is _independent_ of one's saying that one does, or of saying anything at all. Now, in the case of explicit performative sentences, we are dealing with things people _do in_ saying things, "their doing of which is therefore not independent of what they say." The principle fails to apply here: what one says when one utters an explicit performative sentences (namely, _that_ one X-es) is _made true_ by the fact that one says it. Explicit performative sentences make themselves true. 2. It might be objected that if I X _in_ saying _that_ I X, what I say "would substantiate itself and _could_ not be false." But this objection may be met by arguing, (a) that "there is nothing in principle vicious in the idea of a proposition which, while it can be true or false, can't be falsely asserted, or rather, can't be falsely asserted by a particular person," or (b) that since saying 'I X' is not a sufficient element in X-ing, "it is possible to say that one X-es when one does not, just as--or anyway, somewhat as--it is possible to say that one smokes when one does not." This is not the case of, say, a false promise, "in which case one does not promise, not intending to perform." (Warnock is inclined to accept b). 3. It might be objected also that if the explicit performative sentence 'I X' were about what I do, it should be of the form of 'I am X-ing'; and the present continuous is not the verbal form for explicit performatives. The objection is met by arguing: a. that as a matter of fact, explicit performatives occasionally are in the present continuous, and (b) that one resorts to the present continuous when the doings one is engaged in extend beyond the natural boundaries of one's saying. With explicit performative sentences, "one's doing the thing coincides with, does not temporally over-spread, one's saying that one does it." 4. What makes it the case that in saying (happily) 'I promise' or 'I advise you to...,' I promise or I advise, is not a convention in virtue of which to speak so counts as or constitutes promising or advising, but simply the standard, normal <u>meaning</u> of the words I utter. In order to grasp what I do in saying 'I warn you that...,' you need no equipment beyond the understanding of English. 5. Explicit performative sentences are peculiar: "they have the peculiarity that, since in these cases what the speaker says that he does is something that is done in speaking, and indeed is in fact done by himself (if all goes well) in saying the very thing he says, the truth-value of what he says is involved... in a decidedly unusual way with the fact that he says it."

The notion of 'convention' plays a crucial role in Warnock's discussion. He acknowledges a strong sense of 'convention' (convention^1, in what follows) according to which conventions^1 = non-linguistic conventions, i.e., sets of rules, legal provisions, officially recognized practices, etc. Performative utterances presuppose conventions^1. It is by convention^1 (or, better, by a set of conventions^1) that, say, to utter 'Guilty,' in appropriate circumstances, is giving a verdict. It is clear that conventions^1 are not involved in explicit performative sentences, and that Austin took conventions^1 as a "model" for his doctrine about *sayings* that are *doings*. Warnock acknowledges a weaker sense of 'convention' ('convention^2,' in what follows) according to which conventions^2 = *linguistic* conventions, i.e., conventions that essentially enter into all linguistic utterances as such (conventions regulating the meaning of linguistic items; conventions involved in merely speaking a language). Explicit performative sentences presuppose conventions^2 (a feature that they share, democratically, with every sentence of the language). Warnock asserts, further, that "illocutionary forces are not conventionally constituted--which is why they can... be distinguished from meanings." ([1970], 76). Now, Warnock is right in saying this, because 'conventionally constituted' refers, in this context to conventions^1. But, what follows from this? It follows, obviously, that conventions^1 are not involved in explicit performative utterances. It does *not* follow--as Warnock thinks it does--that there are no conventions^2 in virtue of which to say 'I X...' *constitutes* X-ing. And this is a basic point. There is nothing wrong with the idea of a full-fledged language including some sort of conventions^2 to regulate, somehow, *sayings* that are *doings*. Furthermore, there is nothing unusual in the suggestion that a living language may produce devices to make explicit the forces of the utterances, and that these devices may be conventionally^2 regulated in such a way that uttering them, in appropriate circumstances, *constitutes* a doing of a certain sort. Actually, the forces of the utterances being what they are, this is the sort of device we may expect to have in a language. Two objections (at least) may be raised at this point. First, we might be urged to produce clear examples of conventions^2 by which *saying* 'I X...' constitutes X-ing. Second, we might be charged with confusing conventions^2 (i.e., regulations of the *meaning* of performative verbs) with some alleged conventions^2 (obscurely) involved in the utterances of explicit performative sentences. These objections merit rather simply answers. Notice, first, that once it is granted that linguistic conventions are far from being precise, clear-cut, neat regulations concerning the uses of words, it is not difficult to "reconstruct" the eventual content of those conventions^2 which regulate the use of the devices for making explicit the force of the utterances. Secondly, there is no point in suggesting that the "reconstruction" achieved is only of the *meaning* (as opposed to *force*) of such devices. The crucial point here is that *force* is a dimension of <u>meaning</u> (in a broad sense of 'meaning'), and that it differs, as such, from 'linguistic and 'referential' meaning (this is a narrower sense of 'meaning,' close to the Austinian 'sense' and 'reference'). It is true, then, that to grasp what I do in saying 'I warn you that...' you need no equipment beyond the understanding of English. Actually, what we mean by 'understanding of English' includes, among various things, being conversant with the dimension of meaning we are calling 'forces.' Hence it is not correct to assert that what makes it the case that in saying (happily) 'I X...,' I X, is not a convention [i.e., a convention^2] which constitutes X-ing, "but simply the standard, normal meaning of the words I utter." As pointed out before, the exclusion of conventions^2 as constitutive of X-ing, is based on a misunderstanding of the argument about conventions^1: the meaning (broad sense) of X includes conventions^2 which <u>constitute</u> X-ing.

The argument produced so far affects, if valid, contention 4. Contentions 1-3 are intimately related among themselves and call for different comments. According to Warnock, when one utters an explicit performative sentence, 'I X...,' (a) one says <u>that</u> one X-es, and (b), that one X-es is <u>made true</u> by the fact that one says it: "explicit performative sentences make themselves true." Now, the evidence produced by Warnock in support of this thesis is far from being convincing. To the objection that it follows from it that explicit performative sentences will substantiate themselves and <u>could</u> not be false, Warnock answers that saying 'I X...' is not a sufficient element in X-ing. He suggests, further, that we must take into account <u>felicity conditions</u>. But this is a strange thing to say. If the fact that saying <u>that</u> one X-es, makes true that one X-es, it is difficult to see what room is left for felicity conditions to play the role of <u>falsifying</u>, somehow, that one X-es. Again, to the known objection that the continuous present is not properly used in connection with explicit performatives, and the additional point that if 'I X...' were statemental, it should be of the form 'I am X-ing,' Warnock argues that in the case of explicit performative sentences one's doing the thing does not temporarily over-spread one's saying that one does it. That is why the present continuous is not properly used in those cases. But--Warnock suggests--this is not peculiar of explicit performative sentences. This is what happens, normally, when there is a concurrence between words and deeds. "Severing the jugular vein to the patient in the operating theatre I say concurrently and explanatorily to the assembled students: 'I sever the jugular vein.'" Now I agree that the concurrence between words and deeds explains why the present continuous is not available in some cases, but I disagree with the suggestion that there is nothing peculiar to explicit performatives in this respect. In fact, Warnock does not provide further examples to support his claim. And the only example he provides ('I sever the jugular vein') is a case of what Austin called 'suiting the action to the word' (HTD, 81-82); a case which is far from supplying the neat example of a statement about one's deeds that Warnock is in need of. When one suits the action to the words one does not produce a performative, but this does not mean that one produces a statement about one's action. The situations in which one suits the action to the words are variegated and not always, not even normally, allow for a statemental use of language.

There are some additional points, in Warnock's defense of CR II, that deserve attention. But I think that the discussion so far produced gives a fairly acceptable view of a typical statement of CR II, and a sample of objections one might raise against it.

V. What is the outcome of this discussion?

If the points made in Sections III and IV against some versions of CR I and CR II are correct--and I think they are--, there are *prima facie* reasons to think that Austin's doctrine about explicit performative sentences is basically correct. This is not a harmless thing to say in these days. As pointed out in Section II, enemies and partisans of Austin's doctrine have taken explicit performative sentences as a kind of target on which they have projected objections and alleged refinements. The outcome is far from being theoretically fruitful. No one denies the value of Austin's insight, but there is disagreement as to how to cope with the linguistic facts he pointed out so sharply.

To say that Austin's doctrine is basically correct is not to say that Austin's analysis is acceptable as it stands. In the discussion of CR I and CR II, and in the general comments of Section I, I have touched upon several questions that deserve further attention. The distinction between performative utterances and explicit performative sentences, the logical form of explicit performative sentences, the features of the description of utterances expressing force, and the nature of the linguistic conventions involved in such utterances, are, I take it, basic and important problems. I have said something about each of these problems. But I am conscious that this only amounts to a suggestion about the way in which such problems should be approached.

* Philosophica 19 (1977), 89-109.

BIBLIOGRAPHY

Ardall, P. S. "And That's a Promise," Philosophical Quarterly (1968), 225-237.

Austin, J. [HTD] How To Do Things With Words. Oxford: Clarendon Press, 1962.

Cohen, J. L. "Do Illocutionary Forces Exist?" Philosophical Quarterly (1964), 118-37.

Heal, J. "Explicit Performative Utterances and Statements," Philosophical Quarterly (1974), 106-121.

Lemmon, E. J. "On Sentences Verifiable by Their Use," Analysis (1962), 86-80.

Schiffer, S. Meaning. Oxford: Clarendon Press, 1972.

Searle, J. "Austin on Locutionary and Illocutionary Acts," Philosophical Review (1968), 405-424.

Warnock, G. "Some Types of Performative Utterances," in Essays on J. L. Austin. Oxford: Clarendon Press, 1970, pp. 69-89.

Wiggins, D. "On Sentence-Use, Word Sense and Difference of Word-Sense," in Semantics, ed. D. Steinberg and L. Jakobovitz. Cambridge: Cambridge University Press, 1971, pp. 14-34.

Enrique Villanueva

II. PHILOSOPHICAL ANALYSIS IN MEXICO

1. ANTECEDENTS

The present selection of analytical work in Mexico illustrates the variety of forms of philosophical analysis. It collects only works from the recent past which fall into the tradition started at the beginning of this century in the Anglo-Saxon countries.

Among the predecessors of philosophical analysis in Mexico it is necessary to include philosophers who lived during the colonial period--some of their works will be published soon by the Instituto de Investigaciones Filosoficas--as well as the so called positivists, who lived at the end of last century and beginning of the present one. All of them employed philosophical techniques that we could recognize as analytical, even if the topics with which they were concerned and the positions they took were in many occasions considerably different from those in the Anglo-American philosophical tradition.

The immediate antecedents to the practice of philosophical analysis in Mexico derive from the activity of two important philosophers: José Gaos and Eduardo García Máynez. José Gaos arrived in Mexico in 1938 together with many other Spanish philosophers and intellectuals. He had been a student of José Ortega y Gasset in Spain and of Edmund Husserl in Germany. Gaos taught continuously in Mexico from 1938 until his death in 1969, training a large number of pupils.

The way Gaos used to teach could be considered analytic in that he introduced philosophical questions with care and in detail, proceeding then to reconstruct arguments closely tied to the actual texts of the philosophers under examination.

J. J. E. Gracia et al. (eds.), Philosophical Analysis in Latin America, 165–171.

In that way he read and discussed texts such as Aristotle's Metaphysics, the Meditations from Descartes, Kant's Critique of Pure Reason, Heidegger's Being and Time, and many others. We can characterize Gaos' method as a kind of historical analysis. But that characterization does not mean that Gaos did not practice systematic analysis; he did this in some of his works, such as De la Filosofía and Del Hombre.

Something similar could be said about Eduardo García Máynez. His work, as represented by his Filosofía del Derecho and his Lógica Jurídica, for example, should be considered analytic under any reasonable interpretation of that term, even if he was never a part of the Anglo-Saxon tradition.

García Máynez's procedure involved the introduction of philosophical questions and the analysis of the most plausible views. He then evaluated these views, giving various arguments for and against them, and finally offered his own solution to the problem. This technique can be seen in his Teoría de los Tres Círculos, where he offers a solution to the problem of the validity and actuality of law. That thesis appears in his well known Introducción al Estudio del Derecho and in his most recent Filosofía del Derecho.

Moreover, in 1940 García Máynez founded the Centro de Estudios Filosóficos, which is the predecessor of the present Instituto de Investigaciones Filosóficas. He founded it under very difficult conditions, for he had no budget but only a minimum of furniture and his own library which he graciously donated to the Center. It was only in 1952, when the Universidad Autónoma de México moved into the present campus at the Ciudad Universitaria, that he received a complete floor in the Humanities Tower and the necessary financial and institutional support.

Some of Gaos' younger students later became fellows of the Instituto de Investigaciones Filosóficas. Among these, Adolfo García Díaz, Alejandro Rossi, Fernando Salmerón and Luis Villoro are worth mentioning.

2. DEVELOPMENT

These two previous factors, namely, the technique of teaching philosophy of José Gaos and Eduardo García Máynez and the creation of the Centro de Estudios Filosóficos, established the favorable conditions for the development of the philosophical analysis of Anglo-Saxon origin in Mexico. In 1960, Alejandro Rossi went to Oxford and returned the following year to teach courses in Logic, Philosophy of Language and Epistemology. He taught these subjects in a way that was completely new to students. In these courses Rossi dealt mainly with the philosophical theses of Frege, Russell, Strawson, Ayer and Austin. At this time Fernando Salmerón began to teach Ethics as well, while Luis Villoro was dealing with Descartes and Kant in other courses.

The result was that students began to learn the new analytic techniques and later on chose to do post-graduate work in European and American universities where these techniques were practiced. Fernando Salmerón was Director of the Instituto at the time and he sponsored a post-graduate program which made these studies possible.

The fashionable philosophy at those universities was known as Ordinary Language Philosophy. The typical questions examined in this context were: How do we use 'knowledge' in ordinary language? How do definite descriptions work in ordinary language?

In 1965, some scholarships were created for undergraduate students and a supervisory system was established so that the students returning from abroad could work on their dissertations while enjoying the Instituto's facilities. At the beginning those scholarships were few, but year after year they were increased until at present there is no set limit to them.

In 1967, the journal Crítica, Revista Hispanoamericana de Filosofía appeared. Its founders, Alejandro Rossi, Fernando Salmerón and Luis Villoro, expressed in the first issue their preference for a clear, argumentative philosophy backed up by research which would use expository techniques up to the level required by the complex problems of philosophy. The collaboration of the Argentinian philosopher, Thomas Moro Simpson, was of the greatest importance in this enterprise, especially at the beginning.

Before this time several analytic papers had appeared in Diánoia, the Instituto's yearly publication. But the founders of Crítica wanted more unity and frequency.

For the first few years, the topic around which most research was organized at the Institute was the Semantics of Philosophical Logic. The problems of proper names, predication, descriptions, existence and meaning were the preferred ones and, as pointed out before, Frege, Russell, Wittgenstein, Quine, Geach and Strawson were the authors most frequently discussed. Moreover, the philosophical style preferred was not that of Carnap or Quine, but rather that of Austin and Strawson.

It was only years later, from 1972 onwards, after the first generation of post-graduate students had returned from abroad, that other concerns, like the Philosophy of Persons, Epistemology, the Philosophy of Science, Practical Philosophy, and the Philosophy of History emerged and increased.

3. RECENT DEVELOPMENTS

In 1975 Crítica was reorganized and its editorial board revamped with the collaboration of the Argentinian Society for Philosophical Analysis (SADAF). Since that year Crítica has been appearing regularly and has published forty-two issues in forty volumes.

Cuadernos de Filosofía began to appear in 1977. It is a series of articles considered indispensable for the understanding of recent philosophy.

In 1978 a seminar for Research Fellows was begun. This seminar was intended to offer a permanent opportunity to the Fellows of the Institute as well as to other professional philosophers to discuss their ideas in order that their work could be improved. This important seminar has sharpened the practice and technique of philosophical analysis in recent years.

Later, in March 1979, the need was felt for starting a circle of discussion among the students of the Instituto--whose number had increased considerably--so that they also could benefit from exposure to philosophical argument and criticism. The quality of the papers read and discussed made it possible to change the circle into a seminar with regular meetings similar to those conducted for the Fellows.

In October 1976 a seminar on Philosophy and History of Science began with the name "Seminar on Epistemology." Dr. Mario Bunge was given charge of it at the beginning. The purpose of this seminar was to establish a relationship between the work of scientists and that of philosophers and further the understanding of the problems that arise from the relation between science and philosophy. Later, the seminar, under the leadership of Mario Otero and Ulises Moulines, extended its subject matter to other areas; it continues to this day.

After 1978 the advantages of dividing the work of the Fellows into five main areas of research became clear. The five areas are: Philosophy of Logic and Language; Philosophy of Mind and Persons; Epistemology and Philosophy of Science; and Practical Philosophy and History of Philosophy. The Fellows choose the areas of primary interest and are grouped accordingly, so that plans and projects may be developed and realized and eventually all philosophical subjects may be covered.

Conjointly with this reorganization, three additional programs were launched: the program of academic exchange with the main philosophical centers of the world; an ambitious translating and publishing program; and, finally, a program concerned with restructuring and enriching the library of the Institute.

The exchange program includes both inviting foreign philosophers to the Institute and sending our Fellows to the main philosophical centers of the world to give seminars, conferences, etc.

In 1980 the Institute started an annual International Symposium of Philosophy, in which continuous discussion of original papers takes place for a period of a week. This represents the culmination of the Fellows' work for the year and offers a good opportunity to compare their own views with those of leading philosophers.

In 1979 was launched an ambitious publication program; its purpose is to acquire and translate key works of philosophy in order to start new collections and increase those already in existence.

Finally, in 1979 a program of reorganization and acquisition of books and journals was initiated. Its aim was to accumulate all the philosophical classics in their original languages together with all the valuable studies on those classics and other contemporary studies and books as well as important journals from Europe and America.

4. PERSPECTIVES

Mexico, as other Spanish and the Portuguese speaking countries, lacks a philosophical tradition. No classic has ever been written in either of these two languages. All the technical vocabulary is, therefore, framed in other languages. This fact, added to other historic and social facts, makes us vulnerable to imported intellectual fashions and, in most cases, to pitiful caricatures of those fashions which come and go without enriching substantially our philosophical culture. Moreover, any new philosophy is under a permanent danger of becoming one of those fashions.

Thus, it is very important to point out the way in which philosophical analysis could help us to nourish and develop philosophy in Mexico, and the way in which, even when it is no longer fashionable, it could still leave a heritage worth the effort of studying it. The criteria of clarity and vigor which underlie the analytical trend constitute an important method to understand the philosophical problems of Western philosophy.

The Spanish speaking philosophers could thus find a remedy to the historical lack of problems, arguments, and theses they have suffered and could appropriate the fundamental problems, that is, the theses, arguments and fruitful techniques developed by analysis to make possible understanding and creativity in various philosophical fields.

We need to introduce the great traditions of Western philosophical thought so that we may have the appropriate instruments to think about the world and persons. Indeed, it would not be a minor achievement to rescue all what is worthwhile in the thought of all those people who inhabited our countries before the Spanish Conquest as well as the philosophical work done during the Colonial period.

Thus we realize that at least for the time being it is important to follow the _Canon_ of the so called "Analytic Philosophy" as it is practiced in Anglo-Saxon countries.

C. Ulises Moulines

EXISTENTIAL QUANTIFIERS AND GUIDING PRINCIPLES IN PHYSICAL THEORIES*

INTRODUCTION

In the 1969 Postscript to the second edition of The Structure of Scientific Revolutions, Thomas Kuhn set out to reply to several objections that had been made against the first edition of the book (1962). Among those objections, there was one that Kuhn took most seriously. It was the claim that his famous notion of a "paradigm," as a structure underlying the whole of a discipline during a period of "normal science," was irremediably vague and ambiguous. Kuhn tried to make the notion more precise, and for this reason he even changed the terminology: The "paradigm" of the first edition became the "disciplinary matrix" of the Postscript. It seems that the new term has had no good fortune; people continue to talk about "paradigms" and, even worse, they continue to talk about them with the same vagueness which characterized Kuhn's discussion in the first edition of his work. Of course, the issue at stake is not merely terminological, and therefore we can go on using the term "paradigm" instead of "disciplinary matrix" in the present context. What really matters is that Kuhn tried to make the structure of his basic notion more precise and for this reason he identified four components that, in his view, are essential to a paradigm: symbolic generalizations, ontological and heuristic models, methodological values, and "paradigmatic" exemplars of applications. Shortly afterwards, in Structures and Dynamics of Theories, Stegmüller used the formal apparatus of Sneed's Logical Structure of Mathematical Physics to formalize the notions involved in the first and last components identified by Kuhn in a paradigm. According to Stegmüller, Kuhn's "symbolic generalizations" are nothing but the fundamental laws of a theory's so called "structure core."

The present essay deals with this first component of paradigms identified by Kuhn and made precise by Sneed and

J. J. E. Gracia et al. (eds.), Philosophical Analysis in Latin America, 173–197.

Stegmüller. For reasons that will become clearer later on, I propose a new name for the entities that were called "symbolic generalizations" by Kuhn and "fundamental laws" by Stegmüller--even at the cost of terminological inflation. I propose to call the first component of Kuhn's paradigms "guiding principles," a denomination that already exists in the literature. In my view, this label reflects the role of the entities at issue much better than the previous denominations. Of course, the name is unimportant. What really matters is to provide a more complete and detailed (and also more "committed") analysis than the ones offered so far. The conception of guiding principles which I propose purports to be applicable at least to two important cases of scientific paradigms: classical mechanics and phenomenological thermodynamics. Whether the same kind of analysis can be applied to other cases as well I would like to leave as an open question for the time being. I admit the possibility that guiding principles have a different form and function in other scientific paradigms. What I want to stress now is this: If a paradigm's guiding principle had the form I propose, then this would explain some of the peculiar features of paradigms that Kuhn has noted and that have intrigued many of his critics. The features are these:

a) That the content of a paradigm is rather like a <u>promise</u> of future scientific success and not so much a concrete achievement.
b) That paradigms lead to the strange sort of activity Kuhn calls "puzzle-solving."
c) That paradigms are essentially <u>unfalsifiable</u> by experience.

These features of paradigms have been explained in part by Stegmüller in the above mentioned book and in more recent essays (e.g., in his article "Accidental ('Non-Substantial') Theory Change and Theory Dislodgement"). Nevertheless, I consider Stegmüller's analysis to be still too crude; moreover, it does not show clearly the role that guiding principles play (and <u>why</u> they play it) in paradigms. I intend to supplement Stegmüller's analysis with an examination of the <u>logical form</u> peculiar to guiding principles--at least in the two cases of classical mechanics and phenomenological thermodynamics. The axiomatization of classical mechanics offered by Sneed and Stegmüller, which is in turn grounded on the "paradigmatic" axiomatization provided by McKinsey *et. al.*, gives no account of the formal characteristics of the mechanical guiding principle we are interested in now, and in this sense it is incomplete. The same goes for my own axiomatization of thermodynamics in "A Logical Reconstruction of Simple Equilibrium Thermodynamics."

In the present essay, I shall concentrate on the guiding principle of classical mechanics and try to get at the general features of this kind of principles through the analysis of that particular example. The reason for this choice is partly historical (Newton's principle has probably been the most discussed principle in the history of methodology) and partly expository (classical mechanics is the best known physical theory). Toward the end, I shall argue that the same considerations apply to thermodynamics.

1. The Discussion of Newton's Second Law

The guiding principle of classical mechanics is "Newton's Second Law," i.e., the fundamental dynamic equation "$\underline{F}=\underline{m}.\underline{a}$." In "naive" expositions of classical mechanics, this equation always appears as the fundamental law of the theory. It is more general than "Newton's Third Law," which is not taken as true in all applications of the theory (for example, in ballistics or in the cases of charged particles in motion) and it is logically more fundamental than "Newton's First Law," which in present day formulations of mechanics can easily be derived from the Second Law. It goes without saying that it is also more fundamental than more special laws of restricted scope, like the law of gravitation, Coulomb's laws for electrostatic and magnetostatic forces, or the dynamic equations for elastic forces, friction, etc. The central status of the Second Law is not only admitted by "naive" textbooks, but is also clearly present in careful axiomatizations like those provided by Hamel, McKinsey-Sugar-Suppes, and Sneed, and also in Kuhn's and Stegmüller's more philosophical discussions. Some authors, however, have denied, or at least questioned, the fundamental character of the Second Law. Their reasons for doing so have been essentially of a logico-philosophical origin. Let us discuss them.

The history of the methodological controversy around the Second Law is long and dense. It is also the history of the discussion of the Newtonian concept of _force_. A concept and a statement seldom have been so inextricably tied together as in this the case. To have a full understanding of the meaning of this history one has to be familiar with a methodological distinction that has played an important role in modern philosophy of science: the distinction between _statements of fact_ and _definitions_ (or terminological conventions). This distinction may be viewed as an instance of the more general dichotomy "_descriptive-prescriptive_." According to this dichotomy, statements of fact are descriptions of certain states of affairs postulated by a theory, whereas definitions are prescriptions for the use of certain terms in the theory.

Assuming this dychotomy, we are led to ask whether "$F=m.a$" is a statement of fact, that is, a description of empirical situations found in nature, or else a definition. If the second alternative is adopted, the most plausible thing to do is to claim that it is a definition of the concept of _force_, since it will generally be admitted that the concepts of space, time, and mass have a meaning independent of the concept of force.[1]

The two alternative interpretations constitute the core of the controversy around the Second Law. Some say that this principle is an empirical statement of fact, whereas others claim that it is a "mere convention," a definition of the term "force," and it is not an exaggeration to say that the controversy has lasted for centuries. Each of the rival views receives some plausibility from the fact that one can show that the opposite view is laden with serious difficulties. Thus, anyone wishing to maintain that "$F=m.a$" is a statement of fact may note that, if it simply were a definition, the whole of classical mechanics would rest on a mere tautology and therefore would be an _a priori_ discipline, at least in its fundamental part. On the other hand, if someone wants to persuade us that "$F=m.a$" is a definition of "F," he rightly argues that it is difficult to admit that this equation describes a concrete fact (or even a set of facts), since any conceivable empirical situation is compatible with it: One can always construe force-vectors for each concrete system in such a way that the resulting sum be the product of the mass by the acceleration of the bodies involved. Therefore, the principle is not refutable, that is, it is empirically vacuous.

Some authors have tried to find a solution to the dilemma through a sort of "middle way," that is, following an eclectic line. On this view, the Second Law "sometimes" would work like an empirical statement of fact, and "sometimes" like a definition; or else, "from some point of view" it would be one thing, and "from another point of view" it would be something else. Of these eclectic positions, the best articulated one that I know is Nagel's view in The Structure of Science. It is worthwhile to quote Nagel at length. Indeed, his analysis of the Second Law leads him to a position that comes very close to the one which constitutes my starting point:

> ...the second axiom is frequently regarded not primarily as an assertion about the conditions under which accelerations occur but rather as a compact formulation of a specialized guide for research, as a methodological rule which directs the physicist what to look for when he is analyzing the motions of bodies. For in respect to the feature of being incapable of conclusive refutation, the second axiom is very much like such a rule. Any number of failures on the part of a physicist to find what the axiom directs him to obtain, do not necessitate abandoning the search and scuttling the rule. For the rule may nonetheless be a good one because research conducted in accordance with it may have been generally rewarded with success and because even a rule that is only sometimes useful may be better than no rule at all. In point of fact, the second axiom viewed as a regulative principle has been eminently fruitful in guiding the construction of a systematic body of warranted knowledge; and if it continues to be accepted as a rule of procedure, it is clearly not because it is an arbitrary and unfounded rule for investigating the motions of bodies (op. cit., p. 192).

This description of the way the Second Law is actually _used_ seems to be, generally speaking, correct. Accordingly, we could expect that, from this analysis of a particular case, a general and systematic view of this kind of physical principles would follow that really escapes the dilemma mentioned above. However, when Nagel comes to summarize his conclusions about the analysis of classical mechanics in general, he again puts Newton's "three axioms" in the same box (a serious mistake, in my opinion) and takes an uncommitted and ultimately disappointing position.

> It thus becomes evident that no brief and simple answer can be given to the question: What is the logical status of the Newtonian axioms of motion? It is quite certain that the axioms are not _a priori_ truths to which there are no logical alternatives; and it is equally clear that none of them is an inductive generalization, in the sense of a generalization that has been obtained by extrapolating to all bodies interrelations of traits found to hold in observed cases. But beyond these negative characterizations of the axioms, a reasonably satisfactory answer to the question requires reference to the place the axioms occupy in some particular codification of the theory of mechanics, and to the uses to which the axioms are put in various special contexts. What can perhaps be asserted quite generally is that, on the one hand, the Newtonian axioms can often play the role of schema for analyzing the motions of bodies or of stipulations for defining certain experimental notions, and, on the other hand, when the axioms are coupled with additional assumptions (among others, with assumptions concerning force-functions) they can be correctly construed as statements possessing a definite empirical content (_op. cit_., p. 202).

This interpretation has two main shortcomings which seem absolutely hopeless: First, it sets Newton's three axioms on the same level; second, and more important, the talk about "various texts," "different uses," and so on, leads to a most unsatisfactory ambiguity and eclecticism, which fails to meet the standards one expects of a genuine logical reconstruction of a basic aspect of empirical science. In the last analysis, the ambiguity inherent in Nagel's view, in spite of his promising start, comes from the fact that he was not able to free himself of the sharp dichotomy "description-prescription": When facing a case like Newton's Second Law, he must therefore conclude that "in some contexts" it works like a statement of fact, whereas "on other occasions" it plays the role of a definition. To summarize, the starting point in his logical analysis of the Second Law is correct, but his general conclusion is not.

From the preceding remarks, my lack of interest in maintaining that famous dichotomy should already be easy to see. My general attitude is to distrust simple dichotomies: The logical forms of discourse in general, and *also* of scientific discourse, are more varied than philosophers, with their eagerness to simplify, tend to assume. The philosophical investigator, like the detective, has to be cautious and imaginative at the same time. If we follow this general attitude in the present case, and remind ourselves of the obvious shortcomings of the two views controverted and of the shallowness of the eclectic position, we shall be led to look for a new approach to the problem. If possible, this approach should be applicable not only to the particular case of "$F=m.a$," but to other cases as well, so that we might come to a more general view of some structures of science. This is precisely what most logical analyses of the Second Law have lacked: their embedding into a general metatheoretical scheme.

As noted before, the discussion surrounding the Second Law has normally been associated with a discussion of the concept of force itself. This concept is the crux of the problem. To clarify the nature of force is a necessary condition for getting at an adequate conception of the Second Law.

2. The Discussion of the Concept of Force

From the very beginnings of Newtonian mechanics, the concept of force was looked at with much suspicion and even repulsion by some of the most brilliant minds in the history of science and philosophy. This was, in part, but *not exclusively*, because of its association with the notion of "action at distance." Among Newton's contemporaries, it suffices to mention Huygens, Leibniz, and Berkeley as sharp critics of the Newtonian concept of force. In spite of the clear and impressive attainments of Newtonian mechanics through the years, this concept continued to be viewed with suspicion by most authors devoted to the foundations of mechanics. They wanted to see this science "freed from metaphysical obscurities" (following D'Alembert's famous phrase in his Traité de Dynamique), and the main "metaphysical obscurity" certainly was Newtonian force together with the Second Law. From this perspective, the continued efforts made by positivistic or "para-positivistic" authors to eliminate, or at least control rigorously, the concept of force, may be best understood. This tradition begins with D'Alembert in the middle of the 18th century, goes through Kirchhoff, Mach, and Hertz during the second half of the 19th century, and reaches methodologists like Hans Hermes and Herbert Simon in the middle of the 20th century.[2] All these efforts may generally be interpreted as attempts at reducing dynamics to statics or kinematics. They are certainly profound and interesting from a philosophical point of view; however, they changed nothing, the Newtonian concept of force was kept well and alive in all standard textbooks of classical mechanics and in the normal practice of physicists. Famous physicists like Kirchhoff and brilliant logicians like Simon have claimed that "force" is nothing but the abbreviation of "mass times acceleration" and, consequently, that the Second Law is nothing but a tautology; however, they do not appear to have been taken very seriously in the practice of mechanics. The "scientific community" continues to use the concept of force as an entity conceptually independent of mass, space, and time.

In relatively recent times, some philosophically well-versed minds have rebelled against the program of force elimination. The first were Suppes and his collaborators, who, in their already classic axiomatization of Newtonian mechanics, showed by means of a formal argument that the Second Law cannot be taken as a definition of force (nor of mass—of course) unless the structure of the theory becomes substantially modified. More recently, Clifford Truesdell, who is equally versed in the philosophy and the history of science, emphasized in his article "Rückwirkungen der Geschichte der Mechanik" that the Newtonian concept of force, far from being a metaphysical ornament, is Newton's most revolutionary and significant single contribution to the development of mechanics. According to Truesdell, this concept was especially responsible for all great accomplishments of mechanics for two centuries; and this judgment could be applied to the corresponding statement, the Second Law, as well. Both the concept and the principle may be regarded, in Kantian terms, as "the conditions of possibility of all mechanics." Suppes, in the formal aspect, and Truesdell, in historical perspective, rightly appreciate the importance of the concept of force and show its irreducibility in the sense that it cannot be defined by alledgedly more basic concepts, nor can it be directly associated with empirical measurements and observations. (This is the reason why Truesdell calls it an "*a priori* concept.") Nevertheless, neither Suppes nor Truesdell imbed their characterization in a conceptual framework that *explains* (and not only describes) the concept's peculiarities. A deeper analysis is needed to give an account of the role "force" has played in the development of mechanics.

An important step in this direction was made by Sneed in his reconstruction of classical mechanics in the work mentioned above. When he applied his general theoreticity criterion to the particular case of Newtonian mechanics, he came to the conclusion that force and mass are mechanics-theoretical, whereas the kinematical concepts are mechanics-non-theoretical. *T*-theoretical magnitudes in general have some peculiar characteristics as to the way they are used within a given theory *T* to determine "indirectly" the values of the *T*-non-theoretical magnitudes and to establish some correlations between different models of *T*. These features explain, among other things, the predictive power of theoretical concepts and the role they play in providing empirical theories with a holistic structure.[3] In this way, some of the most remarkable peculiarities of the concept of force in mechanics would become justified. This certainly is a relevant result for our discussion. But it is still not complete. *Not all* the peculiarities of the concept of force become sufficiently clarified if this concept is subsumed under the Sneedian category of a *T*-theoretical magnitude. To see this it suffices to note that, in Sneed's analysis, force and mass appear on the same footing as mechanics-theoretical concepts--which means that there would be no essential methodological difference between force and mass. Now, in spite of the impressive power of Sneed's metatheory in so many respects, this point of his analysis has always seemed unsatisfactory to me. My own intuition (and I think the intuition of many others too) on the structure of classical mechanics is that force and mass should not be placed, so to speak, at the same level. If mass is a _T_-theoretical concept, as seems to be the case, then force, besides being _T_-theoretical, is still something else. It is "more abstract." And it is a historical fact that most authors who have examined mechanics with a critical eye have felt much more worried and perplexed by the concept of force than by the mass concept.

Another point worth mentioning is this: The discussion of the concept of force has been inextricably tied, as we have already remarked, to the discussion of the Second Law, and particularly to its alleged tautological or _a priori_ character. Now, Sneed has provided a formal reconstruction of the Second Law in which its apparent tautological or non-empirical nature becomes translated into the following terms: Although the Second Law _cannot_ be taken as a definition, its structure is such that, in Sneed's terminology, _any_ partial (non-theoretical) model can trivially be extended or "completed" so that it becomes a complete (theoretical) model of mechanics, i.e., a structure satisfying the Second Law. In this sense, the Second Law implies no restriction of empirical content and, therefore, it is empirically vacuous. However, for reasons that will be given below, this last characterization is misleading and does not reflect the principle's real content. I would like to anticipate here that it is not "empirically vacuous" in the same sense in which analytic statements like "No bachelor is married" or "If Johnny has eaten the chicken, then the chicken has been eaten by Johnny" are empirically vacuous. In place of the expression "empirically vacuous," I propose a less committed one for the Second Law: "empirically unrestricted." We shall see why this last expression seems to be more appropriate in this context.

The point I wish to emphasize now is this: within Sneed's reconstruction, the fact that the concept of force is meaningful only in terms of the Second Law, and the fact that this principle is empirically unrestricted are, so to speak, two "contingent" features of classical mechanics, historical accidents of the way the theory developed and, therefore, two logically independent aspects of it. This is, in my opinion, the weak point in Sneed's analysis. It is not an "accident" that the fundamental principle of mechanics (in any of its equivalent formulations) is empirically unrestricted (without being a mere definition), nor is it a historical contingency that the discussion of the concept of force has been associated with the discussion of the Second Law. I claim that: (a) It is essential to this theory (and presumably to others as well) that its guiding principle is empirically unrestricted; and (b) the peculiarities of "$\underline{F}=\underline{m}.\underline{a}$" come from the implicit structure of "$\underline{F}$" and it is precisely the structure of "$\underline{F}$" that makes the principle so powerful in spite of its being empirically unrestricted—although this may sound like a methodological paradox. In the following, I shall give the reasons for maintaining this double claim and afterwards I shall show that a similar claim is true of thermodynamics.

3. A Logical Reconstruction of Newton's Second Law

Let us begin with a more detailed analysis of the logical form of the Second Law. Its standard formalization is attributable to McKinsey *et. al*. in the article mentioned above.[4] All later analyses of classical mechanics (with the exception of the "reductionists" Simon and Hermes), including those due to Sneed and Stegmüller, rest on that formalization. Its primitive concepts are those of particle, position, instant, mass, and force, in addition to a numerical index set for distinguishing the different forces that can act on a given particle. The motivation for introducing the numerical indices is that, thanks to this "trick," it is not necessary to quantify over different force functions and therefore it is not necessary to introduce as many functional variables. We shall see, however, that this trick does not attain its goal.

If $\underline{P}$ is a set of particles (= a mechanical system), $\underline{T}$ is the time interval in which the system is considered, $\underline{i}$ is a numerical index for kinds of forces, $\underline{s}$ is the "position" function, $\underline{m}$ the "mass" function, and $\underline{f}$ the "force" function, then the Second Law in the standard formalization would assert that, for any $\underline{p}$ of $\underline{P}$ and for any $\underline{t}$ of $\underline{T}$:

$$\sum_{i} \underline{f}(\underline{p},\underline{t},\underline{i}) = \underline{m}(\underline{p}).D^2_{\underline{t}}\underline{s}(\underline{p},\underline{t}) \quad ,$$

where "$D_t^2 \underline{s}(\underline{p}, \underline{t})$" is an abbreviation for the "second derivative of position with respect to time," that is, the acceleration.

An aspect of this formalization worth emphasizing is the fact that force and mass appear "on the same footing" in the following sense: Both are functions of particles, i.e., of the primitive individuals of the system. The only formal difference between mass and force consists in the first being a monadic function whereas the second is triadic. But, in any case, $\underline{m}$ and $\underline{f}$ are at the same ontological level of functions of particles.

Let us note another feature of the previous formalization. The only bound variables are the individual variables $\underline{p}$, $\underline{t}$, and $\underline{i}$. Therefore, the Second Law could be formalized in a first-order logic. This might seem to be an advantage. However, in this case, it is not, because it would distort its real logical form. Let us say it once and for all: In order to explicate the logical form of the Second Law adequately, it is required to quantify over <u>functional</u>, and not only individual, variables. That is, if we wish to formalize it, we need at least a second-order logic. The reason is not merely that we have to introduce different force functions—since this could be arranged by means of the trick of numerical indices. The crucial reason is that those force functions are <u>functions of functions</u> (or, more precisely, functions of tuples of functions), and not simply functions of particles and instants. In the literature on mathematical logic,[5] the term "functional" is sometimes used for such functions of tuples of functions. By adopting this denomination, we may state our central thesis as follows: Every mechanical system may be described by a certain number of forces, and each one of these forces is a functional. (There are some physicists who also use the term "functional" in a sense identical or similar to the present one. Thus, Mittelstaedt, in his textbook on classical mechanics, describes the force function explicitly as a "functional" of velocity, position, etc.)[6]

Why do we say that forces are functionals and not just functions of particles? Because only in this way can we account for an essential feature of the notion of force.

This feature is apparent in the way this notion is used in concrete applications of mechanics, but it remains concealed in the standard formulation of the Second Law. It is this: For any physical system, the forces considered--often more than one--"depend" on several "additional parameters" that may be specified only in the concrete application of the theory. These parameters are, in turn, functions of the individual variables (particles and instants). The Second Law does not specify either their number or their nature, nor does it tell the specific way in which force is a function of them. For all these reasons, they are "implicit" in the usual formulation of the Second Law--this actually means that we presuppose an indeterminate number of existential quantifiers over the functional variables representing those parameters.

This, by the way, formally corresponds to the intuitive way of speaking of physicists. As a matter of fact, physicists never say that "force is a function of particles and instants," but "different kinds of forces are functions of different parameters," according to the kinds of applications of the theory. These parameters may be, for example, spatial coordinates, instants, velocities, masses, electric charges, magnetic poles, coefficients of elasticity, coefficients of friction, etc. All of them are, of course, functions of particles and/or instants. And, in each case, the functional's specific form will be different. For example, it may be a linear function of distance, or else an inverse-square function of distance, or an exponential function of velocity, etc. By specifying all these different possibilities in concrete applications of mechanics, we obtain different dynamic laws that may be tested empirically. The Second Law implicitly contains the information that all these possibilities _exist_. But it does not say anything specific about them.

If we make explicit the information contained in the Second Law, this may be stated in the following informal terms:

> For any given system _P_ and time interval _T_, a set of vector functionals can be found that are functions of functions of particles of _P_ and instants of _T_, so that the vector sum of the functionals equals the product of the mass times

the acceleration of each particle at each instant considered.

Stated a bit more formally, Newton's Guiding Principle says:

(NGP) For any given $\underline{P}$ and $\underline{T}$: For any $\underline{p}$ in $\underline{P}$ and for any $\underline{t}$ in $\underline{T}$, there exist vector functionals $\underline{f}_i$ (into $\underline{R}^3$) such that, for each $\underline{f}_i$ there exists a number $\underline{n}$ of (scalar or vector) functions g_{i1}, $g_{i2},\ldots,=g_{in}$ of $\underline{p}$ and/or $\underline{t}$, such that:

$$\sum_i f_i(g_{i1}(\underline{p},\underline{t}),g_{i2}(\underline{p},\underline{t}),\ldots,g_{in}(\underline{p},\underline{t}))=\underline{m}(\underline{p}).D_t^2\underline{s}(\underline{p},\underline{t}).$$

A complete formalization of this statement would be feasible, but I avoid it since it would obscure rather than clarify the structure of the principle. The present formulation suffices to show that the logical form of the Second Law is much more complicated than is admitted in the standard version of McKinsey-Sugar-Suppes. This formal complexity corresponds to the real content of the principle.

To be completely accurate, however, it should be noted that the Second Law's implicit content is not yet fully explicated within the present formulation of NGP. This only takes the principle's syntactic-semantic components into account, but not what may be called its pragmatic aspect. Indeed, the use of NGP in physical practice implicitly requires the functionals $\underline{f}_i$ to be both "non-trivial" and "relatively simple." To exclude triviality in this case means that the functional should not amount merely to a surreptitious and concealed repetition of the equation's right member, i.e., "mass times acceleration." Were this to be the case, then NGP would not help us as a guide to look for interesting functions--it would only be an ingenious tautology. In order to give an exact formulation of this requirement, it would be convenient to make use of the formal notion of "essential occurrence of a variable in a formula." However, it is not clear how this notion may be applied rigorously to the present case; besides, it is very likely that it would not be sufficient to immunize NGP against all possible cases of "triviality" about which physicists think intuitively. "Triviality" in physics presumably has a wider scope than "triviality" in logic. On the other hand, the $\underline{f}_i$'s mathematical form should not be so three legs" than when one says, "Juan Pérez González, born on April 16, 1978, in Mexico City, had three legs when he was born.

If the existential quantifiers run over a potentially infinite domain, then it is well known that the statement at stake is empirically unfalsifiable. Let us call this kind of unfalsifiability "first-order unfalsifiability," since it is due to a quantification over the individual variables of a given (infinite) domain. Relative to first-order unfalsifiability, some mathematical and/or empirical strategies may be envisaged that would make the statement--though unfalsifiable in principle--ever less plausible or less probable until it becomes "practically falsified." (Some of these strategies might come from decision theory.) Nevertheless, NGP's unfalsifiability is not first-order, since the problem does not arise for the quantification over particles and instants--these entities being quantified universally. What we have here is a "second-order and third-order unfalsifiability" (a great scandal both for Popperian fallibilists and Carnapian verificationists!), since we quantify existentially over second and third-order functional variables. Therefore, NGP is not only unfalsifiable in principle, but it is also difficult to imagine what kind of empirical strategy or inductive (or verisimilitude) calculus could lead us to determine NGP's degree of implausibility or unlikelihood. Indeed, the principle's characteristic indeterminacy does not come from an existential quantification over empirical entities (bodies, places, times): It rather comes from existentially quantifying over theoretical functions and abstract functionals; that is, what we have here is a quantification over our minds' mathematical capabilities when dealing with empirical problems. It is difficult to imagine, in such a context, that one can meaningfully speak of verification or falsification, or even of degrees of corroboration, confirmation, likelihood, etc. All that methodological apparatus so popular in classical philosophy of science does not seem applicable to the present case.

It may be objected that it is possible to reject NGP because of its logical incompatibility, not with the empirical data, but rather with some other principle having the same or even a greater degree of generality that might have been accepted previously "for good empirical reasons."

complicated as to make calculations impossible. Were this to be the case, the "guide" offered by _NGP_ would be useless—not because of its triviality, but because of its hopelessness. Now, since we do not seem to have a generally accepted _simplicity_ criterion at hand (in spite of Goodman's and others' efforts), the requirement of _non-excess-complication_ for the time being cannot but be relegated to the physicists' intuitions. (As a way of fictitious consolation, this situation is usually described by saying that the concept of simplicity is "pragmatic.")

The two problems I have just discussed (non-triviality and non-excess-complication) make the formal reconstruction offered in _NGP_ still imcomplete. Nevertheless, I do not think this has fatal consequences for our present purposes. In spite of the _lacunae_ just noted, I think the logical form expounded in _NGP_ is much richer and more adequate than any alternative offered so far as an explanation of Newton's Second Law. Moreover, whatever the definitive formulations (if any) of the non-triviality and simplicity criteria for physics might be, it is likely that they will be compatible with the present formulation of _NGP_. As a matter of fact, it is difficult to conceive of any possible reason why they might not be so.

Let us consider the logical form of _NGP_ in greater detail. From a methodological point of view, its most relevant aspect is the great number of existential quantifiers involved. Not only the first-order functions g_{ij} are being existentially quantified, but also their total number $\underline{n}$ and the second order functionals $\underline{f}_i$: Number and nature of both functionals are being existentially quantified, i.e. they are being left indeterminate. Furthermore, here we cannot apply the trick of introducing numerical indices to "compile" all $\underline{f}_i$ in just one $\underline{f}$, since every functional $\underline{f}_i$ possibly depends on a distinct set of g_{ij}, which might differ from other such sets both in cardinality and nature.

Now, it is a well-known methodological fact that the more existential quantifiers one introduces into an empirical statement, the weaker its empirical content and the more difficult its falsification become. Clearly, one expresses a weaker commitment when one says, "There has been or there will be some time at some place some person with

However, it is also difficult to imagine that some other mechanical principle could make us doubt _NGP_'s truth, since _NGP_ does not assert anything, or "almost" anything. And it is quite remarkable that relativistic dynamics, in spite of its radical transformation of the concepts of space, time, and mass, kept _NGP_ as a fundamental dynamic equation _in the same form_.

Our logical reconstruction also explains why the concept of force itself appears to be inextricably tied to the discussion of the Second Law, and why the interpretation of this concept is so elusive. To begin with, the forces are functionals that become fully meaningful only within _NGP_'s framework. Further, our formulation makes clear that the general idea of force synthetizes an indeterminate number of functionals or, rather, of third-order variables for functionals of different forms and content. It is no wonder, then, that Newtonian force apppears in methodological discussion as a "highly abstract" concept whose interpretation is very difficult. It _is_ difficult to interpret because, by its very nature, it has no fixed interpretation--it is the name for an indeterminate series of functional variables. And it appears as highly abstract (and not merely _T_-theoretical) clearly because it is a concept of concepts (a predicate of predicates).

Newton's Guiding Principle is a paradigm's paradigm in a twofold way: First, it plays the most basic role (the most "paradigmatic" role, if you wish) within that paradigm that we call "classical mechanics." Secondly, the analysis of the form and function of _NGP_ may become methodologically paradigmatic in the sense that similar guiding principles may appear in other theories and, consequently, the study of the essential features of _NGP_ may give us the clue for logically reconstructing other guiding principles and for understanding their role within the corresponding paradigms. We cannot expect strict parallels, since every theory has, so to speak, its own "personality." But it is certainly likely that some common features can be found that are relevant to a general scheme for science--e.g., to Kuhn's scheme. To show the plausibility of this thesis, I propose to consider the case of thermodynamics.

4. The Guiding Principle of Reversible Thermodynamics

Let us first examine a particular sub-theory of reversible thermodynamics: the thermodynamics of simple systems. The primitive notions of this theory are: entropy, internal energy, volume, pressure, and moles. In my article "A Logical Reconstruction of Simple Equilibrium Thermodynamics," I provided some reasons for this selection, and also for taking entropy and internal energy as thermodynamic-theoretical magnitudes. Now, the point of interest here is this: Following Callen's *Thermodynamics*, we may identify the theory's fundamental principle as the correlation established between entropy and the extensive (= additive) functions, or alternatively, between energy and the rest. Callen writes down these two (equivalent) "fundamental equations" (as he calls them) as follows:

$$\underline{S} = \underline{S}(\underline{U}, \underline{V}, \underline{N}_1, \ldots, \underline{N}_r) \ ;$$

and

$$\underline{U} = \underline{U}(\underline{S}, \underline{V}, \underline{N}_1, \ldots, \underline{N}_r) \ ,$$

where $\underline{S}$ stands for entropy, $\underline{U}$ for energy, $\underline{V}$ for volume, and the $\underline{N}_i$ for the moles of each chemical substance present in the system.

Of course, this way of formulating the fundamental equations is incorrect (though we understand what it means). The reason is that all magnitudes considered are functions of *state*, i.e., functions of the individuals the theory is about, which are states represented by the variable $\underline{z}$. $\underline{S}$ cannot be at the same time a first-order function (a function of state)—as it is written on the left hand side of the equation—and a second-order function (a functional of functions of state)—as it appears on the right hand side of the equation. The same goes for $\underline{U}$ in the second equation.

It is clear that what Callen wants to express with this *genre* of equations, as many physicists that do not pay attention to logical matters do, is that the numerical values that entropy as a function of states may take *depend* on the values taken by the other functions of state—i.e., those appearing on the right hand side of the equation. The same goes for energy. Now, the formally correct way of expressing this is to say that *there is a functional* (call it f^S) of the extensive functions U, V, N such that the value the entropy takes equals the functional's value (for each given state z). That is: There exists an f^S such that: $S(z)=f^S(U(z),V(z),N_1(z),\ldots,N_r(z))$. And similarly for the alternative principle: There exists an f^U such that: $U(z)=f^U(S(z),V(z),N\ (z),\ldots,N\ (z))$.

Notice that f^S and f^U are functionals in the same sense as the force functions. And, analogously, the functionals are third-order variables existentially quantified--variables that run over correlations of functions of state: energy (respectively, entropy), volume, and moles. The only information which the "fundamental equation" provides about this functional is that it *exists* and that it takes real values. Its specific form is left completely undetermined. When applying the guiding principle to particular cases, an appropriate form for f^S will be sought that provides a more special law of simple systems--e.g., Gay-Lussac's, or Van der Waals', or Stefan's law. It is only at the level of such laws that the empirical check of thermodynamics is possible. The guiding principle itself,

$$S(z)=f^S(U(z),V(z),N_1(z),\ldots,N_r(z)),$$

is not empirically falsifiable.

It is also worth noticing that in this case we quantify existentially not only over f^S, but also over the number r of mole functions and, consequently, in the last analysis over the number of "parameters" appearing on the right hand side of the equation. Here again, we encounter a double indeterminacy in the guiding principle: A part of it is the indeterminacy of the functional and the other part comes from the number of parameters considered. No wonder that the principle is "empirically vacuous" and fruitful at the same time.

On the other hand, it could be noted that the guiding principle of the thermodynamics of simple systems is "less" unrestricted or vacuous--if one would like to speak thus--than Newton's Second Law. Indeed, in the present case we only quantify existentially over *one* functional (f^S, or f^U, respectively), and not over an indeterminate number of them; moreover, the first-order functions covered by the functional are specified almost completely.

Now, the guiding principle of simple thermodynamic systems is, in fact, only a "sub-guiding-principle"--a particular instance of a much more general guiding principle that characterizes the whole paradigm of reversible thermodynamics and is fully comparable in form and role with the guiding principle of classical mechanics.[7]

The formulation of the general guiding principle of the whole of reversible thermodynamics that I offer here rests on the one given by Tisza in his Generalized Thermodynamics, p. 113, although it differs from it in some details. (Tisza's formulation is formally careless, and somewhat misleading, as is normally the case with physicists' formulations). For a full understanding of the following formalization, it should be kept in mind that a thermodynamic system $\underline{Z}$ in general may be regarded as a compound of a series of partial, (relatively) simple sub-systems $\underline{Z}_a$, $\underline{Z}_b, \ldots, \underline{Z}_n$; consequently, any state $\underline{z}$ of $\underline{Z}$ may be regarded as constituted by a tuple of partial states $[\underline{z}_a, \underline{z}_b, \ldots, \underline{z}_n]$. Thus, the general guiding principle of thermodynamics says this:

> (GGPT) For any given thermodynamic system $\underline{Z}$, there exists a decomposition of $\underline{Z}$ in sub-systems $\underline{Z}_a$, $Z_b, \ldots, \underline{Z}_n$ such that, for each sub-system $\underline{Z}_k$, there is a functional $\underline{f}_k^S$ and some extensive functions of state $\underline{G}_1^k$, $\underline{G}_2^k, \ldots, \underline{G}_r^k$ such that:
>
> For any $\underline{z}$ in $\underline{Z}$:
>
> $$\underline{S}(\underline{z})=\sum_k \underline{f}_k^S(\underline{U}_k(\underline{z}_k), \underline{G}_1^k(\underline{z}_k), \underline{G}_2^k(\underline{z}_k), \ldots, \underline{G}_r^k(\underline{z}_k)).$$

The level of indeterminacy of this guiding principle, owing to the use of so many existential quantifiers, is fully comparable with the case of Newton's Second Law. We not only quantify over the functional $\underline{f}_k^S$, but also over the functions of state, over their number, and over the number of sub-systems into which the original system $\underline{Z}$ may be decomposed. The level of "abstraction" of this principle is even higher than that of NGP: In the last one, at least the acceleration explicitly appeared as a $\underline{T}$-non-theoretical magnitude, whereas, in the present case, the only functions explicitly mentioned are entropy and internal energy--both being $\underline{T}$-theoretical. It is not surprising, therefore, that thermodynamics is usually regarded as a "more abstract and general" theory than classical particle mechanics.

5. Conclusion: A Paradigm's Promise

Having concluded our analyses of NGP and GGPT, we may come back to the general discussion which was the starting point of this essay. To begin with, now it should be clear why the simple-minded dichotomy "descriptive-prescriptive" cannot, in good faith, be applied to the guiding principles considered. The generalized use of existential quantifiers, and particularly the quantification over a second-order functional, makes clear why these principles cannot be regarded as descriptions of particular facts, nor as a description of sets of particular facts. To claim that there are some functions of the theory's basic objects and, in addition, that there are some indeterminate functionals of those functions, is not to describe any concrete aspect of the world, unless we use the term "description" in an abstruse and *ad hoc* manner. At most, we could say that they are descriptions of the expression capabilities of our own conceptual apparatus. But this is also an odd way of using the term "description." Clearly, by introducing NGP or GGPT we are not merely aiming at a linguistic or psychological description.

On the other hand, it is obvious that they are not definitions either. (It is relatively easy to show in both cases, by means of Padoa's method, that none of the concepts involved in both principles may be taken as a *definition* with respect to the rest.) Neither of the two principles is intended to settle linguistic rules for the usage of terms. On the contrary, both principles are regarded by physicists as conceptual matrices out of which important empirical laws of nature can be derived. ("To derive" should not be understood here in the sense of a logical deduction.)

The eclectic position, of course, does not work either. It is *not* the case that NGP or GGPT are used *sometimes* as descriptions of facts and *sometimes* as definitions. Both principles are *always* used in the same way, precisely in the way we have tried to explicate here, which cannot be characterized either as a description or as a definition--unless the rationale for these semantical categories is completely destroyed.

Finally, from the analysis proposed it may also be inferred how the guiding principles are "paradigmatic," or more exactly, how some features of paradigms, as revealed in a metaphorical-intuitive way by Kuhn, really come from the logical form of the guiding principles.

In the first place, we have already seen in what sense guiding principles are empirically unrestricted and, therefore, unfalsifiable: To introduce a great number of existential quantifiers into an empirical statement already is a way of substantially reducing its empirical content; but if these quantifiers run over second and third-order variables, the absurdity of looking for counter-examples against the principle becomes apparent. The possibilities of giving an account of any concrete empirical state of affairs by choosing the appropriate functions and functionals are, for strictly logico-mathematical reasons, _indenumerably infinite_.

What are these seemingly vacuous principles good for? Their value is, to use Kuhn's term, that of a _promise_--a powerful driving principle to do science. The promise implicitly contained in the guiding principle is that it guarantees us that, if we adopt the general conceptual scheme propounded by it, in the long run and with enough patience and skill, we will obtain the wanted empirical results. As contained solely in the guiding principle, the promise is very vague and indeterminate; but it defines the general structure within which the required parameters and functionals to deal with _any particular case_ are to be looked for. The search for those parameters and functionals--i.e., from the logical point of view, the search for constants to substitute adequately for the second and third-order functional variables--is the activity that Kuhn called "puzzle-solving." It certainly is a non-trivial and difficult, sometimes even exasperating, activity. But the guiding principle promises us that, if we keep trying, in the long run we will be successful and hit on the adequate substitution of constants for variables.

This is, indeed, the role of Newton's Second Law within classical mechanics. It is the promise that, if we look for appropriate dynamic parameters and the right functional connection among all of them, at last we will obtain a dynamic equation for any specific mechanical situation. This promise may seem to be vague. However, the historical fact is that it was quite stimulating to physicists and engineers, generation after generation, at least for two centuries (and even nowadays--classical mechanics is not "dead"). By including the guiding principle as expressed by NGP, the Newtonian paradigm did what the Cartesian paradigm had been unable to do: To show the way to solve all kinds of mechanical problems, including the most complicated details. This was the task that "Newtonians"—those people who had faith in the original promise--set for themselves: To find out the appropriate number and kind of force functions, not only for the problems of celestial mechanics, but for many other sorts of problems: those of elastomechanics, hydrodynamics, acoustics, electrostatics, magnetostatics, and even (though less successfully) optical and chemical problems. These were the innumerable pieces of the Newtonian puzzle; its solution led to an unprecedented success in the history of physics.

NOTES

* The original Spanish version of this essay appeared in Crítica 29 (1978). I translated it into English without introducing any essential changes. I wish to thank Prof. William Craig (Berkeley) and my friend, Ignacio Jané (Mexico) for their suggestions and criticisms. Of course, I am the only one responsible for the mistakes that might still be present.

1. In some introductory expositions of classical mechanics it is sometimes said that mass "is defined as the quotient of force divided by acceleration." The ground for this characterization apparently is that, in many cases, masses are measured through their corresponding weights in the gravitational field of the Earth. But this alledged "definition" of mass is not only a historical and methodological chimera, but openly contradicts one of the basic "rules" every student of mechanics learns in those same expositions: "Don't confuse mass with weight!"
2. A detailed account of this history is provided in my book, Exploraciones metacientíficas, ch. III.4.
3. The Sneedian distinction between T-theoretical and T-non-theoretical concepts may be characterized as

"functional," "pragmatic," and "relative." A short exposition of Sneed's criterion can be found in my Exploraciones..., ch. II.2.

4. This article is not easy to find. A summary of it can be found in Suppes' Introduction to Logic, though it is somewhat misleading for our present purposes. C. L. García has written a review in Crítica, vol. 28.
5. See e.g., M. Davis, Computability and Unsolvability, p. 124.
6. See P. Mittelstaedt, Klassische Mechanik, p. 59.
7. When I worked on the logical reconstruction of thermodynamics in the article mentioned above,, I was not aware of the existence of this general principle, and therefore my analysis was restricted to simple systems.

BIBLIOGRAPHY

Callen, H. B., Thermodynamics. New York, 1960.

D'Alembert, J. R., Traité de Dynamique. Paris, 1743.

Davis, M., Computability and Unsolvability. New York, 1958.

García, C. L., Review of J. C. C. McKinsey, A. C. Sugar, P. Suppes, "Axiomatic Foundations of Classical Particle Mechanics," Crítica, 28 (1978).

Goodman, N., The Structure of Appearance. Cambridge, Mass., 1951.

Hamel, G., "Die Axiome der Mechanik." In Handbuch der Physik, vol. 5. Berlin, 1925.

Hermes, H., "Zur Axiomatisierung der Mechanik." In L. Henkin, P. Suppes, A. Tarski (eds.), The Axiomatic Method. Amsterdam, 1959.

Hertz, H., Die Prinzipien der Mechanik. Leipzig, 1894.

Kirchhoff, G., Vorlesungen über Mechanik. Leipzig, 1876.

Kuhn, T. S., The Structure of Scientific Revolutions (2nd edition). Chicago, 1970.

Mach, E., Die Mechanik in ihrer Entwicklung. Prague, 1883.

McKinsey, J. C. C., A. C. Sugar and P. Suppes, "Axiomatic Foundations of Classical Particle Mechanics," Journal of Rational Mechanics 2/2 (1953).

Mittelstaedt, P., Klassische Mechanik. Mannheim, 1970.

Moulines, C. U., "A Logical Reconstruction of Simple Equilibrium Thermodynamics," Erkenntnis 9/I (1975).

Moulines, C. U., Exploraciones metacientíficas. Madrid, 1982.

Nagel, E., The Structure of Science. New York, 1961.

Simon, H., "The Axioms of Newtonian Mechanics," Philosophical Magazine 38 (1947).

Simon, H., "The Axiomatization of Classical Mechanics," Philosophy of Science 21 (1954).

Sneed, J. D., The Logical Structure of Mathematical Physics. Dordrecht, 1971.

Stegmüller, W., Theorienstrukturen und Theoriendynamik. Berlin-Heidelberg, 1973.

Stegmüller, W., "Accidental ('Non-substantial') Theory Change and Theory Dislodgement." Erkenntnis 10/II (1976).

Suppes, P., Introduction to Logic. New York, 1957.

Tisza, L., Generalized Thermodynamics. Cambridge, Mass., 1966.

Truesdel, G., "Rückwirkungen der Geschichte der Mechanik auf die moderne Forschung". Humanismus und Technik 13, (1969).

José A. Robles

(SIMPLE) QUALITIES AND RESEMBLANCE*

Several different attempts have been made to solve the problem of resemblance (or similitude or likeness; here I shall use these terms as synonymous) among the so called simple qualities;[1] I shall discuss some of them in this paper. Now, the orthodox view about this problem seems to be to formulate it in the following terms: there are qualities which _are_ simple _and_ similar. But then, if this is so, how can we explain this fact (assuming it to be one) since it seems that, for resemblance to hold among entities, some sort of complexity has to be presupposed in them? Once the problem is put in these terms, it seems that for its solution one has to postulate the existence of an equally simple resemblance relation (it is clear that the relation cannot be characterized in terms of a partial identity of some sort), i.e. the one linking those simple entities.

But once one agrees to the above, some difficulties arise when attempting to give an explanation of what that likeness might be. Apparently it is not possible to appeal to an ostensive definition of it without running the risk of begging the question, since, after having been shown several cases of resemblance among simple qualities, it is no good for the definer to conclude his task with the remark "...and any other similar cases." It seems clear that there is an assumption that we _already_ know the meaning of the term which is being explained to us. At this juncture it also seems that we cannot appeal anymore to some kind of direct awareness of the relation (as a universal) which is present in each and every case of resemblance. For, what is the constant element in all of them? Assuming that the simple qualities _A_, _B_, _C_ are different from each other (three different shades of a certain color, say) and such that _A_ is similar to _B_ and to _C_, what is the identical element which is present in both cases? It does not seem that it could be _the_ relation of similarity, since we have no criterion which would let us explain what is common to _A_, _B_ and _C_, (if there is something which is common to them all). Again,

J. J. E. Gracia et al. (eds.), Philosophical Analysis in Latin America, 199–217.

assuming—as we are doing—that *A*, *B* and *C* are different and simple, it is not clear why the resemblance between *A* and *B* should be the same as that which holds between *A* and *C*. There would be some similarity between *the fact* that *A* resembles *B* and the fact that *A* resembles *C*, but this resemblance could be explained by saying that both facts have a common ingredient, i.e. *A*, but from this it clearly does not follow that the relation holding between *A* and *B* has to be the same as that holding between *A* and *C*. Besides that, in case terms *A*, *B* and *C* had something in common, this again would go against their assumed simplicity (I am assuming here that, in this case, if some of those terms had something in common, they would have everything in common). It seems, thus, that the road is closed to any type of explanation by means of universals. But, it could be argued, even if those terms have nothing in common, it is exactly the same relation, formally at least, which holds between *A* and *B* and between *A* and *C*. What is, however, the sense which can be given to "same relation" in this context? We can perfectly well assume that each member of a pair of relations, R^0 and R^1, have identical formal properties without being the same. *To be at most the same height as* is a binary relation (defined on the set of human beings, say) which cannot be formally distinguished from the relation *less than or equal* (defined on the set of natural numbers). We can perfectly well understand what it would be for two terms of the former relation to shift their respective positions at a certain time *t* even though no such change is admissible in case of the latter (is this so *just* because of the different nature of the relata? I am not sure that this completely explains the situation). Wouldn't this be the case among resemblance relations among simple entities? What guarantee do we have that all their properties, whether formal or otherwise, would be and remain the same? At this point it seems that a possible answer might be that at least the resemblance relations among simple qualities should be similar; if all of them are to be resemblance relations, they must be, at least, reflexive and symmetric. But in so saying we are importing an identity criterion which holds for resemblance relations among *complex entities*. In this case we could give a definition of "is similar to" in more or less the following terms:

> _A_ is similar to _B_ if _A_ and _B_ share (at least) a property in common,

and assuming we had an idea of what a property is, a relation so defined would have the relevant formal properties. But for the case we are dealing with, any definition of "resembles" ("is similar to"), in which it is implied that its terms are simple entities, is completely irrelevant. Our problem is that of formulating an identity criterion for resemblance relations among simple entities and thus we are not justified—not until we have it—even to talk about similarities among resemblance relations among simples. Here it seems that an appeal to an extreme nominalism would be in order. This, however, is a solution difficult, if not impossible, to accept. But if it is rejected, one has to appeal to some heroic measures of some other kind, e.g. to state that resemblance is an unanalyzable, basic relation of which we come to be aware through some sort of intuition,[2] or else that its notion is part of our natural baggage (innatism), etc.

Some authors, like Price,[3] have pointed out precisely that resemblance is a basic, undefinable relation. It is impossible to put it on the same footing as almost any other relation. Price comes up with a very complex explanation of what resemblance is, but I have the feeling that it is somehow circular. In a different vein, Quine advances the view that our own ability to grasp similarities is the result of the fact that we have an "innate quality space" or else "an innate standard of similarity or innate spacing of qualities."[4] It seems to me that answers of this sort just circumvent the problem without solving it. Questions such as "Why are two (simple) shades of green similar?" or "why are they more similar to one another than to a shade of red?" have no answer. Or, if you like, the answer is given by saying: "The shades _are_ simple _and_ similar. That's a fact of life."

What I want to do here is to study a different way out which, I believe, goes deeper into this matter. This answer to the problem consists in pointing out that the problem originates in attempting to bestow an ontological status to certain epistemic characters of (simple) qualities; more explicitly: the simplicity of the qualities is just an epistemic character of them, while their similarity is an ontological character of those qualities in their relation with the object (objects) of which they are qualities, and this character might be very complex; in other words, that which appears (epistemically) is not (ontologically) as it appears. What is important in this situation is that, eventually, this ontological complexity will allow us to explain the similarity among qualities in such a way as to avoid introducing several types of relations: one to explain similarity among complex entities, a different one for resemblance among simple entities. And, on the other hand, it will also let us dispense with an appeal, as a solution to this problem, to so problematic a notion as intuition.

The problem with which I am dealing here, as I understand it, has to be viewed within the context of a certain theory of perception. Positions of an idealistic or phenomenalistic type (in which the main ontological burden is placed on sense-data) seem to turn this into an insoluble problem or, else, if from such viewpoint the problem is not taken to be insoluble, the solutions to it will turn out to be extremely artificial.

A theory of perception which, I assume, brings us closer to a satisfactory solution to our problem is a realist theory such as, e.g. the direct realism of the Armstrong type,[5] or some sort of critical realism as that proposed by Mandelbaum.[6] In these cases there will also be difficulties for finding a satisfactory solution to our problem but, as I just pointed out, I think that the way to solve it will be clearer and more adequate.

In what follows I shall discuss and reject some of the attempts which have been made to answer the problem of resemblance of (simple) sense qualities. In a forthcoming paper I shall try to substantiate my contention that a better solution is afforded by basing it on a realist theory of perception.

I. I shall start by quoting the locus classicus for this issue. Hume, in his Treatise (I, 1, 7), in a note in the Appendix, says:

> 'Tis evident, that even different simple ideas may have a similarity or resemblance to each other; nor is it necessary, that the point or circumstance of resemblance shou'd be distinct or separable from that in which they differ. _Blue_ and _green_ are different simple ideas but are more resembling than _blue_ and _scarlet_, tho' their perfect simplicity excludes all possibility of separation or distinction. 'Tis the same case with particular sounds, and tastes and smells. These admit of infinite resemblances upon the general appearance and comparison, without having any common circumstance the same.[7]

It seems that for Hume there is no problem in considering this situation. A possible account one can give of it is as follows: ideas are such as they appear; ideas of secondary qualities appear as being simple, hence they _are_ simple.[8] Since some of them appear as being similar to one another, they _are_ similar. In this note Hume points only to a fact and does not question the description he gives of it. The described fact seems to be an unanalyzable simple fact.

Against this way of considering things, Passmore has replied:

> ...if blue and green are more resembling than blue and scarlet, this can only be because they are not simple; they are dull or intense or light or vivid as well as blue or green; or else we see the blue in the green. If all that can be said of blue is that it _is_ blue and of green that it _is_ green, there is no possible way in which they could resemble one another.[9]

In his reply, Passmore clearly states how he wants to take "resemblance." He wants resemblance to be taken as a relation having as its terms complex entities and so he wants this relation to be accounted for as Bradley did,[10] in terms of partial identity. Now, it might be true that resemblance is a relation which needs not be explained in the same way in every context. But, at least in the present case, the explanation suggested by Passmore seems to be the one required. Many familiar facts are more easily explained in this way. E.g. the well known dexterity by means of which paint mixers can make a color shade of paint indistinguishable from a given one: they mix their dyes, tints, pigments, etc. to obtain the desired effect and painting restorers attain the goal of making their labor go unnoticed by reproducing the shades of the original master's painting which is being restored. At this point I would insist that the colors so obtained are complex, since they have been obtained from differet mixtures. On the other hand, in colors one can distinguish another kind of complexity, i.e. saturation, hue, brilliance, which are different characteristics present in color shades. But, is not each one of these characteristics something simple? Hence, the critic would go on, even if one could agree that the problem of simplicity is not present at the level of colors themselves, because of their assumed complexity, it can be encountered at the level of these color characteristics. But against this another question might be asked: Is it a necessary truth that in the knowing process we have to come ultimately to entities which in a certain absolute fashion are simple? And besides, even if we came to find epistemically simple entities, are we to conclude that they are ontologically simple as well? Well, this will depend on the relations we are willing to accept as existing among (simple) qualities and their objects. For the time being I would add that what I have said about colors holds also for qualities perceptible by means of other senses: an experienced sommelier can distinguish flavors and shades of flavors which go unnoticed by the average wine drinker, an orchestra conductor (a perceptive musician, in general) can distinguish... etc.

At this stage in my argument, somebody might point out that in the above discussion there has been a confusion of two quite different aspects which have to be sharply distinguished, i.e. the qualitative-perceptual, according to which qualities are simple, and the complexity of the objects to which those qualities are somehow related. We can perfectly well assume that even a sommelier with a most delicate palate will have a limit in his detection of shades of flavor and that some of them will present themselves to him as being simple and similar to others; the same can be said in the case of painters, etc. We can also assume that the object of which flavors are flavors or colors are colors is a complex one without being able to appeal to this complexity to explain the resemblance noted. A reason which can be given, and one to which we shall return, is that the relation among qualities and objects is a contingent one; in Hume's case, there might be a different one: that we are ignorant of the sort of relation which holds among qualities and objects. After talking about our instinctive belief on the independent existence of the objects we perceive, he tells us thus:

> But this universal and primary opinion of all men is soon destroyed by the slightest philosophy, which teaches us, that nothing can ever be present to the mind but an image or perception, and that the senses are only the inlets, through which these images are conveyed, without being able to produce any immediate intercourse between the mind and the object.[11]

What we find in Hume, then, is the following solution to the problem of resemblance among simple qualities (ideas): sensible qualities are presented to us as they really are. These qualities present themselves as being simple, hence they are simple. On the other hand, these qualities have no relation of any kind whatsoever with any complex entities by means of which we could explain their similarity to each other, so this similarity has to be explained in terms of the simplicity of the former. But similarity is a relation which is given immediately to us; we come to be aware of it by means of some sort of intuition. Hence, our way of knowing that two simple qualities, _A_ and _B_, are similar, is immediate; their similarity presents itself to us. For this to happen there is no need of any sort of proof whatsoever.

Some of the problems which face us in giving a solution of this kind have been rehearsed at the beginning of this paper; I shall come back to them later on and shall try to be more precise about their scope.

Now let us go back to the first text of Hume's quoted at the beginning of this section. It is pointed out in it that blue and green have a _greater_ similarity to each other than blue and scarlet. It is very important here to note this comparative in the case of simple qualities. If these qualities are _really_ simple, we are faced with the problem of how to explain their resemblance and now, as seems clear, we shall have a bigger problem if we have not only resemblance, but degrees of it as well. For, of what do blue and green have <u>more</u> that puts them in a closer relation to each other than blue and scarlet? According to the premises used in stating the problem, one cannot say of colors, which resemble more each other than something else, that they do have (quantitatively) more of a certain generic color, since it would seem that such qualities would be complex in some sense. Hume himself states the following:

> [Simple ideas] resemble each other in their simplicity. And yet from their very nature, which excludes all composition, this circumstance, in which they resemble, is not distinguishable nor separable from the rest. 'Tis the same case with all the degrees in any quality. They are all resembling, and yet the quality, in any individual, is not distinct from the degree.[12]

Then, according to Hume's view, it would seem that we would have to accept that

> (i) by their very nature, different simple qualities are or are not similar to one another,
>
> and
>
> (ii) taking into account the degree of the quality one can say, if some qualities are similar to each other, that some of them are more similar to each other than to others.

Briefly, one can say that, acording to Hume, it is because of the same graded feature of a quality, a feature which is nothing else but the quality itself, that the quality is more or less similar to and different from other qualities. This is so because of the very nature of <u>simple</u> qualities. This is an ontological point. Our way of coming to know this is by means of an intuition which shows us the simplicity of those qualities, the degree of their resemblance to others and the complex network of interrelations into which they enter. This on the epistemic side. Hume does not tell us if one intuition is enough for us to be aware of the complexities of simplicity. What is clear is that, for him, resemblance and quality degrees are, in a way, distinct intuitible features of simple qualities which do not alter the simplicity of the latter.

Some authors, writing after Hume, have tried to alleviate the burden imposed on simple qualities, arguing that the dyadic relation "*x* is similar to *y*" is definable in terms of the triadic relation "*x* is more similar to *y* than to *z*," where this relation, for the case of simple qualities, is explained in different ways: by an appeal to degrees of *difference*, by comparisons steming from a certain subject, by means of difference in behavior to certain given stimuli, etc.[13]

On the other hand, to take similarity as a triadic relation is no accident since, as it has been plausibly argued, to take it as a dyadic relation is to take it as something incomplete[14]: if *A* and *B* are similar, in what respect are they so? That simple qualities are no exception to this way of viewing resemblance among them is made clear by Hume himself, who in the passage last quoted says that by their simplicity all qualities are similar. So, if *A* and *B* are similar, this similarity must be so in a certain respect, introducing in this way a third term in the relation.

In the view we have presented as Hume's, the following points are somehow included. Appealing to the degree of qualities, one can easily define the triadic relation to which I have alluded. But the following problem has to be met. If it is stated that *A* has a bigger resemblance to red than to *B*, and this is said *because* there is more of red in *A* than in *B*, then it would seem that the simplicity of these qualities is questionable. Following Passmore at this point, we could say, e.g., that we *see* the blue in the violet, etc. The notion of simplicity, with all these complications, becomes rather obscure.

In the following section we shall see that Russell tries to eliminate, from the simplicity of qualities, the excess burden which Hume had put on it. Gradation, for Russell, will not be something that qualities *per se* will have; it will be located in *the difference among qualities*. There will be greater or lesser differences among qualities and these will have, so to speak, a greater simplicity than

Hume's. Let us now see these points in Russell.

II. In _The Principles of Mathematics_, Russell states:

> O ruddier than the cherry,
> O sweeter than the berry,
> O nymph more bright
> Than moonshine light,
>
> are lines containing three comparatives. As regards sweetness and brightness, we have, I think, a genuine quantitative comparison; but as regards ruddiness, this may be doubted. The comparative here—and generally where colours are concerned—indicates, I think, not more of a given colour, but more likeness to a standard colour. Various shades of colour are supposed to be arranged in a series, such that the difference of quality is greater or less according as the distance in the series is greater or less. One of these shades is the ideal "ruddiness," and others are called more or less ruddy according as they are nearer to or further from this shade in the series. The same explanation applies, I think, to such terms as _whiter_, _blacker_, _redder_. The true quantity involved seems to be, in all these cases, a relation, namely the relation of similarity. The difference between two shades of colour is certainly a difference of quality, not merely of magnitude; and when we say that one thing is redder than another, we do not imply that the two are of the same shade. If there were no difference of shade, we should probably say one was _brighter_ than the other, which is quite a different kind of comparison. But though the difference of two shades is a difference of quality, yet, as the possibility of serial arrangement shows, this difference of quality is itself susceptible of degrees. Each shade of colour seems to be simple and unanalyzable; but neighbouring colours in the spectrum are certainly more similar than remote colours. It is this that gives continuity to colours. Between two shades of colour, _A_ and _B_, we should say, there is always a third colour _C_; and this means that _C_ resembles

> A or B more than B or A does. But for such relations of immediate resemblance, we should not be able to arrange colours in series. The resemblance must be immediate, since all shades of colour are unanalyzable, as appears from any attempt at description or definition. Thus we have an indubitable case of relations which have magnitude. The difference or resemblance of two colours is a relation, and is a magnitude; for it is greater or less than other differences or resemblances.[15]

This passage by Russell, I think, can be summed up in the following six points:

> (i) color shades are simple;
> (ii) two color shades are completely different (if they are different); their difference exhausts their simplicity;
> (iii) the difference among shades is susceptible of gradation. Between any two different shades, A and B, there can be a greater or lesser difference than between two shades A and C;
> (iv) the gradation of difference generates a serial ordering of colors. The series so generated is dense or saturated, i.e., between any two given different shades, A and B, one can always locate a different shade C;
> (v) to say that A is more similar to B than to C is to say that the difference between A and B is lesser than that between A and C;
> (iv) the similarity relation among color shades is an immediate relation.

Now, a comment on these points: (i) simply states the thesis of qualitative simplicity; (ii) comes as a corollary of (i). As we pointed out, at the end of section I, Russell takes the thesis of simplicity more in earnest than Hume. For Hume two different shades could, at the same time and for the very same reason of their difference, be similar; he considered that to have degrees was a feature of the quality itself. For Russell, instead, "degrees of a quality" does not seem to be a meaningful expression; he substitutes it, as pointed out in (iii), by "degrees of difference." Coming back to (ii), we can say that in it is expressed a necessary condition for there to be a difference among _simple_ qualities. However, a brief note has to be added about the qualifier "completely" in (ii). This expression has to be understood in the sense of "completely in the relevant respects" so as not to invite the trivial objection that two simple entities are identical in their being simple. This can also be put as follows: once the relevant domain is selected (color shades in the case we are considering), form the subclasses of elements of such a domain that have something in common. Russell would say that, in this way, we would just get singletons, i.e., classes with just one element, since for him there are no two different shades which would partake of a feature, besides their being colors. This seems to be a correct consequence (even though it seems to go against some natural ways of viewing colors) once simplicity is taken seriously. Another consequence (or another way of viewing the above consequence) is that if two simple entities have a feature in common, they are the very same simple entity. This is something which does _not_ seem to follow from a position like Hume's, since it can be argued that two different shades might partake of the same qualitative degree and, in Humean shades--from a Russellian viewpoint--this would serve to distinguish two different features in the quality: the degree _and_ the quality itself, since they might vary in an independent way from each other. Russell is more careful at this point.

What Russell does is to adopt an extreme nominalism: "red" is a term applicable to different color shades, _not_ because they have something in common, since if they differ they differ completely, but because these shades are similar:

> Consider for example two shades of colour. It seems undeniable that two shades of red are more similar to each other than either is to a shade of blue; yet there is no common property in the one case which is not found in the other also. _Red_ is a mere collective name for a certain series of shades, and the only reason for giving a collective name to this series lies in the close resemblance between its terms. Hence _red_ must not be regarded as a common property in virtue of which two shades of red resemble each other. And since relations are not even conceivably

> divisible, greater and less among relations cannot depend upon number of parts.[16]

In (iii) difference among shades is qualified as being a magnitude. (iv) follows from (iii), but the point about the density of the series introduces some problems. If what Russell says about density is taken psychologically, i.e., given any two color shades, _A_ and _B_, discernibly different for a subject _s_ at a time _t_, we can _always_ introduce between them a third shade _C_ such that _s_ will be able to discern _C_ as different from _A_ and _B_ and _A_ closer (more similar) to _C_ than to _B_, then this is false. We have a limited (finite) capability for discerning differences among color shades and once that limit is reached, between shades _A_ and _B_, say, any other shade _C_ which (theoretically(?)) could be located between the former, will seem, to _s_, either identical to _A_ or to _B_. Here we have another problem related to the nature of the shades we see, since we must ask whether the color shades Russell is talking about are taken as perceived by us or as attributes of a colored object. In _The Principles of Mathematics_ Russell expressly denies our second proposal, identifying himself at this point with Hume:

> On the view which regarded matter as the subject of which qualities were attributes, one piece of colour was distinguished from another by the matter whose attribute it was, even when the two colours were exactly similar. I should prefer to say that the colour is the same, and has no direct relation to the matter in the place. The relation is indirect, and consists in occupation of the same place.[17]

But then, if this is so, color shades have to be given to the perceiving subject as they really are. That is, we have once again an identification of the epistemic and the ontological respects, but this goes against the possibility that the series of shades be dense, as I argued before. For density to obtain, we would have to assume a difference between the epistemic and the ontological aspects of qualities and to take color shades as somehow related, in a close way, to features of the colored object. In this way one could argue that to an (ontologically) infinite set of color shades there would correspond (epistemically) one and the same perceptual shade.[18] If such were the case, one could give a clearer sense to the statement about the density of the series, even though some _different_ problems would then come to the fore. But, as we have just seen, Russell himself excludes this as a possible interpretation of his view.

In the above discussion many problems crop up which I shall not be able to deal with in this paper. However, I want to underline a difficulty pretty closely related to what I have just said in the previous paragraph. Let us consider a subject _s_, who at time _t_ perceives a uniform red, _r_. It might happen that at a later time, _t_', and without any important change in the perceiving conditions, _s_ perceives _r_ not as a red uniform, but instead as a mosaic of different shades of red. A way of describing this situation would be by saying that _s_, at _t_', has perceived complexity in what at _t_ showed itself to him as simple. Russell would have to point out[19] that the object of perception has changed, i.e., that _r_-at-_t_ and _r_-at-_t_' are two different objects and so, reject the above description. But there will be cases in which the perception of these differences will give us greater and better information about our environment, say. How are we to reconcile this point with that which, following Russell, states the separation between color shades and those objects of which they are colors?

Let us now come back to the points into which I have summarized Russell's view. According to (v) Russell offers an explanation of the triadic relation (among simples), "_x_ is more similar to _y_ than to _z_," in terms of (graded) difference, and (vi) tells us how to understand what resemblance (or graded difference) is among color shades. This, for Russell, is an "immediate" relation and, I assume, this can be taken to mean that resemblance among shades is a simple unanalyzable relation which presents itself to us in an immediate way (we become aware of it intuitively). Here we have, again, the identification of the epistemic and ontological aspects. That this is so I take it to follow from (ii): given that the terms of the relation are simple and different, this difference exhausts their simplicity and so, any other relation among them has to hold (if it does) without our having elements which would let us offer an analysis of it. On the other hand, it must be pointed out that the relation itself, as well as the terms it relates, is one in which, once again, the epistemic and the ontological aspects are identified, the relation is as it appears. The lack of ontological economy is clear at this point: besides the enormous amount of unrelated color shades, which originate in Russell's account, owing to the complete lack of ontological links among those one would (intuitively) take to belong to one and the same family, we will have, for each case of resemblance among shades, a different relation which will necessitate a different intuition for it to be accounted for by a subject.[20] The general argument used to conclude that each relation must be different, was given at the beginning of this paper.

Besides the lack of economy, there is, in all this, a more serious problem. Since according to (ii), as we have seen, different color shades are completely different, in the shades themselves there is no element whatsoever which will let us order them in a series. But if it is not by the shades themselves that the series is formed, what is the founding element for the relation of graded difference to hold? Why is it that not any ordering of shades is a serial ordering of them? In other words, if different color shades are completely different (cf. Russell's own words above) there seems to be nothing in them which would allow the grounding of a relation among them and, even less, a relation like resemblance which has to be anchored, so to speak, in the very same nature of its terms (quite different from some other more "external" type of relation like to be to the right of, to be under...). As we have seen, Russell himself acknowledges this much; what he does as an attempt to remedy this situation is to appeal to a mysterious immediate relation of resemblance. This relation, though, has nothing to support it; its terms cannot serve this purpose since their difference exhausts their simplicity. I think that one has to conclude that Russell's proposal is as problematic as Hume's, although for different reasons. Russell does indeed retain the simplicity of color shades by taking away from them the Humean degrees, but then he introduces a relation with no support, a supportless bridge which cannot span the gap among simple and resembling qualities. There is no intuition which will create a relation that, I take it, is impossible in principle, as the problem is stated; as is impossible, by means of an intuition, to eliminate the paradoxical complexity of Humean simple qualities.

NOTES

* Published in <u>Crítica</u> 9(1977), 91-111.

1. Simple sense qualities, to be more precise. These are, basically, Lockean secondary qualities: colors, tastes, smells, etc. In this paper I shall concentrate on colors, although I think that what I shall say can be generalized so as to apply to the others.
2. Thus, for example, Hume says: "...[similarity] fall[s] more properly under the province of intuition than demonstration. When any objects <u>resemble</u> each other, the resemblance will at first strike the eye, or rather the mind; and seldom requires a second examination." [2] (ii), p. 70. On Section I of this paper more will be said about Hume's position.
3. [3], pp. 25-6 and Ch. I, <u>passim</u>. Hume tells us that "...this [resemblance] is a relation, without which no philosophical relation can exist; since no objects will admit of comparison, but what have some degree of resemblance." [2] (ii), p. 14.
4. Quoted in [4], pp. 13-4. Quine presents these points in his paper "Natural Kinds," collected in his <u>Ontological Relativity and Other Essays</u>. In [4], Shoemaker suggests that causality is a notion which is basic for the analysis and explanation of what genuine similarities among things are. In a footnote, and referring to work done by Quine and Goodman on this subject, he says: "...yet it is <u>via</u> this notion [causality], I think, that the notion of inductive projectibility is linked with the notions of natural kinds, genuine similarity, and genuine properties. I believe that the identity conditions for genuine properties, those the sharing of which contributes to genuine similarity, can be stated in causal terms..." p. 12. I basically agree with this formulation. However, in case of the problem dealt with in this paper, I think there is no <u>clear</u> causal relation between the secondary qualities of an object and its genuine properties. I shall come back later to these points.
5. See, for example, "Colour-Realism and the Argument from Microscopes," in Robert Brown and C. S. Collins (eds.), <u>Contemporary Philosophy in Australia</u> (London, 1969), as

well as Ch. 5, "Concepts and Ideas," Sect. II, "Simple and Complex Concepts and Ideas," in his _Belief, Truth and Knowledge_ (Cambridge, 1973). In Brown and Rollins see also the interesting paper by K. Campbell, "Colours."

6. _Philosophy, Science, and Sense Perception_ (Baltimore, 1964).
7. [2] (ii), p. 637.
8. Cf. [2] (ii), p. 190. There, Hume says: "...every impression, external and internal, passions, affections, sensations, pains and pleasures, are originally on the same footing... For since all actions and sensations of the mind are known to us by consciousness, they must necessarily appear in every particular what they are, and be what they appear."
9. _Hume's Intensions_, p. 110, quoted in [1], p. 32. In this paper, Hawkins puts forth an argument against Passmore's position. The interested reader will find in this paper a sharp interpretation of Hume's views on the topics pointed out in its title.
10. Bradley defended his position on resemblance among simples in a polemic he had with James. Bradley's paper, which originates the discussion, James' answer and a couple of additional notes by both contenders, are found in _Mind_ 2 (1893), 83-8 (Bradley), 208 (James), 366-9 (B), 509-10 (J), 510 (B).
11. [12] (i), p. 152.
12. For an enumeration of "all those qualities, which make objects admit of comparison," cf. [2] (ii), pp. 14-5. In this place it seems that Hume takes degrees of a quality as a relational property of a quality. We become aware of a degree of a quality once this is compared with another, similar but with a different quality degree. It is possible to understand thus the last quotation of Hume in which he says that the degree of the quality is _not_ different from the quality itself. Another way in which this might be read is to consider that Hume takes degrees as something inseparable from qualities. For a discussion of this point, cf. [1], pp. 34-6.
13. Examples of the first two ways of taking resemblance are Russell and van Steenburgh, respectively. A representative of the last view pointed out, is Smart in "Sensations and Brain Processes" (_Philosophical Review_ 68 (1959), 149-50) and later on in _Philosophy and_

Scientific Realism (Routledge, 1963), pp. 75-84.
14. For a presentation and discussion of this cf. [3], pp. 20 ff.
15. [5] (i), Sect. 159, pp. 170-71.
16. [5] (i), Sect. 153, p. 162.
17. [5] (i), Sect. 440, p. 467.
18. See this point in [5] (ii), p. 119, where Russell presents it, however, in considering the extension of a perceptually simple object in its relation with the object located in "real" space with geometric properties.
19. In fact, he does point this out later on. Cf. [5] (ii), p. 119.
20. Here there seems to arise a problem for Russell within his own system in The Principles. In [5] (i), Sect. 55, p. 52, Russell argued that "...relations have no instances, but are strictly the same in all propositions in which they occur." Even if this point might seem to be founded for the relation of difference which is the one that Russell discusses in this Section (cf. my own discussion of this point in "Teoría de relaciones y universales en B. Russell," Diánoia (1974), 86-97) it is not clear what one means by saying that the resemblance between A and B is the same as the resemblance between C and D; that is, that the degree of difference of the first pair is the same as the degree of difference of the second. Russell tells us that the degree of difference between two shades is measured by the intermediate shades between the former ([5] (i), Sect. 160, pp. 171-2). Now the problem here is not only that the subjective appreciation of distance can vary for different subjects or for the same subject at different times, but that, according to the premise of complete difference among shades, we have no guarantee whatsoever that we are dealing with the same concept of shade in making a survey of the spectrum and not even among different members within the same family of shades.

BIBLIOGRAPHY

[1] Hawkins, R. J. "Simplicity, Resemblance and Contrariety in Hume's Treatise," Philosophical Quarterly 26 (1976).
[2] Hume, D. (i) Enquiries Concerning the Human Understanding and Concerning The Principles of Morals, 2nd ed. L. A. Selby-Bigge (ed.). Oxford: Clarendon

Press, 1902.

[2] Hume, D. (ii) A Treatise of Human Nature, L. A. Selby-Bigge (ed.). Oxford: Clarendon Press, 1888.

[3] Price, H. H. Thinking and Experience. London: Hutchinson University Library, 1969.

[4] Shoemaker, S. "Phenomenal Similarity," Crítica 7 (1975), 3-34.

[5] Russell, B. (i) The Principles of Mathematics. London: G. Allen and Unwin, 2nd ed. 1937.

[5] Russell, B. (ii) "On the Relations of Universals and Particulars," in R. C. Marsh (ed.), Logic and Knowledge. London: G. Allen and Unwin, 1959, pp. 103-24.

Alejandro Rossi

THEORY OF DESCRIPTIONS, MEANING AND PRESUPPOSITION*

The aim of this article is to stage a confrontation in broad terms between Russell's Theory of Descriptions and the critique which Strawson made of it, so as to set out, with the minimum of precision required, the diverse levels of the objections; we shall not attempt to criticize, from our own point of view, the doctrines of Strawson. We shall begin with an exposition of the theory of descriptions which will be, in length, considerably greater than the part of the paper dedicated to Strawson. The reason for this is simply that Russell's theses are more complex and developed than those of Strawson and the latter's criticisms cannot be posed if the theory of descriptions is not carefully presented. Furthermore, our exposition of Russell will pass over many problems and difficulties for his theory, since they are not considered to be important given the aims of this paper. We also wish to make it clear that the confrontation will be staged from the viewpoint of ordinary language. By this we do not mean to say in any way that the theory of descriptions is restricted to this area. But it is there that the dispute with Strawson arises. Whether that is an injustice or not is a separate problem. In any case, there is enough evidence that at least Russell himself does not exclude ordinary language from the scope of application of his theory. And, finally, we wish to point out that Strawson's doctrine about presupposition will be considered only in relation to definite descriptions, leaving aside its applications, for example, to the square of opposition of classical logic.

1. It is, perhaps, no exaggeration to state that the arguments which Russell opposed with his theory of definite or singular descriptions constitute a commonplace within contemporary philosophical literature. We shall not delay, therefore, with a detailed analysis of them, nor with a historical presentation which does justice to the diverse shades and to the variants of the central thesis. We shall simply note those basic features which are essential for

J. J. E. Gracia et al. (eds.), Philosophical Analysis in Latin America, 219–242.

formulating the problems that concern us.[1]

In general terms, the arguments purport to establish that it is necessary to accept certain entities if we wish to explain certain indisputable facts. The facts would be the following: First, expressions such as 'the square circle,' 'the golden mountain,' 'the present king of France,' 'the present king of Italy,' etc.--that is, expressions lacking reference--are perfectly meaningful considered in isolation; and, secondly, they can be the subjects of propositions in which predications are made of them and which are too meaningful. We find ourselves, then, faced with expressions and propositions of ordinary language which are legitimate in point of meaning. Now, it is added, if these expressions and the propositions of which they are subjects are meaningful, it is necessary that they refer to or denote something--since otherwise they could not be meaningful. Nevertheless, as we have just pointed out, the expressions and the propositions of which they are subjects do not in fact refer to any existing thing: neither in France nor in Italy, for example, are there actually kings, since these countries are not monarchies. If we wish, therefore, to explain the indubitable fact of the meaningfulness of these propositions, it will have to be admitted that nonetheless they refer to something; and it is impossible not to recognize that that to which they refer is an entity. In the interests of brevity we can say that it is at least an object--whose _ontological status_ will later be characterized as ideal, logical, etc. The argument, then, concludes thus: either they refer to--or denote--something or they are not meaningful. Since they are meaningful, it is always necessary to assume some _denotatum_ which, given the nature of the case, will have to be of a logical type, an ideal type, etc.

Another argument consists in pointing out that if the propositions whose subjects are definite descriptions are meaningful--and _ex hypothesi_ they are--then, in virtue of the principle of excluded middle, they are true or false. But, it is added, a proposition of the subject-predicate form is true if the subject possesses the relevant attribute and false if it does not. In both cases, the truth or falsity of the proposition implies the existence of the entity to which the subject of the propositions refers. Therefore, if the principle of excluded middle applies--and it does since the propositions are meaningful--it is necessary to admit always the existence of that to which the subject of the proposition refers. But let us suppose that someone holds that the proposition 'The golden mountain does not exist' is a proposition of the subject-predicate form--we shall concede this by way of example--and moreover that it is true because there does not exist spatio-temporally a golden mountain. If that person holds, as he ought to according to this line of reasoning, that if the proposition is true then that to which the subject refers has to exist, he immediately finds himself faced with a problem. In order that the proposition be true the subject has to exist, yet if the subject exists then the proposition 'The golden mountain does not exist' is not true but false. So, how can it be held at the same time that the proposition in question is meaningful and, in this case, true, because no golden mountain exists spatio-temporally, without falling into contradiction? The only way of reconciling these facts is by recognizing that the subject 'the golden mountain' refers to an ideal or logical entity of which it is said (predicated) that it does not exist spatio-temporally--which is true. In this way it is possible to hold that the proposition is meaningful, that its truth implies the existence of its subject, and that it is true because no golden mountain is found in the world. And all this without incurring any contradiction. Now, even in the case in which it is replied that the said proposition is in no way of the subject-predicate form, that 'existence' is not a predicate, the same difficulty would arise if the thesis were accepted which claims that the expression 'the golden mountain' is meaningful so that of necessity it denotes or refers. In effect, it is not possible to hold that that proposition is true because there is no mountain with those characteristics in the world, and to hold that its denotation is a golden mountain found somewhere in the world without falling into a contradiction. Hence, the postulation of an ideal or logical _denotatum_. With regard to other examples, the argument relying among other things on the principle of excluded middle seems to apply more clearly. If 'The present king of France is wise' is meaningful, then it is true or false: which implies, in both cases, that that to which the subject refers exists. But since at the present time nobody reigns in France, it is not the case that the truth or falsity of the proposition implies the spatio-temporal existence of an individual.

Therefore, if we want to maintain that the proposition is meaningful and hence true or false, we shall have to recognize that the subject refers to an ideal or logical entity.

In general terms it will be said, then, that to deny all reference for that class of expressions seems absurd. It will be admitted of course that they do not refer to anything existing spatio-temporally; but it seems evident that, to the extent that they are meaningful, they must refer to some particular object, and they must name that object since otherwise they would not form part of the language. So there will arise the need to characterize that object from an ontological point of view; and it will then be possible to consider what kind of object it is, what its properties are, etc. Hence, even while accepting this model of explanation, there will remain a wide area for possible agreements and disagreements.[2] But however diverse the answers to those questions are, all such accounts accept some entity as the inevitable end-point of an explanation of the meaning of at least those parts of the language. Thus in both arguments the necessity of admitting ideal or logical entities is purportedly shown as following from the meaningfulness of the expressions and of the propositions of which they are subjects.

2. Let us move on, then, to the Theory of Descriptions.[3] Speaking in general terms, we can say that Russell tries to account for the meaning of those expressions and propositions in such a way that it is not necessary to admit ideal entities. But there are diverse ways of refuting the previous arguments. If the postulation of ideal entities is founded upon a thesis about the meaning of those expressions and propositions, one could object to those arguments in the first place by showing that that theory of meaning is mistaken, and, secondly, by rejecting the thesis that if a proposition is meaningful then it is true or false. It is clear, of course, that this is not the route that Russell follows in the Theory of Descriptions. Another way would be to accept the validity of the thesis about meaning, but to hold that the mistake is to be found in the analysis of the logical form of those expressions and of the resulting propositions--which implies, clearly, that if they had had the logical form supposed by the arguments that the Theory of Descriptions opposes, the explanation of their meaning

should have been in accordance with those theses. This, it should be noted, would have one of two immediate consequences: either ideal entities are accepted or the expressions concerned lack meaning. Further--and this is an important point--Russell does not reject the principle of excluded middle: propositions whose subjects are definite descriptions will always be true or false, as the case may be. In a broad sense, this is the route that Russell follows.

Consequently, the first thing that has to be established is that the analysis of the <u>logical form</u> of those expressions is mistaken. That is equivalent to demonstrating, to put it with maximum brevity, that, on the one hand, descriptions cannot be analyzed as names and, on the other hand, that the propositions of which they are grammatical subjects are not of the subject-predicate form, since in such a case the subject would have to be a name. But doubtless this requires explanation. In one of the arguments in favor of ideal or logical individual entities at least three elements are found: (1) A general thesis about meaning which could be formulated thus: In order for an expression to be meaningful it has to have a <u>denotatum</u> and this constitutes the meaning of the expression; (2) An interpretation of definite descriptions as expressions whose function is to denote an <u>individual</u> or a <u>particular object</u>; (3) The recognition, <u>ex hypothesi</u>, that the propositions in which such expressions figure as grammatical subjects are meaningful. If one accepts these three points one will have to admit that the universe contains previously unsuspected entities. Russell's position is the following: he accepts (1) and (3), but denies (2), by which he escapes the unwanted conclusion. Now, to refute (2) is precisely to call into question the logical form of definite descriptions and of propositions in which they appear as grammatical subjects. But in order to reject (2), that is, in order to show that definite descriptions are expressions that do <u>not</u> denote individuals or particular objects, it is necessary to define the conditions under which a symbol denotes an individual. More precisely, how must we think of the relation of denotation between symbol and individual object <u>if</u> we accept the general thesis expressed in point (1)?

The answer is the following: in accordance with that thesis, a symbol will denote only in the case in which it is limited to indicating an object, to pointing it out without predicating of it, explicitly or implicitly, any property. The symbol would be a species of linguistic 'representative,' like a simple sign which indicates without predicating any property. That the general thesis would commit us to this could perhaps be explained in this way: only when the symbol is a mere 'indicator' can it be held that the denotatum constitutes the meaning of the symbol, or, that is, that if the object picked out is specified, there is nothing more to clarify with regard to its meaning, which depends completely upon the indicated object. On the other hand, if the function of the symbol reduces to that of indicating, it will be meaningful only when the object or individual which it picks out exists. If it were not to exist, the symbol would fulfill no function, it would lack meaning. From which it follows that if an expression is a symbol which denotes an individual, then it can be inferred that the denoted individual exists. And now let us notice what happens when we have two denotative symbols. If both denote, two possibilities arise: that the two denote the same object or individual or that each denotes a different object. If the first occurs, then they will have the same meaning and, therefore, a sentence which affirms an identity between them will have the same meaning as those sentences that affirm the identity of each of them with itself. That is to say, in both cases the propositions will be tautologically true. If, on the contrary, the symbols denote different objects, the sentence affirming an identity between them will mean something different from the sentences affirming the identity of each one with itself and moreover will be of necessity false. So if a symbol does not behave in this way it will not be a symbol denoting individuals. Now, Russell calls symbols or expressions which denote objects or individuals 'logically proper names,'[4] it being a further question whether proper names in ordinary language are denotative symbols or disguised descriptions. For the subject that concerns us this problem has no great importance, even though it is clear again that, when Russell, in the context of the theory of descriptions, uses ordinary proper names and contrasts them with descriptions, he conceives them as if they were purely indicative symbols. Now, Russell's thesis about what is a denotative symbol will obviously enable him to interpret the undeniable differences in logical behavior between proper names and definite descriptions as showing that descriptions are not symbols that denote individuals. Further, the differences in logical behavior are shown clearly in the following example. Suppose, that in the proposition 'Cervantes is the author of _Quijote_' any proper name symbolized by the letter _c_ is substituted for the definite description. Then we would have the following: if _c_ is the name of someone other than Cervantes, the proposition obviously is false; if, instead, _c_ denotes the same person as 'Cervantes' names, the proposition expresses a trivial truth, it becomes a tautology. But it happens that the proposition 'Cervantes is the author of _Quijote_' is neither false nor tautological and hence does not have the same form as the proposition which results when a proper name is substituted for the description.[5] So that the reading Russell will give of these differences between proper names and definite descriptions will be to the effect that the latter do not denote objects or individuals. This interpretation, as we have pointed out _en route_, is founded upon: (a) acceptance of the general thesis about meaning; (b) the conclusion that, if that thesis is accepted, symbols that denote individuals must be thought of as purely indicative; (c) the use of ordinary proper names as denotative symbols. It is clear, then, that the difference will be seen as that between a symbol that denotes and another that does not. It is worthwhile insisting, on the other hand, upon another feature of the question, that is, that the indubitable possibility of offering some other interpretation of the differences between a name and a description is ruled out if the general thesis about meaning is accepted. Let us suppose, in order to show what we mean, that someone interprets the differences between names and descriptions not in the manner of Russell but as if two different forms of denoting an individual or an object are at issue; one of them, the proper name, would do it in a direct way--a purely indicative way--and the other, descriptions, in a predicative or descriptive way. But both would _denote_ the same object. That this is a possible reading of the situation is something, we repeat, that is not disputed. But if it is adopted, it cannot then be held that the meaning of the definite description _reduces_ to the _denotatum_ and, therefore, this reading would go against the general thesis. In effect, only if the denoted object _is_ or completely constitutes the meaning of the symbol--in this case of the description--can it be maintained that when that object does not exist the expression lacks meaning. On the second reading, however, it is held first, that the name as much as the description denotes the _same_ object, and, secondly, that one of them also describes it. This implies that the object--which is the same--does not explain the difference between a description and a name. That is to say, the meaning of the description is not totally reducible to the object, since, if it were, no difference could be admitted between two expressions which denote the same object. In other words, that reading would be equivalent to admitting that descriptions do not conform to the general thesis. But if the meaning of the description does not depend completely upon the object, it is not possible to move from meaning to existence. And the force of Russell's argument consists, precisely, in making us see that if we accept one of the chains of reasoning in favor of ideal entities we have to rely upon the general thesis and that if this is the case we shall also have to accept that descriptions do not denote, so that we cannot then postulate any type of entity taking those descriptions as our starting point. In sum: if one wants to defend the argument leading to ideal entities, it is necessary (A) to support the general thesis and (B) to accept, moreover, that definite descriptions are expressions that denote individuals. But if we accept (A), we ought to reject (B)--which is what, according to what we have said, is done by Russell. Thus the fundamental error would not reside in the general thesis about meaning--whose epistemological justification is left aside in this paper--but in the persistent confusion of supposing that descriptions have certain semantic characteristics which would be found only in names. Given that propositions like 'The present King of France is wise' are meaningful, Russell argues, either the expressions that figure as grammatical subjects are interpreted in accordance with the model of names, it then being necessary to postulate an object or individual, or else that model is abandoned and then the meaning of those propositions, which remains to be explained, does not imply the existence of an entity.[6]

The distinction between names, or denotative symbols, and descriptions has at least the advantage of having limited the validity of the arguments in favor of ideal entities to the particular case of names and, further, to a specific theory--which Russell shares--about their meaning. We find ourselves, then, in this situation: descriptions are not symbols with the function of 'representing' objects or individuals within a proposition--the task exclusively of names--and yet propositions whose grammatical subjects are descriptions are used in some way in order to talk of individuals or particular objects. How is it possible to explain this without committing ourselves to the model of the name? How is it possible to analyze those propositions without their meaning implying the existence of the object even while they can at the same time be true or false? In order to answer these questions we are going to examine the analysis which Russell offers of propositions whose grammatical subjects are definite descriptions.

According to Russell, a proposition like 'The present King of France is wise' would assert, prima facie, two things: that an individual exists which is characterized by a determinate property--being King of France--and that that individual is, in this case, wise--that is, has a further property. The first thing that we must note is that in this formulation the concept of a propositional function is used.[7] In fact, the paraphrase just given would be equivalent to saying that an *x*--individual variable--exists that has such and such properties, being King of France and being wise in the previous example. And now the hidden propositional function comes to light; the use of the concept of a propositional function emphasizes a basic fact about descriptions, namely that they refer to an individual through characteristics, through predicates which, according to the case, may or may not have application. A name, on the other hand, consists, as we have seen, in a simple indication. And this would be, in the last analysis, the reason why it is possible for a proposition whose subject is a definite description to be comprehensible even in the case in which we are unaware of the individual so described. Continuing with the analysis we see, then, that 'The present King of France is wise' is analyzed thus: 'The present King of France exists and is wise.'[8]

Let us now examine how the affirmation that an individual exists that possesses the property of being King of France is analyzed: that is to say, how the statement 'The present King of France exists,' or any other of that form, is analyzed. When in relation to a description we ask about existence, what we would be asking in the first place is whether an individual exists who possesses these properties; we would be asking a question relative to a propositional function.[9] So the analysis of 'The present King of France exists' becomes an explanation of what is meant when existence is affirmed in relation to a propositional function. How is the question about existence formulated when a propositional function is under consideration? In the following way: whether an individual exists such that when we substitute the name of the individual, or an individual constant, for the individual variable in a propositional function, that function becomes a true proposition. So that when we affirm that 'The such and such exists,' we would be affirming that the propositional function '*x* is such and such' is true *at least* with respect to one *x*--*x* being an individual variable. Another way of expressing this is to say that the propositional function is sometimes true.[10] All of which would be summarized in the formula $(\exists x)\, \phi x$, ϕ being a symbol which designates a property in general.

But the assertion of existence has another aspect, that is, that of the singularity of reference or denotation which in ordinary language is expressed by the use which the definite article has in these cases.[11] When we affirm that 'The victor of Jena is Corsican' or that 'The author of *Quijote* is Spanish' we are implying that there only exists one individual satisfying these descriptions. So that it is not just affirmed that an individual exists with those attributes--the first aspect of the assertion of existence--but, further, that it is not the case that there are, for example, two individuals satisfying the description. Because it is obvious that the explanation of the first part of the assertion of existence is compatible with the existence of various individuals which satisfy the propositional function--as is illustrated by the fact that it is symbolized with an existential quantifier. Russell summarized this second aspect by saying that the propositional function has to be true *at most* with respect to one *x*--*x* being an individual variable.[12] This is equivalent to saying that any other individual, that satisfies the propositional function will be identical to the first individual of which it is affirmed that it satisfies the propositional function. If we symbolize the two aspects of the analysis we obtain the following: $(\exists x)[\phi x . (y)(\phi y \supset y = x)]$. With this ends the explanation of the first component into which the statement 'The present King of France is wise' is analyzed. At this level it is easy to understand how the second component, that asserting in this case that the present King of France is wise, will be interpreted; we would be saying that the individual of which it is said that it, and only it, possesses the attributes mentioned in the description also has this other attribute. This is equivalent to affirming that there is nobody that is King of France and is not wise. So the complete symbolization of a proposition of the form of 'The present King of France is wise' would be the following: $(\exists \underline{x})[\phi\underline{x}.(\underline{y})(\phi\underline{y} \supset \underline{y} = \underline{x}).\psi\underline{x}]$--where ψ denotes a property in general.[13] Another way of putting this would be: in the proposition 'The present King of France is wise' we affirm the existence of someone who possesses the property or properties present in the description and in the final predicate; that is, we affirm that there is at least one value of the complex propositional function $(F\underline{x} . S\underline{x})$ which converts it into a true proposition. In equivalent form: the propositional function '$\underline{x}$ is King of France and $\underline{x}$ is wise' is sometimes true. Nevertheless, if we were to leave the matter here the condition of singularity with respect to the first predicate would not be guaranteed since if we formalize the preceding we would have $(\exists \underline{x})(F\underline{x} . S\underline{x})$ and an existential quantifier does not guarantee singularity; in relation to the second predicate this condition is not of interest, given that it is not affirmed that no other wise individual exists. We have to add, then, that the propositional function 'if $\underline{x}$ is King of France and $\underline{y}$ is King of France, then $\underline{y} = \underline{x}$' is always true, with which the condition of singularity would be assured. If we now bring together both formulations we obtain the following as the final result: the propositional function '$\underline{x}$ is King of France and $\underline{x}$ is wise' is sometimes true and the propositional function 'if $\underline{x}$ is King of France and $\underline{y}$ is King of France, then $\underline{y} = \underline{x}$' is always true. In short, then, a proposition whose grammatical subject is a definite description would affirm, in the final analysis, three things--since, as we saw, the affirmation of existence subdivides into two--which, formulated in ordinary language without recourse to logical language, would say this in relation to the example 'The present King of France is wise': (1) at least one person is King of France (or if one prefers: there is someone that is King of France); (2) at most one person is King of France (or if one prefers: there is no more than one person that is King of France); (3)

whoever the person is that is King of France that person is wise (or if one prefers: there is nobody that is King of France and is not wise).[14] According to Russell, (1), (2), and (3) are implied in the initial proposition.

3. The foregoing concludes the analysis of this kind of proposition. Let us see, now, how we are placed in relation to the initial problem. We believe that in Russell's contextual analysis two fundamental aspects can be discerned: In the first place, the elimination of the grammatical subject, which is analyzed into predicates and quantified variables; in the second place, the conversion of the original proposition into a kind of complex existential proposition. As regards the first aspect of the analysis, its merit resides in the fact that the symbols or words which are used in order to express its meaning do not imply, through the mere fact of being meaningful, the existence of any object or individual. In fact, the burden of reference, as Quine would say,[15] is now carried by the individual variables, or more exactly in the final analysis by bound individual variables, which, as is known, correspond in ordinary language to words like 'something,' 'anything,' etc. Now, it is evident that these words are not names of individuals or particulars, they do not name an individual object; which is equivalent to saying that their meaning does not imply the existence of a determinate individual.[16] One could say at most, and this is still arguable, that they refer to entities in general: '...with a kind of studied ambiguity peculiar to themselves.'[17] The essential point is, in any case, that they are not names of particular objects. As regards the predicates which occur in the analysis, it is not necessary to argue that their meaningfulness does not in any way assume the existence of a particular since their function is not that of indicating individuals; this argument in favor of ideal entities would not apply to predicates. So none of the symbols which Russell employs, in his paraphrase of what is affirmed in propositions whose grammatical subjects, are descriptions can be taken as naming an individual. It is therefore not possible to apply the argument from meaning so as to postulate ideal entities. With regards to the second aspect, this allows for propositions of this type always to be true or false. In fact, if we are, with a proposition of this type, really affirming what is said by (1), (2) and (3)--in which no symbol or word is found whose meaning

implies the existence of a particular object--it is then obvious that, for example, the proposition 'The author of *Quijote* is Spanish' is false in the case in which nobody wrote *Quijote*, or in the case in which more than one person wrote *Quijote*, or in the case in which one, and only one person wrote it, but this person is not Spanish. So that when that which is asserted in point (1) is not realized--these being the cases of interest to us--the proposition turns out to be false. In symbols we would have it that the truth of $\sim(\exists x)\,\phi x$ implies the truth of $\sim(\exists x)[\phi x.(y)(\phi y \supset y = x).\psi x]$. Hence, it will not be possible to argue that if the proposition, by being meaningful, is true or false, it is necessary to admit the existence of that to which the grammatical subject supposedly refers. For, according to this analysis, the truth or falsity of the proposition does not only depend upon the application or non-application of the predicate 'Spanish' in the previous example, or of 'wise' in the case of 'The present King of France is wise,' as would be the case if they were propositions of the subject-predicate form. Hence, simple recognition that the proposition, through being meaningful, is always 'true or false,' does not require us to admit ideal entities. In order to settle that matter it is necessary to find out whether condition (1) is fulfilled or not. In this way it is possible to maintain that propositions of this class are meaningful and always true or false without the necessity of accepting the ideal existence of an entity. Russell's paraphrase sacrifices neither the principle of excluded middle nor the meaningfulness of propositions like 'The golden mountain does not exist' or 'The present King of France is wise.' Thus is refuted the purported 'necessity' of the argument in its two versions. *Entia non sunt multiplicanda praeter necessitatem*.

Let us return to the example 'The golden mountain does not exist.' If someone holds that this proposition is true *because* no golden mountain exists spatio-temporally, he is not thereby admitting, according to Russell, that that to which the subject refers exists, thus falling into all the known complications; he will be maintaining that (1) is not satisfied. That is, that the formula preceded by $\sim(\exists x)$ is true. The proposition is meaningful and true with no need for another entity. If someone holds that 'The present King of France is wise' is false *because* at the present time nobody reigns in France, then he too will not be obliged to postulate an entity through having assigned a truth-value. He will be able to say that it is false and meaningful because condition (1) is not realized. He will be saying that $\sim(\exists \underline{x})[F\underline{x}.(\underline{y} \supset \underline{y} = \underline{x})S\underline{x}]$ is true.

But let us suppose that someone admits the correctness of Russell's paraphrase and yet nonetheless holds that 'The golden mountain exists' is a true proposition. That is to say, he holds that there is one and only one value of the individual variable which converts the complex propositional function into a true proposition. Let us suppose also that he is shown that no golden mountain exists nor has ever existed spatio-temporally; and suppose he then replies that he accepts that, but that he is referring to the ideal or logical mountain. It is clear that, in such a situation, the Theory of Descriptions is impotent. If in postulating that logical or ideal entity he does not invoke the arguments based upon the thesis about meaning and the assignment of truth-values, the Theory of Descriptions is completely neutral. The only thing that remains, then, is to ask him for other arguments in favor of that ideal entity, and to object to them, if possible, using other methods. Now, if the Theory of Descriptions is neutral in a situation of this kind, that shows that its validity as an analysis of what is affirmed in propositions whose subjects are definite descriptions does not presuppose in any way acceptance of one unique mode of existence, for example, the spatio-temporal; the existential quantifier must not be interpreted as indicating spatio-temporal existence. Since if someone, using arguments other than those we have examined, postulates ideal entities, he will be maintaining that the value of the individual variable is not spatio-temporal, and in that case the proposition will be true for him even accepting Russell's paraphrase. And the basic reason for this is that for Russell to affirm that something exists is simply to affirm that the propositional function is sometimes true. If the existence is affirmed, to take an example of Quine's, of the square root of 27, existence here will not have a spatio-temporal connotation, yet nonetheless we can apply Russell's analysis in order to make the meaning of the proposition clear.[18]

It is confirmed, then, as was said some pages back, that the way in which Russell refutes those arguments is neither by objecting to the principle of excluded middle nor by questioning the general thesis of the theory of meaning. Far from rejecting the latter, he uses it in order to explain the meaning of names, implying, as we have seen, that if the grammatical subjects of those propositions were interpreted according to the model of the name we would find ourselves defenseless when faced with the first argument in favor of ideal entities. It is clear, then, that the fact that Russell subscribes to that theory of meaning provides at least _one_ of his essential motives for putting forward the Theory of Descriptions; part of the plausibility, so to speak, of the Theory of Descriptions stands in direct relation to the correctness of those theses about the meaning of names. For if those theses are presumed correct, the Theory of Descriptions presents itself as _the_ alternative for refuting that argument. It is likewise obvious that if it is possible to invalidate that argument, by showing the thesis about meaning to be incorrect, the Theory of Descriptions will lose its importance, or will in fact be unnecessary, at least as an instrument for blocking the route which leads to ideal entities. It is in this sense that we have said that part of the plausibility of the Theory of Descriptions depends upon the correctness of the thesis about meaning. But if it is shown that in this respect the Theory's plausibility is reduced, it does not follow that the model of translation offered by the Theory of Descriptions has been shown to be mistaken independently of the ontological problem. A little more precisely, following what has been said, the essential features of Russell's analysis are the analysis of the proposition into predicates and quantified variables and, consequently, the conversion of the original proposition into a complex existential one. Now, Russell _could_ have simply announced that with the grammatical subjects of those propositions we make reference by means of predicates; in order to arrive at that conclusion it would not be necessary to employ a special thesis about meaning. The truth of the thesis that in those cases we refer _via_ predicates does not depend upon the truth of a particular thesis about meaning. As regards the existential aspect of the analysis, this is still more obvious; to hold that in these propositions condition (1), for example, is affirmed does not imply a general thesis about meaning of the type to which Russell subscribes. We see, then, that if criticism of this theory of meaning--the theory Russell accepts--were correct, the only thing that would have to be abandoned would be, first, the idea that the Theory of Descriptions is <u>the</u> alternative for refuting the first argument, and further, clearly, the thesis about the meaning of Logically Proper Names. Things being thus, no inconsistency would be incurred by accepting criticism of the thesis about meaning while subscribing to Russell's analysis in its two essential aspects. Now, with Russell's analysis it is <u>also</u> possible to object to the second argument in favor of ideal entities; this is not possible relying only upon criticism of the thesis about meaning. The reason for this is simply that the second argument is not directly founded upon that thesis. The fundamental points in the second argument are the unrestricted application of the principle of excluded middle to meaningful propositions and the recognition that the relevant propositions are of the subject-predicate form. This is very different from simply saying that if they are meaningful they denote something. In the second argument, no explicit appeal is made to an explanation of the way in which those expressions are meaningful; appeal is made to a consequence of the fact that they are meaningful--without putting forward any explanation of this--and to a thesis about their logical form. So the objections must be different. Russell, by showing that the subjects must not be interpreted according to the model of names, denies that they are of subject-predicate form, while the existential aspect of his analysis shows that, nevertheless, they are always true or false--thus refuting <u>also</u> the second argument. It would therefore be logically possible to reject the first argument by means of criticism of the theory of meaning--thus reducing the plausibility of the Theory of Desriptions. But if one wants to reject the second argument in a way other than that done by using the Theory of Descriptions--by holding, for example, that the principle of excluded middle does not always apply, for whatever reasons--one will have to modify, in such a case necessarily, Russell's model of analysis. If those reasons, whatever they may be, were correct, the Theory of Descriptions would be inadequate as a general analysis of propositions like 'The present King of France is wise.'

4. Strawson's critique[19] of the arguments in favor of ideal entities consists in attacking, in the first place, the implicit theory of meaning, and, secondly, the thesis that sentences are always true or false. Strawson begins by

drawing the following distinctions: On the one hand (A1) a sentence, (A2) a use of a sentence, (A3) an utterance of a sentence; and, on the other hand, (B1) an expression, (B2) a use of an expression, (B3) an utterance of an expression. If we consider once more 'The present King of France is wise,' we see that it is possible that it was said or uttered in France during various reigns; it will be said, then, that the same sentence was uttered or said in different circumstances. It is in this sense that Strawson uses (A1). Now, the same sentence can be used on different occasions to refer to different individuals: if two people used it, for example, to refer to Louis XV, they would make the same use of the same sentence; while if one person used it during the reign of Louis XIV to refer to that King and another used it during the reign of Louis XV to refer to this other King, they would make different uses of the same sentence. In the first case it will be said that they have performed two distinct acts of uttering the same sentence with the same use thereof.[20] From this two things emerge: in the first place it is obviously possible that a determinate use of the sentence results in a true proposition or statement while a distinct use of the same sentence results in a false proposition or statement. In the second place--and as a consequence of the preceding--it is absurd to hold that the sentence refers to one person in particular, since, as Strawson writes, '...the same sentence may be used at different times to talk about quite different particular persons.'[21] Only of a particular use of the sentence can we say that it refers to one person in particular. Something similar, but not identical, can be said of an expression (B1) to refer to a particular person and, naturally, can have diverse uses, referring to different people. Is is, therefore, also incorrect to say that the expression refers: "'mentioning' or 'referring' is not something an expression does; it is something that someone can use an expression to do."[22] Leaving to one side minor points, the foregoing shows that it would be an error to mix affirmations about sentences and expressions, (A1) and (B2). According to Strawson, it is only possible to speak, for example, about meaning in relation to a sentence or expression; about reference, in relation to the use of the sentence or of the expression; about truth or falsity, only in relation to the use of the sentence. Thus, to give the meaning of 'The present King of France' is more or less equivalent to giving the general rules governing its use to refer to particular objects or persons.[23] It does not consist, then, in speaking of the person or object to which it perhaps refers in a specific use. So the meaning of an expression is not identical with any object to which it can refer on a determinate occasion.[24] As for 'The present King of France is wise,' to give its meaning also consists in part in stating the general rules for using it in assertions which can be true or false. So, neither should its meaning be identified with that which is affirmed in some particular circumstance.[25] In general terms, then, the meaning of an expression or sentence of this type is the conjunction or habits, conventions and rules for using the expression referentially and for using the sentence in affirmations.[26]

Now, the use of a sentence can be genuine or spurious (or secondary).[27] It will be genuine if, for example, a person during the reign of Louis XV had said 'The present King of France is wise'; spurious if in 1964 someone had said the same sentence. That it be spurious (or secondary) has to do with the use of the sentence and does not call into question its meaningfulness; in the case in which reference fails, it will be said that the sentence has been used spuriously and not that the sentence lacks meaning, since in order that it be meaningful it is sufficient, Strawson tells us, that it be possible to describe circumstances in which its use will result in a true or false statement.[28] This is obviously possible in the case of 'The present King of France is wise,' In sum, the meaning of a sentence of this type does not require that, each time that it be used, there be something to which it refers.[29] It is not possible, therefore, to argue that if it is meaningful it is necessary that it always refer to something in particular, leaving one of the ways open for ideal entities. The definitive error common to Russell and to those favoring such an argument would be that of having mixed meaning with reference--of not having distinguished (A1) and (B1) from (A2) and (B2). The Theory of Descriptions, in this point, tries to solve a pseudo-problem.

Now, when a sentence is used spuriously or secondarily, Strawson claims it to be neither true nor false. This, in his scheme of things, is the same as saying that if 'The present King of France is wise' is meaningful, it does not follow of necessity that uses of the sentence will always result in true or false propositions or statements; its use will decide whether it is true or false or neither of the two. By this is countered the second argument in favor of ideal entities without having to invoke one objection that the sentence under consideration is not of subject-predicate form. Note, nonetheless, that the distinction between the meaningfulness of a sentence and a particular use of it in no way implies, on its own, that when reference fails the sentence is neither true nor false. From the one the other does not follow. One could accept the distinction and yet maintain that when in a particular use reference fails what results is a false proposition. The only thing that the distinction commits one to holding is that it is not necessary to assume some other reference to something ideal in order to explain the meaningfulness of the sentence: if we accept the distinction, the sentence is meaningful without the supposition of something to which it refers. But we remain free to say that it is false, so that if Strawson thinks that in a spurious use the sentence is neither true nor false, he will have to adduce other reasons. To accept the distinction, in sum, is to concede that the outcome of a use of the sentence does not affect its meaningfulness--which does not mean that there are uses which are neither true nor false. So the fact that a sentence is not susceptible of being true or false---truth-values not applying to the sentence--should not be confused with the fact that it is neither true nor false in a determinate use. Now, in order to maintain that in a spurious use the sentence is neither true nor false, one has to reject the thesis that uses of the sentence affirm the unique existence of the subject, that is to say, points (1) and (2) of Russell's analysis. But before considering Strawson's reasons it is useful to note that, if Strawson limits himself to the first distinction, that about meaning and reference, Russell would still be free to accept that distinction and to retain at the same time and without falling into any inconsistency his analysis making all the propositions true or false. In that case, as has already been pointed out, his theory would be unnecessary in order to counter one of the arguments in favor of ideal entities. But the validity of his analysis would remain untouched. And, related to it, Strawson's first distinction could be correct without his reasons being correct for thinking that when a sentence of that type is used elements (1) and (2) of Russell's paraphrase are not affirmed.

Strawson holds that in place of being affirmed they are *presupposed*. According to this, when one asserts 'The present King of France is wise' one presupposes that one, and only one, man exists who reigns in France. The translation of his definition of presupposition says this: '*S presupposes* S^1' is defined in the following way: 'The truth of S^1 is a necessary condition of the truth or falsity of *S*.[30] So that if S^1 is not true, *S* is neither true nor false. If in 1964 someone asserts that 'The present King of France is wise,' the sentence will express neither a true statement nor a false one since the presupposed statement is not true--even when it is believed that something true is being said.[31] So if S^1 were a necessary condition only of the truth of *S*, it would be contradictory to assert *S* and the negation of S^1--as would be the case according to Russell's analysis, while to assert *S* and the negation of S^1, if S^1 is a necessary condition of the truth or falsity of *S*, will not result in a contradiction but in another kind of logical absurdity. Strawson thinks that this way of putting the matter reflects with greater fidelity the way these sentences are used in ordinary language; and, in reality, this is the essential reason (if not the only one) that he offers in support of his Theory of Presupposition in relation to this type of sentence. Because if this were not the reason, the debate with Russell would be set up problematically within a vacuum and one would not know the criteria by which to choose between them. For we have seen that the two theories are able to counter the second argument so that this cannot be used as a criterion of choice. The advantage which Strawson claims for the Theory of Presupposition is that, beyond solving all the problems which the Theory of Descriptions tries to solve, it better reflects ordinary language. In order to support the idea that the Theory of Presupposition is a more faithful account of what happens in ordinary language, Strawson relies, naturally, upon examples. Let us examine one of them. If someone in 1964 told us in all seriousness that the present King of France is wise, Strawson thinks that we would not say that it is not true; and that if we were asked if what he had said was true or false, we would answer that neither the one nor the other, that the question of truth or falsity does not arise--precisely because at the present time nobody reigns in France. In place of assigning truth-values, we should try to explain to him that he is mistaken. So that in saying that at present time nobody reigns in France, we should not be contradicting the statement, but rather giving a reason why truth and falsity do not arise.[32] Thus we arrive at the situation which the definition of Presupposition tries to encapsulate. The correctness of the Theory of Presupposition, inasmuch as it reflects ordinary language, is always proven by Strawson in analogous situations: if someone asserts S and another person knows that S^1 is not true, the second person, if asked whether S is true or false, will refuse to assign a truth-value to S.

It would be absurd to deny that there are contexts which exemplify Strawson's Theory; it would be equally mistaken to think it the most adequate in every situation, since he himself has recognized in subsequent writings that there are exceptions.[33] Nonetheless, it has to be pointed out that even if Strawson were completely right—and this is not the case--in the sense that his Theory of Presuppposition always 'reflected' with greater precision the use of those sentences in ordinary language, the analysis Russell applied to affirmations of existence would not be affected. That is to say, if Strawson were right, 'The author of Quijote exists' would not form part of what is affirmed by—or of the analysis of, if one prefers--'The author of Quijote is Spanish'; but not one reason would have been adduced against the way Russell interprets the affirmation of existence. When existence is asserted, Russell's analysis will be the appropriate one. Now, the Theory of Descriptions does not just apply to propositions like 'The author of Quijote exists'; so we would be making a mistake if we held that acceptance of the Theory of Presupposition involved a wholesale repudiation of Russell's analysis. This limits considerably the scope of Strawson's critique. Finally, it is worth noting that the criterion--correspondence to ordinary language--by which we are invited to choose between Presupposition and the Theory of Descriptions--in relation only, let us not forget, to sentences like 'The present King of France is wise'--will be reasonable, of course, but is in no way the only one. For other kinds of considerations it may be convenient that all propositions be true or false. Then the Theory of Descriptions would be an appropriate instrument. But that would be to enter into another problem. In any case, what it is not possible to reject is that the theory of meaning Strawson proposes is, in broad terms, true.

NOTES

*This article was first published in Diánoia in 1964. A modified version appeared in Lenguaje y Significado (Mexico: Siglo XXI, 1969). This translation by Prof. Mark Platts is from the latter version. I wish to thank him for his effort.

1. I do not intend to state precisely to what extent Russell interprets Meinong correctly. For an exposition on Meinong see J. N. Findlay, Meinong's Theory of Objects and Values (Oxford: Oxford University Press, Second Edition, 1963).
2. See, among others, Max Black, "Russell's Philosophy of Language," originally published in A. Schilpp (ed.), The Philosophy of Bertrand Russell (Tudor Publishing Company, Third Edition, 1951). Also published in M. Black, Language and Philosophy (Ithaca, NY: Cornell University Press, 1949).
3. Henceforth I shall use "Theory of Descriptions" for Theory of Definite Descriptions and "Descriptions" for Definite Descriptions.
4. B. Russell, Introduction to Mathematical Philosophy, p. 174; Logic and Knowledge, p. 244.
5. B. Russell, Introduction..., pp. 174-75; pp. 245-46; see also Principia Mathematica (Cambridge: Cambridge University Press, 1962), p. 67.
6. P. F. Strawson, "On Referring," in Essays in Conceptual Analysis (Macmillan, 1956), pp. 24-25.
7. B. Russell, Introduction..., pp. 155-56.
8. B. Russell, Logic..., p. 250; Introduction..., pp. 177-78.
9. B. Russell, Logic..., p. 232.
10. Ibid., p. 249; also Introduction..., p. 177.
11. That this is not the only way to use a sentence of this type is evident. "The whale is a mammal" expresses a universal proposition. See Stebbing, A Modern Introduction to Logic (Methuen, 7th Edition), p. 149.
12. B. Russell, Logic..., p. 249; Introduction..., p. 177.
13. It seems nevertheless misleading to say that it is only a matter of assigning another predicate, for this would make us suppose that they all serve the same purpose. This is confusing. The common trait is that we are really dealing with predicates in both cases and in relation to the same individual; the distinction

consists in the fact that the last property is predicated of someone who has been identified by predicates. We find an indication of this distinction in the fact that the condition of singularity is established only in relation to the first predicate; the one in the definite description.

14. B. Russell, Introduction..., p. 177.
15. W. O. Quine, "On What There Is," in From a Logical Point of View (Harvard University Press, 1953), p. 6.
16. That is the reason why, for Russell, it is meaningful to affirm, deny or question existence only with respect to descriptions. If at a certain moment we ask, for instance, if something exists--and the question is a genuine one--then the grammatical or apparent subject of the question is certain to be a description for, in accordance with his thesis, if a word is a name, that of which it is a name exists. Therefore, if it is a name, there is no room for doubting or questioning the existence of that to which it refers. If we did such a thing we would fall into an absurdity similar to asking: " That which exists, does it exist?" See B. Russell, Introduction..., pp. 178-79; Logic..., pp. 252 ff.
17. W. O. Quine, op. cit., p. 6.
18. Ibid., p. 3. It is impossible to examine this problem more closely here, for that would take us into the theory of types. See Stebbing, op. cit., p. 162 and M. Black, op. cit.
19. P. F. Strawson, op. cit., Introduction to Logical Theory (Methuen, 1952), chapter 6, esp. pp. 184 ff. Naturally in what follows I do not pretend to consider everything Strawson says in his paper or his book about this problem. I shall only take into account some of his theses.
20. P. F. Strawson, "On Referring," pp. 27 ff.
21. Ibid., pp. 28-29.
22. Ibid., p. 29.
23. Ibid., p. 30.
24. Ibidem.
25. Ibidem.
26. Ibid., p. 31.
27. Here we will consider only one spurious use: the one in which we pretend to be seriously referring to something or someone. I will not touch upon the problem of these uses in, for instance, fictional contexts.
28. P. F. Strawson, "On Referring," p. 35; Introduction...,

p. 185.

29. P. F. Strawson, Introduction..., p. 185.
30. P. F. Strawson, "A Reply to Mr. Sellars," Philosophical Review 63 (1954), 216; Introduction..., p. 175. Relative to the difference between Frege and Strawson on this point and in relation to certain difficulties of Strawson's definition, see Max Black, "Presupposition and Implication," in his book Models and Metaphors (Ithaca, NY: Cornell University Press, 1962), pp. 48 ff.
31. Strawson explicitly states that the relation of presupposition should not be interpreted in terms of 'belief.' On this point and in relation to the role assigned to the belief that S^1 is true, see "A Reply to Mr. Sellars," pp. 216-217.
32. P. F. Strawson, "On Referring," p. 34.
33. P. F. Strawson, "A Reply to Mr. Sellars," pp. 225 ff.

Fernando Salmerón

ETHICS AND THE LANGUAGE OF MORALITY*

1. CONCERNING THE STUDY OF MORALITY

What is sought here is only to point out the relation of ethics, understood as a branch of philosophy in its strictest sense, to the language of morality. But it is taken for granted that in these few pages no precision is possible upon the meaning of "philosophy"—although by the end of this article some light may be shed on it in an indirect way. The same will be the case with the concept of morality, the definition of which it will also not be possible to advance here, although at times we shall have to imagine some of its essential characteristics. Within these limitations, these notes would like to take the form of an analysis of ethics, that is to say, a philosophical exercise related to questions concerning the language of morality.

The approach to the complex phenomenon of morality can be made in different ways. It is possible to study morality as a social fact, to speak of its function as a harmonizer of interests, to relate it to other social realities and to explain, in relation to these, its structures and changes throughout history. It is also possible to elucidate upon its origins in an individual psychology, its functions as a guide to action in the service of a moral attitude, its relation to impulses and beliefs, thus making the conduct of an individual or a group more comprehensible. In all these cases, it is clear that the task remains in the hands of those scientific disciplines, i.e. psychology, history, anthropology, and sociology, to answer our questions concerning actual moral behavior in a determined community and in a certain period in history. However, although these answers can contain ample information about moral conduct—of trials of value and in general of the reactions which inspire and stimulate this conduct; of standards and principles really in force; and of the convictions and doctrines of educated people—they do not, strictly speaking answer moral questions.

J. J. E. Gracia et al. (eds.), Philosophical Analysis in Latin America, 243–253.

The answers to moral questions--attitudes, words, actions--are given in our everyday moral life and, together, constitute morality itself, but do not originate from scientific investigation. Although science, with its advances, can contribute to modify our beliefs and our moral ideals, it is by itself incapable of giving us other new ideals in exchange. The transition from considerations concerning what is reality in our experience to conclusions of what behavior should be, remains an impossible leap from a scientific point of view. On the other hand, all the logical operations--to describe, to theorize, to explain, to foretell, etc., require a certain distance between the investigator and his subject, in relation to his practical commitment.

Even in the case where the subject matter to be investigated is morality in any one of its aspects, the commitment remains part of the object of the study. On enunciating a rule, for example, or stating an obligation of a moral recommendation, the investigator quotes an expression and on writing it in quotation marks takes it out of its original context and leaves it in the field of description and of theory. Scientific operations, in this sense, are neutral from the point of view of moral appraisal --which does not mean that, in determinate contexts the results cannot have important moral consequences and even form part of a moral argument--even though science itself as such, as a human task, is not neutral from the point of view of morality and moral attitudes.

Does the middle class of Mexico consider abortion to be good or worthy of praise? How is the diversity of rules which govern the forms of matrimony in the towns of the coast of Guerrero explained? What part does school education play in the formation of moral and political ideals of Mexican youth? Questions such as these should be answered by social scientists, but whatever their answers may be, they do not constitute a pronouncement either on the morality of abortion or on the value of marriage or the political ideals of the dominant classes in relation to other ideals. Pronouncements of this kind are not the result of scientific work and operate, in practical life, as guides to action.

Nor can philosophy--ethics as a philosophical discipline --give answers to questions regarding the moral life, although it can contribute to clarify them. What philosophy faces are questions arising from morality, but which are not examined as psychological or as social phenomena, but rather as expressions of these same phenomena. It amounts to an approximation to reality itself, but at another level and by another route: that of the analysis of moral language. Philosophy inquires about the significance of gestures of rejection and expressions of praise; the significance of key expressions of the moral language; the relations between the standards which govern conduct and the principles which serve as a basic pattern; the basis of these principles and the ultimate ideals of life; the logical structure of moral reasoning and the truth value of the most complete systematic moral doctrines, which compose a universal view of the world and assume this or that characteristic of human nature. But to make hypotheses explicit, clarify significations, examine logical ties and question attempts at developing foundations, does not primarily offer guides to action, although it can at times contribute to their change or the loss of their effectiveness in our conviction.

The characteristic feature of morality, in contrast, is exactly the essential character of its practical element: its commitment to action in circumstances which cannot be eliminated however much we may accumulate reflexive power in the service of this practice. The traditional distinction between customary morality and reflexive morality, which appears to divide in two the history of Western thought, is a relative distinction, debatable as historical happening--and requiring greater precision from the point of view of the examination of morality. In a way, reflection has intervened in the formation of a traditional system based on custom, and it is equally important to recognize the importance of habits and social usage even in doctrines which are offered as rational.

In the examination of the language of morality it is possible to note greater distinctions, each of which can be shown as a different level of discourse. The determination of these levels makes it possible to manifest the complexity of moral reasoning, the difficulty of the most frequent logical steps, and the truly theoretical and critical character of ethics as a philosophical discipline.

2. THE FOUR LEVELS OF MORAL DISCOURSE

The first level of moral discourse can be registered from expressions of satisfaction or disgust, rejection or praise, apparently spontaneous, which by their own nature do not raise questions of truth or validity. Taken seperately, and not as part of a moral attitude, which is by definition a complex phenomenon, these expressions can, however, seriously influence the feelings and behavior of others. Its evocative and influencing capacity is what has been put into evidence by the analysis of emotivist philosophers, who see in this characteristic trait the essence of moral discourse. This does not cease to be an exaggeration difficult to accept. For, even accepting that any kind of expression can have an influence upon someone's conduct, and therefore moral significance, it seems prudent to admit that this extensive area of the isolated expressions of feelings cannot be presented as an example of moral discourse. Questions relevant to ownership or justification of expressions do not fit in here and when these are raised in connection with a disagreement, for example, the discourse has been changed to another level.

At a second level of the language of morality, what chiefly counts is not the mere influence upon conduct, but the guide. An expression which acts as a guide to action should present itself as intelligible from all angles and, for that reason, has to have a rational accessibility which can be discussed in terms of correctness or propriety. It is offered as a reply to a practical question and presents itself in the form of a prescription--independently of its emotive force. Prescriptivist philosophers have studied the simple imperative as the elemental and basic form of moral discourse. But this should not be understood as meaning that to express a moral judgment is always prescribed, and that all statements of this kind invariably reveal themselves in the imperative form. There can be a great variety of grammatical forms in the expression of choices and decisions, of evaluations and criticisms, of admonitions and advice--the decisive factor is that all of them function as guides to action and, one way or another, presuppose, imply, or lead to an imperative.

The difference with the level of emotive expressions is exactly in the recourse to the imperative which functions as a rule of conduct and that, by definition, is safe from the allure of changing inclinations and personal preferences. Even that norm which orders the pursuit of the call of passions as a rule of conduct, as paradoxical as it may seem, would be presented as a principle of justification with pretensions of universal validity.

In our daily judgments concerning the government of our conduct and the criticism of that of others, we frequently trust in a set of norms which may be called the moral code accepted by the community. At this level, our reasoning does not question norms or institutions, nor does it doubt the possibility of future changes, but simply relates an accepted principle to the description of a particular event. An argument which justifies a particular behavior demonstrates its agreement with a given body of norms and has the same logical structure as any validation analysis in the realm of scientific knowledge. When it measures the practical results of an action, it makes inductive inferences; when it places a particular case under an ordinary moral rule, it makes deductive inferences. And the procedure ends by making explicit the norm applicable to the case. Something else happens if we come across a conflict among norms or if we discover that the people reject a certain description for ruling behavior. The conflict leads to a criticism of norms, to adjustments among them, to a search for principles of a more general order which could harmonize them and support their justification. But the search for reasons, completely subordinate to the demands of practice, does not submit itself to special methodological requirements--although it already represents a new level of moral discourse.

Moral reasoning prolongs, in this third level, the normative discussion in a search for the relation between ultimate principles and norms; it tries to make explicit the reasons which are behind the justification and it even organizes them systematically, but does not exceed the limits of the systems themselves. All its reflective strength remains, as it were, within the primary moral discourse, integrally rooted in practical life in such a way that the systematic and doctrinal display does not seek the objectivity of scientific theories, even though it may assume decisions of attitude, comparable in many ways to scientific requirements. This is fittingly the field of moral reasoning, where the normative systems, codes, and great doctrines come to life. All the problems of morality have a place here and before any new proposal, the system will find flexibility and resources for the consequent reply --without excluding even the understanding of other moral parallel doctrines. But this understanding does not eliminate the conflict between doctrines nor the diversity of attitudes and, above all, it does not reach the extreme point of this diversity, which is the questioning of morality itself as such. The question: "Why should I be moral?" is a complete challenge which has no answer within ordinary moral discourse.

The challenge of the egoist who asks for specific norms of justification for morality as such, in so far as it guides human action, has no place in the realm of morality. The question goes much further than moral discourse, including the treatment of the doctrinal and metaphysical elements of moral discourse, in a useless attempt to exact the establishment of attitudes and ideals of life. Moral reasoning reasons from principles, but the final questions are answered with decisions concerning the value of the basic principles of a moral attitude. The most systematically elaborated moral doctrine still requires the vindication of a fundamental norm, of a final principle or of an idea which, in an inevitable way, is connected with a universal view of the world.

If, in connection with emotive expressions, from which the first level of the language of morality arises, a rigorous formulation of questions of truth or validity were not possible, neither would it be possible in connection with the final level, which is that of the expression of moral attitudes. And even though these expressions do not escape completely from analysis, they maintain an irreducible point of decision which signals the rational limits of discourse.

In reality, in the second and third levels, the stated moral argument is formed. Here the logical steps are located, as well as the descriptions and definitions which are possible to control using the same procedures used to control the discourse of science. But ordinary moral discourse cuts across these two levels and goes beyond them —although rarely in an explicit way. The references to imperatives and to formulas of universal claim is established from attempts at the justifications of emotive reactions. In this way, isolated reactions demonstrate that their simplicity is only apparent, make explicit their dependence upon moral traditions, and in the background point toward very complex conceptions, which are expressions of fundamental attitudes. The usual moral arguments connect norms and principles with descriptions of conduct and events of all kinds, relating statements of diverse grammatical forms and diverse idealogical ranks; they determine prescriptions, indicate the pertinent value of facts, logically relate norms and events, and defend the articulation of all of them and the soundness of the whole. But in doing the latter they move to the fourth level of moral discourse where use is made of statements whose content is impossible to corroborate.

3. THE LEVEL OF ETHICS AND ITS FOUR LEVELS OF STUDY

Ordinary moral discourse, however, is only exceptionally explicit in all its steps. The complexity and dynamics of moral life, at the service of which this discourse arises, maintains such an index of demands that it remains condemned to form elliptical shapes: to manifest itself in incomplete reasonings, to silence premises and presuppositions, to allude hardly to, when not ignoring altogether, the conditions resulting from circumstances and cultural heritage.

To rebuild the path of discourse in its integrity, to make explicit cultural conditions and doctrinal hypotheses and to clarify the significance of expressions and key terms, is something which is beyond systematic moral doctrines and reflexive morality. It concerns an undertaking of a different nature which requires some distance from the exigencies of practical life, a critical interest, but above all a theoretical interest which distinguishes it from moral discourse in all its forms. This represents another different level in the language of morality: exactly that which corresponds to ethics or moral philosophy, understood as a philosophical discipline in the strictest sense.

Properly speaking this is a meeting point of moral thought, the nucleus of the origin of philosophy as wisdom or universal view of the world with philosophy understood in the strictest sense. Here an attempt is made to show that the indicated levels of difference do not represent mere opposition--so common in the philosophical literature of recent years--between ethics and metaethics, but a more radical distinction between two very different meanings of the term "philosophy." There is no objection in treating it as a distinction of levels of moral language, so long as all that follows is taken into account.

In the first place, it must be made very clear that the level of philosophical analysis maintains as the object of study the language of morality, and that the use of expressions and moral terms throughout the discourse is a form of mention which neutralizes its nature and its practical function. The neutrality of philosophical investigation, in the face of moral doctrines and systems, does not signify anything else. But it is obvious that the results of the investigation can have indirect consequences in moral life, and it is also true that the investigator can be led to the formulation of philosophical problems as a result of preoccupations arising in practical life.

The best way to make clear the distance which prevents ethics from identifying itself with its object--which is the language of morality, whether in spontaneous form or as a reflexive morality--is to point out within ethics itself a few investigative undertakings, parallel to the levels indicated for the language of morality. For example, opposite the first level, relative to the most spontaneous expressions of our moral life, it corresponds to the ethical theory the task of taking into account the significance of these expressions and of clarifying the role of the key terms that appear in them. At the second level, that of moral reasoning which relates particular cases to expressions in the imperative form, norms or principles, it corresponds to ethics the logical structure of moral reasoning and the critical analysis of the ways of confirmation. If we consider the third level, in which moral discussion systematically organizes norms and principles of justification, we find in ethics the parallel task of the analysis of these logical connections, of a theory of norms and normative systems--all of which leads to what is called deontic logic. Finally, faced with the basic decisions of an attitude and its final expressions, that is to say, faced with individual and collective ideals which are connected in a universal view of the world, ethics has to respond to the most complex of its tasks: the theory of action, liberty and responsibility, and the examination of the truth value of systematic moral doctrines. In this field questions are not only logical or semantic, but frankly epistemological.

Secondly, one must take into account that the four levels of the language of morality mentioned above can function in principle without any methodological restriction, and from fundamental decisions of attitude--which do not prevent it from attempting the formulation of a moral doctrine with rational and even scientific claims. On the other hand, within the field of ethics and its four levels of study, there are no alternatives: philosophy in the strict sense proceeds in the same way as the sciences, with established method and from a theoretical framework justified by pragmatic reasons. However, some very brief considerations have still to be added with regard to these.

4. ETHICS AND MORAL ATTITUDE

The question "Why should one be moral?", without a reply in the field of morality, can be presented from the point of view of philosophical analysis as a question without any meaning. At the same time, it can prolong its intrusion so as to arrive at philosophical questions: Why should one reflect philosophically about morality? Now it is not only a question of the openly accepted possibility of unreflecting moral positions or even of irrational doctrines. It is a question of asking philosophy itself to account and give reasons for its existence, regardless of the acceptability of its methods and of the pragmatic reasons which may support the theoretical framework of its work of investigation. It is a question which has no place within ethics, but rather one which has a clear moral sense and returns the discussion to the field of practical problems. It is a terrain in which ethics cannot give answers, because its task is not to dictate norms, nor to justify principles or behavior. When it is to characterize the general procedures of validation, by which means the imperatives can be justified, ethics does not offer rules of conduct nor models of behavior, but only empty formulas which allow any rule to be put to the test. These are formal criteria which fulfill a regulating function, which remain open to discussion and which can be changed the moment they appear inadequate.

The question indicates the limits of philosophical activity and, on the other hand, the critical and corrective character of ethics discovers that formal criteria assume methodological decisions which are not self-sufficient from the point of view of its justification as a human task. Philosophy remains as a second order activity, but the restriction to which its proceedings are subjected are not foreign attitudes and clearly define its theoretical and critical functions.

This is the other point of contact between moral thought, the nucleus of the origin of philosophy as wisdom or universal view of the world with ethics understood as a philosophical discipline in the strictest sense. The all embracing and comprehensive character of universal views of the world offers itself as the only way to begin the moral justification of ethics as a human task. And it is quite clear that at this point of contact there is no deductive or logical connection--it is simply a practical connection, a relation of moral order. Once again the language of moral deliberation recovers its first task and philosophical analyses remain out of place. All neutrality has ended.

The neutrality achieved at the levels of philosophical investigation of morality has, as has been noted, the same limitations as the neutrality of science: those which originate from its methodological options. Imperceptible limitations in the first analysis of moral expressions or of the fundamental notions, but conspicuous in the treatment of systems and in the concept of morality itself. If it has been accepted as a methodological rule, for example, that all procedures of analysis should be subject to revision and in no case lead to statements which in turn cannot be subjected to rational criticism, a proposal has been accepted--which must be discussed in the light of results, but which still remains as a proposition of dubious neutrality in philosophical practice. On the one hand, it presupposes attachment to a point of view which is not independent of the rationalistic tradition and criticism of western thinking; and on the other hand, it qualifies in advance the concept of morality and the soundness of moral systems.

* From _Man and His Conduct: Philosophical Essays in Honor of Risieri Frondizi (El hombre y su conducta: Ensayos filosóficos en honor de Risieri Frondizi)_, ed. Jorge J. E. Gracia, (San Juan: University of Puerto Rico Press, 1980), pp.305-315.

Enrique Villanueva

THE PRIVATE LANGUAGE ARGUMENT*

In the past twenty one years the possibility or impossibility of a private language has been discussed with a certain degree of obsession. The name "Private Language Argument"[1] covers a multiplicity of different issues. Some philosophers want to see in the PLA a paradigm of philosophical argument with devastating consequences for the Cartesian Thesis on the Philosophy of Mind and the Metaphysics of Persons;[2] others only see obscure statements that in no case make an argument or a definite point.

In the first part of this paper I shall present one line or argument. In a second part I shall pursue this line as it appears in several contemporary philosophers.

FIRST PART

1. A. J. Ayer's Interpretation of the PLA

In a joint Aristotelian Society's symposium with R. Rhees (1954), Ayer held that for an isolated man like Robinson Crusoe it is possible to have and use a PL. Crusoe introduces the words of that PL by private ostensive definitions.[3]

According to Ayer, there is no difficulty for Crusoe to invent a PL for his private sensations and, even if a certain technical difficulty is possible, there is no logical difficulty or difficulty of principle. The main technical difficulty is that of verifying or checking Crusoe's use of his words. But Ayer held that to be also a difficulty for our public or shared language.

It is far from clear what the meaning of Ayer's thesis is. According to J. J. Thomson, Ayer holds two quite irrelevant theses, namely:

J. J. E. Gracia et al. (eds.), Philosophical Analysis in Latin America, 255–276.

(1) That there is a language which, in fact, has not been understood by any other person besides the one who speaks it.

(2) That sensation reports are not unintelligible for other persons besides the one who utters them.

Medlin, instead, thinks that Ayer distinguished two senses or 'private,' namely:

Private^1: A language used by one person to refer to his private experiences.

Private^2: A language whose expression could convey indirectly some information to other persons but would not mean the same for them as it means for the one who has the experiences. (Here we have a logical impossibility).

According to Medlin, there could be a PL which is Private^1 while not Private^2; that is, Ayer would hold that Private^1 does not imply Private^2 and thus it could be perfectly possible to have a PL which is not Private^2 and therefore not absurd or unintelligible.

But then, against J. J. Thomson's interpretation, (1) and (2) are relevant in the following way:

(2) Breaks the implication from Private^1 to Private^2.

(1) Is a counter-example to the interpretation according to which to report a sensation is not different from reporting the expression of that sensation (this is the thesis of <u>Aeusserungen</u>,[4] according to which there is a non-contingent relationship between a sensation and its manifestations or expressions). Ayer's retort consists in stating that there is something else besides the expression and that additional thing is tied only contingently to the expression: This additional thing is what 'private' is meant to express.

It is important to observe two things:

(I) Ayer does not commit himself to the doctrine of private objects (a doctrine that many people who use the PLA think must be held together with the idea of a PL). Ayer holds only an empirical view of sensations or else a proof must be offered that Ayer's empirical view implies the idea of private objects.

(II) Ayer holds that Aeusserungen is a mistaken view if taken as a general account of all self-ascription; there is something else beyond the natural expression, but this additional thing is only an empirical item.

But then, according to this interpretation of Ayer's views by Medlin, Ayer does not go against the PLA and his views are perfectly compatible with it. According to Medlin's interpretation, the PLA states that Private^1 reduces to Private^2 and thus not even the user of PL could check or verify his use of, i.e., the sign S. Against this contention Ayer gives the following reply:

> My argument is that since every process of checking must terminate in some act of recognition, no process of checking can establish anything unless some acts of recognition are taken as valid in themselves.[5]

That is, Ayer says that it is arbitrary to suppose that a private^2 statement cannot--because it is Private^2--be verified or corroborated by another Private^2 statement.

Let's set aside the matter concerning who has to prove the implication from Private^1 to Private^2; Ayer does not discuss it but instead tries to refute his thesis (2) whereas he should prove that there could be re-identification even if the sensation is completely private, that is, with no possible expression. Ayer offers a very poor argument for this last point[6] which cannot eliminate the arbitrariness which befalls the applications of S and thus cannot show that S is a private name.

More important for the later development of the dispute was the diagnosis Ayer made of why S cannot become a private name. Ayer thought that there are two reasons why S is not a name, namely:

> (A) For a person to be able to attach meaning to a sign it is necessary that other people should be capable of understanding it too.
>
> (B) It is... logically impossible to understand a sign unless one can either observe the object it signifies, or at least observe something with which this object is naturally associated.[8]

Thus Ayer fuses the nominative theory of meaning with the use theory of meaning. (Such heterodoxy was what spoiled his efforts according to the defenders of the PLA.)

Concerning (A) Ayer argued that, even if psychologically relevant, there is no logical absurdity in granting that no one understands the meaning of a word or sentence apart from its user; this much agrees with the PLA. What Ayer does not make clear--and is relevant according to the PLA defenders--is whether it is coherent to hold the existence of a meaning that nobody could possibly understand apart from its possessor. Later, Ayer accepted that this was a mistake and said:

> In any language which allows reference to individuals there must be criteria of identity which make it possible for different speakers to refer to the same individual. This would not prevent the language from containing private sectors, but it would mean that my idea that these private sectors could be made to absorb the public sectors was not tenable.[9]

In addition, the quotation allows us to see the factual connection between a defense of PL and holding a phenomenalistic program.

Concerning (B) Ayer ascribed the nominative view of meaning to the defender of the PLA, bringing with it great confusion. For, from then onwards, it was held that the reason why S cannot be a private name was because other people cannot observe what S designates or refers to and Crusoe's observation cannot help, as was stated above. Ayer thought this thesis was too strong, since it also questions the meaning of sentences about the past. According to Ayer, once we eliminate (B) there is no difficulty in observing indirectly and thus checking uses of S and so granting a PL for Crusoe.

2. Strawson's Version

P. T. Strawson introduced the notion of 'criteria'[10] into the PLA discussion, in such a way that the validity of the PLA implied philosophical behaviorism.

According to Strawson, what makes a language private is the fact that the referents of its words are sensations and these are private. Thus the language with which Strawson is concerned is in fact private. Then, he says that there are two theses about PL, namely, one strong and one weak. According to the strong one, no words name sensations; the weak one says that certain conditions must be satisfied if it is to be possible to ascribe sensations to the persons who have them.

But the weak thesis becomes the strong one when the thesis of criteria is introduced; it is at that moment that the horror for the private makes the PLA something excessive and implausible.

According to Strawson, the reason for not having a criterion of correctness for the application of names to sensations is that the defenders of PLA bring scepticism into the use of memory. Just as Ayer, Strawson thinks that this reason would cancel a public language, e.g., color language. On the other hand, Strawson sees no difficulty in introducing a private name: It is introduced from some recurrent characteristic in the sensations.

Strawson believes that there is another argument against the PL: If 'it hurts' cannot be doubted, nor is it an identification by criteria, then it is equivalent to a scream (uttered only after a certain teaching). Then 'pain' is neither the name nor the description of a sensation. Names or descriptions require identification or recognition and those in turn demand criteria. But if in self-ascription there are no criteria, 'pain' will not be a name in those cases, and so, there will be no private names.

In order to get rid of private names and thus of a PL, it is stipulated that self-ascriptions are not names but 'expressions' (Aeusserungen), and it is concluded that there are no names for sensations: The weak thesis has turned into the strong one.

Strawson thinks that this is excessive and must be corrected. There are names for sensations--this is a common empirical fact--and the condition for them is that sensations express themselves in the behavior of persons. To recognize the existence of that condition is something different from stating that pain is only an exclamation. Obviously, pain is something more than an exclamation, but without the latter it would be impossible to identify and name the former. Pain needs criteria and criteria lie in the behavior of persons, but it is a mistake to substitute pain for pain behavior. It is because there are criteria that philosophical doubts about other persons' pain or the knowledge we have of these experiences cannot be introduced. This is all we can conclude from the PLA.

Now, if anyone thinks the opposite, namely, that a PL is possible, Strawson thinks that such belief is not senseless or unintelligible, but only vacuous or pointless. This is enough; any other demand goes too far and is an exaggeration.

It will be clear that Strawson never considers the philosophical privacy that the PL seems to imply. Probably he thought that no philosopher in his right mind would assume such a thesis.

In any case, Strawson made a general and comprehensive interpretation that prompted discussion. The answer to Strawson and Ayer assumed two forms: one essentialist and the other pragmatic. Malcom and Rhees were the champions of each of these. In what follows, I shall be concerned with the 'essentialist' answer.

3. Malcom's Answer

Defenders of the PLA thought that the Strawson-Ayer interpretation was full of common sense but in some ways naive. Malcom thought that in the *Philosophical Investigations* there is a serious, deep and conclusive argument against a PL and on the basis of it the Cartesian is left without the means of expressing his own private dualist theory of persons. Strawson and Ayer commit the mistake of thinking only of a language that no one else apart from Crusoe understands, but the issue is with a language that no one else could understand. The idea of a PL is an extreme idea whose error had escaped philosophers thus far.

Britist Empiricists, Cartesians and those who believe in the idea of the construction or foundation of the 'external world' or justify our ascription of mental states to other minds, phenomenalists, sense-data theorists, all of them assume the premise that self-ascription is meaningful and true but doubtful when other-ascribed. An epistemological primacy is granted to self-ascription: it is certain to a degree that there cannot be other-ascription. In this sense it is assumed that in self-ascription a PL is used, a PL that other people are prevented from understanding: other people cannot understand, i.e., S because they cannot know if S designates something or is a mere noise.

The Cartesian construes mental items as private objects. It is a mistake to think that sensations are philosophically private: It is the Cartesian's sensations that are private. But the language for sensations is not private. It is the language for the Cartesian's private sensations that is a PL. Strawson and Ayer err on both counts, according to Malcom.

The Cartesian stipulation is this: S is a private name because it refers to a private object. Then this idea of a private object becomes one with the nominative theory of meaning. But then a paradox arises: In what sense is it possible to talk of language or meaning if these have been absorbed by the private object or reference?[11]

Once Malcom has cleared the ground from the Strawson-Ayer mistaken interpretation, he goes on to present his own view of the PLA. He presents two forms of argument, one internal and the other external. We could summarize his internal attack in the following way:

(I) A sign becomes a word when it follows a rule.

(II) The rule should be independent from the sign.

(III) The sign could be used correctly or incorrectly. In order to decide this, appeal has to be made to the criteria for the application of the rule.[12]

(IV) In the PL there is no correct-incorrrect use of the sign.

(V) Inductions or appeal to memory have no place in the use of sign in a PL.[13]

(VI) Thus in a PL there are no rules and so there is no language.

According to the external attack, it is impossible to get the mental concepts privately, that is, for one's own case and then start applying them to other people, because the idea of such a transference is a different one.

Thus Malcom declared the idea of a PL senseless and unintelligible. A tremendous controversy broke out and 'rules,' 'criteria,' 'memory,' 'senselessness,' etc., were the words which came into question. Many philosophers thought that Malcom presented the PLA in a weak form for either (i) it involved a _petitio_ supposing as proved what he had to prove, namely, that in the PL there cannot be correct-incorrect or worse, (ii) was question-begging, for he was neglecting something he had the right to introduce, namely, private rules.

The growing dispute aligned philosophers into two main groups: The Wittgensteinian group, defending the PLA, and the group that defended the idea of a PL and thus dismissed the PLA as a new wording for the old physicalistic behaviorism of the thirties. The main characteristics of the two groups were the following:

On the Wittgensteinian side:

(I) Strong defense of the _a priori_ condemnation of a PL as senseless.

(II) Stubborn repetition of Wittgenstein's texts in order to defend the ideas of a new philosophical semantics in terms of 'rule,' 'criteria,' etc.

(III) Construction and additional refinement of the PLA.

On the side of those attacking the PLA:

(I) Attack of the 'senseless,' 'unintelligible,' etc. condemnations.

(II) Ingenious search for alternatives to each one of the premisses of the PLA.

(III) Determination to introduce 'experiences' back into the philosophy of mind.[14]

(IV) Attempts to argue differently against the Cartesians.[15]

4. Castañeda, Chappell and Thomson Symposium

Castañeda went straight against Malcom's contentions and with a daring effort distinguished as many as six senses of 'private object' and seven postulates of pain. Then he developed the idea of language--one with logical connectives, quantifiers, etc.--and concluded that there are perfectly intelligible private languages.

Castañeda interpreted Malcom's premiss that in the PL there cannot be an independent test of correctness as the impossibility of self-correction. To this Castañeda objected that self-correction is a clear possibility and the speaker of the PL could have recourse to other experiences, inductions, laws he discovers among experiences, memories, etc. Mental states and processes are objects, and one could make substantial mistakes about them but also correct himself and come back to the truth.

According to Castañeda, it is not that Malcom's premiss is weak or unargued, but that it is straightforwardly false and that the PLA is an outright invalid argument.

Castañeda introduced into the interpretation of the argument that it cannot be said 'I know (am certain) I have a pain' as a short way—very short indeed--of stating the incoherence of a PL. The idea is that if 'I know (am certain)' cannot be applied in self-ascription, such as 'I have a pain,' then this last one cannot be an object which could be known, and thus the question of a language for one such object does not arise. If that were true, the defender of the PL would have to find another candidate for the role of private object.[16]

The co-symposiasts of Castañeda, J. F. Thomson and V. Chappell, expressed doubts concerning the existence of a PLA. According to Thomson the premiss that only the possessor of the sensations could understand what S means is too strong. And since Castañeda's reconstruction of the PLA and others are unlikely, he remained sceptical about the existence of a PLA.

V. Chappell went further than Thomson and accused Castañeda of Ignoratio Elenchi for not discussing the language--one that no one apart from his possessor could understand--Malcom and others were discussing. According to Chappell, Castañeda got it wrong, for he supposed that there was no trouble with the idea of a PL and thus he proceeded quickly to construe it. But the situation is very different, for the question is: given the thesis of the philosophical privacy, could there be something like a language? The idea is that of providing a soil for the Cartesian intuition, so that the vacuous character of that intuition could be appreciated through the PLA.

Chappell thinks that the candidates Castañeda offered for distinguishing correct from incorrect uses of S are not good. First, he rejects in general the appeal to something external or non-subjective. Then he goes on to eliminate the other candidates. I shall consider only Chappell's rejections of other private objects. Let us suppose, says Chappell, that we could scrutinize closely the use of S. Even so, what could establish such scrutiny or closer examination?

> A closer look might convince me that the object really is as I first thought it was, and hence

> that this is the right word. But that only touches my _conviction_; the question is whether I am _right_, which is to say, whether this _is_ the word that I have used for this sort of object in the past.[17]

Thus, recourse to other private objects becomes useless, and so, memories, other words and inductive generalizations made from those private objects become useless. Lastly, recourse to logical connections to other words of the public language would destroy the private character of that language.

In that way Chappell concludes that Castañeda's strong effort fails to prove the existence of a philosophical private language.

In his rejoinder, Castañeda replied that if Chappell is right, then the distinction correct-incorrect cannot be made good in a public language. Later he developed this idea and concluded that he had provided straight refutation of the PLA even if the question concerning PL remained undecided.

The conclusions by Chappell and Castañeda clearly indicate an _impasse_ and thus the need for further, deep work in the whole issue. On the other hand, the Castañeda, Chappell, Thomson symposium left as a gain a way of taking the PLA, a way that has continued until A. Kenny. I shall pursue it in what follows.

5. The Clarification by J. W. COOK

J. W. Cook tried to explain both Castañeda's confusion and Thomson's perplexity by making clear the character of the privacy that is involved in the PLA. Cook made explicit that conception of privacy and its absurdity. In order to do this he constructed the following argument:

> P1 No one can feel (experience, be acquainted with) another person's sensations.
>
> P2 The proper and necessary means of coming to know what sensation another person is having is to feel that person's sensation.

P3 Anyone who has a sensation knows that he has it because he feels it, and whatever can be known to exist by being felt cannot be known (in the same sense of "known") to exist in any other way.

(C) No one can know what sensations another person is having.[18]

According to Cook, from (C) it follows that the names of sensations cannot be thought of other persons. Also it follows that nobody except for the owner can understand those names, for he will be prevented from knowing the referents of the names. And as this is a necessary prevention we face a case of PL.

Cook's strategy is to show that this conception of the knowledge of sensations is unintelligible and it eliminates not only the other's case, but also one's own.

To begin with, Cook clarifies the mistakes involved in the treatment of sensations. There are two main mistakes:

(I) Sensations are private and thus there are no names for them (even if apparently there are).

(II) Sensations are private but they have names; so, our language of sensations is a private one.

(I) Is Strawson and Pitcher's interpretation; (II) is Castañeda's. Accordingly, it is concluded that (I) is the price paid if PL is eliminated; (II) is the price paid by reducing sensations to private objects. But according to Cook the above diagnosis is completely mistaken and in no way are (I) or (II) the PLA's target.

What Cook's argument shows is precisely this. Let us see how he does it.

According to a Cartesian, P1 expresses a 'logical impossibility,' but according to Cook it does not express any such thing. If P1 seems to say something it is because it is dressed in the wrong logical form, that is, it seems similar to 'nobody could have Caesar's robe.' But it is not similar, for whereas one could identify the robe and then ask who the owner is, that cannot be done in the case of 'pain.'[19]

P2 and P3 express a necessary condition for the knowledge of a sensation and say that unless one feels the sensation one cannot know it. But--Cook argues--there is no such necessary condition, for in order for 'know' to be an expression of certainty in first person present tense statements, it is necessary for the person who makes the statement to be mistaken, and this is not possible with 'I have a pain,' for it is incorrigible.

The only way to prove such incorrigibility is to offer examples of uses of language, or to show that the only mistakes that could be made are mistaken descriptions of pains, but not mistakes about the pain itself.

Hence, P2 and P3 lack sense; the same happens to P1. What seemed an argument vanishes as a hallucination leaving no traces. The only thing left are words whose sense exists only in other language games.[20]

But why a strong conclusion such as lack of sense or unintelligibility? Why not restrict it to falsity? Cook seems to think that Wittgenstein's ideas--the ones he handles in his argument—rest on a kind of super-semantics which demarcates very strongly different kinds of words and kinds of possible combinations. According to that, if one makes the wrong combination the result is senselessness. This is an essentialist _a priori_ view that stands in need of justification or so it is thought by those who have suffered such attacks.

It is worth noticing that Cook never bases his condemnations of 'senseless' on the existence of rules or criteria as some do. But this fact only makes Cook's condemnations appear worse to the eyes of those who reject such strong talk.

6. The Dispute on Private Rules

In the entry on PL in Edwards' _Encyclopedia of Philosophy_ Castañeda[21] went on to attack the idea that one cannot obey a private rule. He objected that it is not a requisite of rules that there be the possibility of disobeying them. There are rules such that thinking that one is obeying them is to obey them.

J. J. Thomson[22] argues in a similar way. He provides the following examples:

> (I) Mummy writes 'whenever you feel the least bit gloomy, think of your mummy (it will cheer you up)'
>
> (II) 'If you are in pain, think you are in pain'

In both of these cases understanding the rule involves obeying it. Thus, there is no problem with private rules.

Carl Ginet objected that both Castañeda and Thomson slippered before a common case of disobeying rules, namely, even in the case of rules easily understood it is possible that someone <u>thinks</u> he understands the rule and yet does not do so.[23]

Thus in case (I), contrary to the mother's intention, the daughter could be repelled by the idea of her mother or motherhood in general and so, in misunderstanding the intention of the maternal advice--even if she thinks she follows the advice--she could arrive at worst results than the initial situation.

In this way Ginet could save the PLA's premiss according to which every rule should offer the possibility of disobedience in order to count as a rule.

7. <u>PLA and the Charge of Verificationism</u>

J. J. Thomson developed the suspicion held by many that the key premiss of the PLA rests on verificationism. She asks questions like the following: Why is it not possible for S to be a name in a PL? Why cannot the user of PL check his own applications? Why, in short, cannot a private rule be followed? She answers: Because the user of the PL cannot find out, discover or become aware that his current mental occurrence is not S or F or some other name he wanted to give it. From this she concludes:

> That it <u>is</u> conclusive or strong finding out which is required--either because good reasons are not

> enough, or because where there can be no strong finding out there cannot be good reasons either.[24]

Thus, according to her, if Malcom and the others think that there is no sense for common or kind names in a PL, they base their case on a verificationist premiss. Furthermore, they hold that the only kind of mental ascription that can be verified is that which refers to something physical, that is, to behavior. In his view they do not go from verificationism to phenomenalism--as was usual--but to physicalism.

Thomson thinks that all the fuss Malcom and others were making was produced only by a tricky change of words, but when looked upon closely, it only appears to be the old verificationism of the thirties.

Thomson concludes by reminding us of the vice of circularity that befalls all verificationism. She also concludes that Wittgenstein is poorly served by ascribing to him such a controversial philosophical thesis against his own expressed warnings.

I think it is possible to argue against the circularity Thomson believes she finds in her PLA. On the other hand, it is possible to argue <u>ad hominem</u> in the following way: Verificationism could be taken as a general line of demarcation of sense from non-sense, i.e., of scientific language from metaphysical language. But Thomson seems to think of it in another way, namely, as giving internal necessary and sufficient conditions of meaningfulness. No clear view about this is offered, however. When uncovered, it might be that such verificationist theory of meaning could be a serious contender which cannot be dismissed as easily as she takes it to be.

8. <u>The PLA and the Conditions of Propositionhood</u>

Anthony Kenny thinks, like Malcom, that the idea of a PL is implied in the epistemology and philosophy of mind of those that assume private experiences. The PLA departs from that implication and deduces the absurdity that in a PL nothing could be asked or said. The PLA uncovers two basic mistakes, namely: The mistake of conceiving experiences as private objects and the mistake of thinking that words are parts of language because they get their meanings by private ostention.

Let us see first the mistake about language. The idea according to which S refers to pain sensations by private ostention is only a simplification. In order to make 'reference to...' a complex basis of customs, circumstances, etc., is necessary. Mere experience is not enough, as it is not either mere behavior. If emphasis is laid on criteria, it is only to stress the insufficiency of experience to provide for meaning, but it is not meant to say that criteria are enough.

The philosophical thesis of privacy pretends to derive itself from the 'fact' of incommunicability or from the 'fact' of inalienability. The first is epistemological, viz., 'I cannot know your pains.' The second is of ownership, viz., 'Nobody else could have my pains.' But privacy could be derived from both 'facts.' 'I cannot know your pain because I do not have it.' Kenny elminates the three possibilities, calling them 'senseless.' The only privacy remaining is the one coming from secrecy, and that does not cause any philosophical trouble.

That shows that 'pain' is not something philosophically private, and thus that our sensation-language is not a PL. But, could there not be a pain similar to ours but with the characteristic of incommunicability? That would be a pseudo-pain. But then, could I not define S in a way that would refer to that pseudo-pain by means of an association between S and the pseudo-pain?

Kenny rejects that possibility but without assuming scepticism on memory or verificationism. According to Kenny, there are three possible answers to the question: What do you mean by S? These are the following:

(1) By S I mean to say <u>this</u>.
(2) By S I mean the S impression which I remember.
(3) By S I mean an X physical correlate.

Kenny dismisses the three answers for the following reasons: (1) is no good because:

> 'This is S' is not a genuine proposition capable of being true or false; for what gives it its content is the very same thing as gives it its truth: the significance of the predicate is supposed to be settled by the reference of the subject.[25]

(2) is no good either because:

> He must call up a memory-sample of S and compare it with his current sensation to see if the two are alike. But of course he must call up the <u>right</u> memory. Now is it possible that the wrong memory might come at his call? If not, then 'S' means whatever memory occurs to him in connection with 'S' and again whatever seems right is right.[26]

(1) and (2) cannot thus eliminate arbitrariness and make S a vacuous name. (3) seems to provide a way out, for its use relies on something other than the current sensation, namely, the rising of my blood pressure; that certainly works but at a high cost, namely, that the name is no longer private but a public one.

Now, Kenny thinks Thomson is wrong, for the reason to reject S as a name is not verificationism, but the thesis that propositions must be bipolar. It is not that a strong testing or checking is being demanded but that the space where verification could take place has been cancelled when the possibility of the poles of truth and falsity have been taken away.

The immediacy of experience as well as the idea of private rules cancel the logical space where judgment could take place and thus make it impossible to talk of true or false.

Kenny points, furthermore, to the connection between that theory of the bipolarity of propositions and the doctrine that self-ascriptions are <u>Aeusserungen</u>, that is, are not descriptions and cannot tolerate an epistemic operator as 'I know.' What it is meant when it is said that 'It hurts' is not a description is that the sentence is not read off from observation or facts. On the other hand, 'It hurts' is part of my pain[27] because it is part of my pain-behavior and thus is not something independent from the pain. For that reason it cannot be true or false.

Something similar occurs[28] with 'I know it is hurting me': In order that 'I know' have an epistemological use it must be possible that it could be a true description of a fact, and in order to do this it must be possible for the one who utters it to be mistaken. But that is not a possibility in the above sentence, and 'I know' adds nothing epistemological to 'It is hurting me.'

A general thesis emerges thus: When the logical space is cancelled between a name and its referent or between a description and a fact or between a knowledge-claim and the object to be known by that very fact 'name,' 'description' or 'knowledge' are cancelled. This is Descartes' mistake against the sceptic, a mistake that is found also in the theories of the logical proper name and the immediate experience.

I think that the line of argument initiated in Strawson and Ayer comes to an end in the works of A. Kenny. According to it, both defenders and attackers of the PLA demand philosophical arguments based on philosophical theses. Kenny uses all those concepts involved: 'meaning,' 'criteria,' 'rule,' 'language,' 'experience,' etc., and in a new way provides an answer to the objection to the PLA.

SUMMARY OF SECOND PART

In the first part of this paper, the orthodox line of the PLA was pursued. In this second part, another line of argumentation is introduced.

Strawson's reconstruction of Kant's objectivity arguments is presented as an argument having a similar objective as the PLA. Cartesian contents of consciousness cannot be identified or re-identified and thus there cannot be a language for them. These are pseudo-objects that keep their appearance of objects only because of an illegitimate trading with our public-notion of objects.

Shoemaker's argument against private similarities also keeps in view the same target as the PLA. Shoemaker claims that the notion of private similarities turns out to be circularly empty. As a result, there would be no kinds, no inductions, etc. These private objects would then keep their denomination only as a gracious gift.

Saunders and Henze have brought up the PLA in relation to the idea of a private ego. If the notion of such an ego cannot be introduced, the idea of a world and that of a language collapses. Hacker also likens the PLA to a doctrine of Kant, according to which 'experiences without concepts are blind.' Hacker argues that the PLA indicates that there cannot be private concepts and, consequently, no private language.

Pears claims that there is a weak interpretation of the idea of a private language which is not obviously mistaken. He uses the sense-reference distinction in order to claim that in a private language the sense would be public, whereas the reference would remain private. However, the question arises as to what extent the sense could be left indeterminate by the reference. It seems that under this hypothesis there would be nothing private in such a language.

Another interpretation of the PLA was introduced by R. Rhees. In contrast with Malcom, Rhees does not rest his argument on notions like 'rules,' 'criteria,' etc. Rhees' strategy is different, namely, he points to the use that Ayer's Crusoe would like to give to notions like that of 'giving meaning' and observes that a number of other notions are lacking and, because of that, Ayer's Crusoe cannot introduce his 'meaning.' Rhees advances his criticism in a persuasive manner; he would express himself in this way: "Look, naming brings such and such consequences in a language, why do you want to refer to names and language when none of these consequences follow?"

Castañeda has made an important observation concerning the PLA. He points out that the PLA goes against the strong essentialism assumed by the idea of a private language, according to which the private particulars are supposed to wear their own identity and their own names. The PLA, according to Castañeda, would oppose this view, adopting nominalism and an extreme conventionalism.

Rorty, under the influence of Quine, has tried to rescue the PLA from the Wittgensteinians. First, he has rejected essentialist notions such as 'senseless' as necessary to attack the idea of immediacy which is precisely what the PLA is against. Next, he has rejected the charge that the PLA has to assume verificationism; according to him, the PLA assumes rather a holistic view of reference.

Marks thinks that the notion of private language rests on the notion of a private object and this notion is empty, since it is impossible to sort natural kinds out of these private objects. He qualifies his objection by saying that so far no private kinds seem available and it is up to the privatist to show that there could be private kinds. However, Marks feels confident he can show the illusory character of these kinds each time. One is presented.

In the final remarks it is pointed out that the PLA remains questionable. It is still unclear to what extent current philosophies and philosophers hold a view as extreme as that of a private language; on the other hand, the task of bringing the PLA to take sides on central questions of the metaphysics of persons and on current issues like the 'Identity Thesis' remains unaccomplished.

NOTES

* Originally published in <u>Crítica</u>, <u>Revista Latinoamericana de Filosofía</u> 7 (1975). The first part is translated in its entirety (with minor alterations). For reasons of space, only a summary of the second part is given.

1. PL is short for 'private language' and PLA for 'private language argument.' PLA should be understood as against the idea of a private language. Even though Wittgenstein originated the dispute on PLA, this is not an exegesis of Wittgenstein's work, but rather a survey of the different interpretations currently held around the common denomination PLA and thus goes beyond what Wittgenstein actually said to what, probably, he meant to say or what he should have said.
2. Different authors have worked on specific classical philosophers to prove this. Thus Kenny (1) states that Descartes held the idea of PL. Hacker does it for Locke and Frege. The charge is that Cartesian philosophers

from Descartes to Husserl and Russell suppose that the idea of a PL is a coherent one and has a primary place for our theory of the world and persons. An enormous bibliography has developed on this issue, viz., Saunders and Henze, _The Private Language Problem_ (New York: Random House, 1967).

3. The idea of a private ostensive definition could be specified like a sort of pointing with the mind's eye to what cannot be shared or is expressible, i.e., the pain-quality that each one feels and can point to himself but cannot communicate to the others. That is the private reference or private object.
4. The _Aeusserungen's_ thesis opposes description to expression. According to it 'it hurts' does not describe a pain but rather expresses it and thus is more on the side of behavior than of theory and language. The thesis, then, is that those expressions, just because they do not describe, lack a truth-value.
5. A. J. Ayer, "Can There Be a Private Language?" in O. R. Jones, ed., _The Private Language Argument_ (New York: Macmillan, 1971), p. 277.
6. _Ibid_., p. 56.
7. _Ibid_., p. 55.
8. _Ibid_., p. 55.
9. _Ibid_., p. 277.
10. The notion of 'criterion' is one of the most controversial (cf. A. Kenny, "Criterion," in _Encyclopedia of Philosophy_ (New York, 1967). According to it "'X is a criterion for Y' if X is non-inductive evidence for Y." Thus, for example, it is said that screaming, contorting, etc., are behavioral criteria that a person is in pain.
11. Cf. P. T. Geach, _Mental Acts_ (London: Routledge and Kegan Paul, 1964), p. 4. Ginet, "How Words Mean Kinds of Sensations," _Philosophical Review_, 1968, agrees that there could be an absolutely private reference but in such a case there would not be a kind-name for it and thus no language to express it. Some think that S is not a _flatus vocis_ but that neither qualifies as language.
12. Malcom accuses Strawson of another confusion on the thesis of criteria, for he thinks that the PLA confuses criteria with the mental state, i.e., pain, and thus falls into behaviorism; Malcom thinks that pain and criteria for pain are different things but that there

would not be pain without pain-criteria (that is the reason why there cannot be a private pain).

13. Premisses (IV) and (V) are usually interpreted as a mixture of verificationism (as we shall see later)--and scepticism on memory.
14. Very important for this purpose was H. P. Grice's "The Causal Theory of Perception," in _Proceedings of the Aristotelian Society_ 35 (1961).
15. The influence of Quine's thought is the most relevant in those attempts.
16. H. N. Castañeda, "The Private Language Argument," in Jones, p. 140. See the thesis by Cook and Kenny below.
17. C. Chappell, "Comentarios a Castañeda," in Jones, p. 163.
18. J. Cook, "Wittgenstein on Privacy," in G. Pitcher, _Wittgenstein: The Philosophical Investigations_ (New York: Macmillan, 1968), p. 289-96.
19. _Ibid_., p. 302.
20. _Ibid_., pp. 293-5.
21. H. N. Castañeda, "Private Language Problem," in _Encyclopedia of Philosophy_, p. 461-2.
22. J. J. Thomson, "Comentarios a Castañeda," in Jones, p. 191.
23. C. Ginet, "Wittgenstein's Argument That One Cannot Obey a Rule Privately," _Nous_ (1970), p. 363.
24. Thomson, p. 198.
25. Kenny, p. 219.
26. _Ibidem_.
27. _Ibid_., p. 224.
28. _Ibid_., p. 226.

Marcelo Dascal

III. PHILOSOPHICAL ANALYSIS IN BRAZIL

Traditionally, the teaching of philosophy in Brazil has been associated with law and theology. Scholastic philosophy is taught to this day in many universities, both public and private (especially religious), throughout the country. In such institutions, all turns around Thomism, and philosophy is conceived mainly as the history of philosophy. To be sure, logic is taught too, but mainly Aristotelian logic--with the help of such textbooks as Maritain's Logica minor, which had its fourth Portuguese edition published in 1962. When not purely historical and scholastic, the interest of philosophy teachers used to be mainly addressed to moral and political philosophy, with eventual incursions into metaphysics, following the vagaries of its development in Europe. Epistemological questions were rarely dealt with, and when they were, only insofar as they were considered relevant to other problems. Institutionally, the most prominent Brazilian philosophers of the past were faculty members of the best law schools in the country, at Fortaleza, Recife, São Paulo, etc. To this day, the "Brazilian Institute of Philosophy", which publishes the Revista Brasileira de Filosofia, has its seat at the law school of the University of São Paulo. Lawyers and professors of law have always been very active in national politics. Hence, it was natural that they should attempt to employ the philosophical ideas that were fashionable in Europe in order to justify one or another political ideology. For instance, the issue of the theoretical justification of a republic as opposed to a monarchy, as well as the connected issues of the nature of liberty, of the relationship between individuals and society, of the foundations of a legal system, etc., monopolized philosophical attention for many decades, and occupies the majority of the pages of many books and journals.

Among the imported philosophical doctrines employed in the local political debate, at the end of the 19th and

J. J. E. Gracia et al. (eds.), Philosophical Analysis in Latin America, 277–284.

beginning of the 20th centuries, the most influential was, no doubt, positivism. This is the only philosophical doctrine that can boast having been able to dominate the Brazilian philosophical scene for several decades. Its influence went beyond ethics and politics, and was felt also in metaphysics, epistemology, aesthetics, philosophy of religion, and even in the very conception of the history of philosophy. Beyond philosophy itself, a "positive mentality" developed in Brazil, which was instrumental in shaping the attitude of Brazilian intellectuals toward science, in particular toward the emerging social sciences. Positivism has been so influential in Brazil, that it is not an exaggeration to describe a great deal of Brazilian thought after it as a reaction—not entirely negative —to positivism.[1] Central figures in Brazilian philosophy, such as Tobias Barreto, fit perfectly well into such a description.

It would be natural to expect that the influence of positivism would be transferred to one of its philosophical heirs, namely to logical positivism, and thence to philosophical analysis in general. The course of history, however, has been quite different. Philosophical analysis had to fend its way in Brazil by its own means, many years after the decline of positivism. It had to fight not only against positivism and its remnants, but also against Marxist tendencies which were the direct beneficiaries of the "positive mentality" and of the privileged position assigned to politics among the philosophical disciplines.

Only in 1934, with the creation of the Faculty of Philosophy, Sciences, and Letters of the University of São Paulo ('USP'), a department of philosophy, independent of both law and theological studies, appeared. France sent professors, and young students went to France to pursue doctoral studies. At first, the orientation was mainly historical. The influence of Martial Guéroult, the great historian of philosophy, who taught in São Paulo in 1948, was decisive. Slowly, the department began to enlarge its philosophical perspective. Quine visited São Paulo in the beginning of the forties, and published in Portuguese his *O Sentido da Nova Lógica* ("The meaning of the new logic"; São Paulo: 1944). Although he did not leave any direct disciples at the university, after his visit modern logic became an obligatory part of the philosophy curriculum.[2]

Quine called attention not only to the importance and power of symbolic logic, but also to the recent progress made in the philosophy of science and in the philosophy of language. In 1947 and 1948, Gilles Granger, a former student of Gaston Bachelard, taught logic and philosophy of science in São Paulo. His course notes gave birth to a book, Lógica e Filosofia das Ciências (São Paulo; 1955), which offered an overview of the recent developments in both fields, and included some chapters with advanced material, which discuss some of the main questions of interest to analytic philosophers. Both in its attitude toward philosophical problems, and in its method, Granger's book--which was originally written in Portuguese, is analytic in spirit and, although out of print for many years, is still influential today.

The seeds planted by Quine, Granger, and others have certainly borne fruits in São Paulo. Yet, one cannot say that the analytic orientation became dominant, or even important within the department of philosophy of USP. From the two chair-holding professors that have directed the department for many years, one--Lívio Teixeira--was essentially a historian of philosophy, and the other--João Cruz Costa--devoted himself to the study of the history of ideas in Brazil, and was mainly interested in political philosophy. Although he had a clear preference for Marxism, Cruz Costa, who was in charge of the chair of general philosophy, adopted an open-minded and tolerant attitude toward the new philosophical ideas. He acknowledged the importance of logic, of philosophical analysis, and of the philosophy of science, stimulated his students and teaching assistants to specialize in these fields, and included courses on these topics in the curriculum. Such an attitude had to face the permanent opposition of teachers and students who were interested exclusively in ethical and political issues. According to them, to devote any time to abstract subjects such as logic or philosophical analysis, in the socio-political circumstances of the country, was a dangerous form of "alienation." To this day the same kind of objection is raised against those who want to pursue these themes in Brazil.

The 1964 coup d'état, with its consequent exorcization of the "leftist" trends in Brazilian universities (which reached its paroxysm in 1967), has led to a sudden interruption of the development of philosophy in general, and of philosophical analysis in particular, in the country. Many of the best teachers and researchers were forced to retire. The departments of philosophy of the public universities in São Paulo, Porto Alegre, Brasília, Rio de Janeiro, and other cities lost many of their young members in this way. Such a situation, rather than inhibiting, stimulated the process of ascension to power, in most departments, of the defenders of the ethico-political orientation, thus leading to a most complete neglect of the logico-analytic field. (The history of philosophy, however, continued to be a "legitimate" field of investigation accepted by everybody). This situation is mirrored in the pages of the journal _Discurso_, published since 1970 by the department of philosophy of USP. It rarely includes a paper of analytic character and, although published by a department of philosophy, it devotes much space to material related to literary and artistic criticism. The more philosophically minded articles are either historical or ethico-political. Such a distribution is not the result of a deliberate editorial policy, I presume, but of the kind of research interests that guide the members of the department and those associated with them. As a reaction to such a situation, João Paulo Monteiro, a specialist in Hume who is also interested in the philosophy of science, began to publish in 1979 a new journal, _Ciência e Filosofia_. Its editorial policy defines it as "an interdisciplinary journal devoted to epistemological and philosophical themes: the theory of knowledge and the philosophy of science, the history, the sociology and the methodology of science, as well as those aspects of logic and philosophy of language directly connected to epistemology." This journal, which, as _Discurso_, is a publication of the department of philosophy of USP, may perhaps become a catalyzer for the analytically inspired work done in São Paulo, as well as other Brazilian centers of philosophical research.

Oswaldo Porchat Pereira, a specialist in Aristotle who has also studied logic in Berkeley, has been for many years a very active member of the department of philosophy in São Paulo. He brought to the department visiting professors from the U.S. and from Europe, who delivered advanced seminars and courses in the philosophy of language, logic, and philosophy of science. In 1975, Porchat left the University of São Paulo, and founded a new "Centro de

Lógica, Epistemologia, e História da Ciência" ('CLE'), at the State University of Campinas ('UNICAMP'), sixty miles from São Paulo. The creation of this Center was, no doubt, a decisive step in the development of philosophical analysis in Brazil. To be sure, the Center's concerns are not purely analytical, but the kind of interdisciplinary interaction fostered by it, the selection of visiting professors and researchers, and the research performed in Campinas under the sponsorship of the CLE, all indicate a general orientation that favors the analytic methods. Some of the doctoral dissertations recently prepared in Campinas clearly demonstrate this trend: "Frege's Logic" (Luiz Henrique Lopes dos Santos), "Behaviorism in Quine's Philosophy" (Antonio Trajano M. Arruda), "The Construction of the Physical Object in Carnap" (Rejane M. M. Carrion), "The Theory of Cognitive Simulation" (Maria Eunice Quilice Gonzales). No doubt, one might point out that these are, for the most part, historical theses--following in this respect the Brazilian tradition mentioned above. Yet, for the first time, the philosophers whose doctrines are investigated clearly belong to the analytic tradition.

The CLE, which does not provide regular courses, but is entirely devoted to philosophical research, has led to the creation of a department of philosophy at UNICAMP, which has a master and doctoral program in the area of "logic and the philosophy of science", where students can specialize in the philosophy of language, in epistemology, in the history of science and of philosophy, and in logic. But the main contribution of the CLE--which is unique in Brazil--lies in the organization of periodical seminars and colloquia with the participation of researchers from Brazil and abroad. Among the topics of such meetings, analytical themes or themes treated in an analytical manner are prominent. Here are some of them: "philosophy and common sense," "logical positivism," "the concept of truth." There are also yearly encounters on the history and philosophy of science, as well as national meetings of logicians (the CLE has also sponsored a Latin-American Congress of Logic). Recently, the Center, together with the Institute of Language Studies of UNICAMP, organized a major International Encounter on the Philosophy of Language (August 1981). Participants included well known American and European philosophers (such as Donald Davidson, John Searle, Kuno Lorenz, Gilles Granger, David Harrah) as well as scholars from all over Brazil and

Latin America. In all, more than seventy participants, of sixteen countries, attended and read papers before an audience of about three hundred registered persons. Many of the contributions presented had an analytic character, and gave those who practice philosophical analysis in Brazil the opportunity to meet each other, and to submit their work to serious international criticism.[3]

The Center is also active in the area of publications. Since 1977 it has published the journal Manuscrito, the first international journal of philosophy published in Brazil (with articles in Portuguese, Spanish, English, and French). Its board of consulting editors includes many philosophers, from all over the world, who are identified with the analytic tradition and, in the published volumes, the proportion of analytic articles is high. A recent issue devoted to the philosophy of action, contains original papers by R. Chisholm, H-N. Castañeda, A. Goldman, J. Searle, L. Apostel, and others. The CLE also publishes Cadernos de Filosofia da Ciência, and is about to begin the publication of another international journal devoted to non-standard logics, whose editor is the logician Newton da Costa, a member of the CLE.

The last years of the seventies witnessed an increasing interest in philosophical analysis throughout the country. One of the reasons for this was certainly the development of linguistics as an independent scientific discipline, and to the popularity it achieved by virtue of the initial success of transformational grammar. As linguistics became an obligatory topic in the curriculum of language and literature studies--those with the greatest number of students in the country--, the demand for courses in the philosophy of language, logic, semantics, and more recently, pragmatics, was also up. To meet this demand, the volume of translations in these and related areas has dramatically increased in recent years, and students can read classics of analytic philosophy, such as Carnap, Schlick, Russell, Moore, Wittgenstein, Popper, and Frege, in high quality and relatively inexpensive Portuguese versions.

Although the state of São Paulo is still the main center of activity, there are also groups of philosophers working within the tradition of analysis in other regions. The return to Brazil of teachers trained in the U.S.,

Germany, Belgium, and England has contributed to the development of such nuclei. In Rio de Janeiro, the Federal University and the Catholic University have formed a group of this kind, which includes, among others, Oswaldo Chateaubriand, a former student of Quine, Raul Landim Filho, Danilo Marcondes Filho. Gama Filho University (Rio de Janeiro) has a graduate program in the philosophy of science, and so does the Federal University of Rio Grande do Sul (Porto Alegre). The Fluminense Federal University (Niterói) has a good program in mathematical logic, which is connected more with the department of mathematics than to the department of philosophy. There are also isolated researchers and small groups in other places, such as Brasília, Belo Horizonte, Curitiba, etc. The geographic diversification under way is illustrated by a list of the Brazilian participants in the Campinas Encounter on the philosophy of language, who have presented analytically oriented contributions: A. M. Mora de Oliveira (São Paulo), Balthazar Barbosa Filho (Campinas), Danilo Marcondes S. Filho (Rio de Janeiro), Glória M. Vilhena de Paiva (Rio de Janeiro), Guido de Almeida (Belo Horizonte), Julio R. Cabrera Alvarez (Santa Maria), Luiz Henrique Lopes dos Santos (Campinas), Marcelo Dascal (Campinas), Oswaldo Porchat Pereira (Campinas), Raul Landim Filho (Rio de Janeiro), Rejane M. M. Carrion (Rio de Janeiro), and Vera Lucia Caldas Vidal (Rio de Janeiro). Furthermore, the influence of philosophical analysis is felt also in neighboring fields. The Brazilian Association of Catholic Philosophers has devoted a few of its International Philosophical Weeks to analytic subjects (philosophy of language, philosophy of science); in the law school of the University of São Paulo, Tércio Sampaio Ferraz Jr. pursues research on the pragmatic and semantic analysis of juridical discourse; at the Instituto Tecnológico de Aeronáutica in São José dos Campos, an elite engineering school, Leônidas Hegenberg practices philosophical analysis, and contributes, with his numerous translations and textbooks, to its diffusion; and, of course, the courses on linguistics, speech communication, semiotics, and the like, let their students at least know about the existence of philosophical analysis.[4]

NOTES

1. The role of positivism in Brazil is described by João Cruz Costa in his Contribuição à História das Idéias no Brasil (São Paulo; 1960; published in English as A History of Ideas in Brazil: The Development of Philosophy in Brazil and the Evolution of Natural History (Berkeley: University of California Press, 1964). The following claim exemplifies a rather common attitude toward positivism: "One has to point out that the need to face positivism--as well as the scientific trends in general--marks the shift toward the contemporary era in Western philosophy. Once its purposes are assessed, and its possibility is denied, metaphysics begins a renovation cycle, whose salient phases are neo-Kantianism and neo-Thomism. In the Brazilian case, positivism ascended to a position of hegemony, throughout the republican period, becoming, on the one hand, a political philosophy of uncommon vitality, and, on the other, the diffuse phenomenon called the 'positive mentality'" (A. Paim, "Momentos del pensamiento filosófico brasileño," Revista de la Universidad de Medellín 22 (1976), 27-62).
2. Vicente Ferreira da Silva had published his Elementos de Lógica Matemática in 1940, and courses on this topic were given at the Colégio Pedro II, in Rio. Nevertheless, in 1950, Martial Guéroult, in his preface to Granger's book, still regretted the nonexistence of a satisfactory textbook of modern logic in Portuguese (in G. Granger, Lógica e Filosofia das Ciências, São Paulo, 1955, p. 17). Nowadays, mathematical logic is a well established discipline in Brazil, with internationally known researchers.
3. Some of the papers presented in this Encounter are forthcoming in M. Dascal (ed.), Dialogue - An Inter-disciplinary Approach (Amsterdam: John Benjamins).
4. The information contained in this introduction is, of necessity, fragmentary and incomplete, since there is, so far, no systematic source on current philosophical research and teaching in the country. I apologize to those who have not been mentioned for my ignorance of their work.

Marcelo Dascal

PHILOSOPHY, COMMON SENSE, AND SCIENCE*

> To live without philosophizing is to have one's eyes closed without ever trying to open them.
>
> Descartes

> This strange doubt or fiction, according to which we have been eventually created for being mistaken, even in the most evident things, need not disturb anyone, since it contradicts the very nature of evidence, and the experience and events of life prove its contrary.
>
> Leibniz

Traditionally, philosophy begins when Thales begins to 'open his eyes'—thus anticipating Descartes' advice--in order to try to see things *beyond their appearance*. Apparently, there is little or nothing in common to all the innumerable individual things in the world, which seem to differ *radically* from each other. Nevertheless, Thales claims that, ultimately, they must have--all of them—a shared foundation which has to be found. Though the solution he himself proposed did not gain wide and lasting support, his question, once asked, could not be set aside any more. His followers replaced water by the *apeiron*, by the *pneuma*, by the *nous*, and later on, by the atoms. But they remained faithful to the basic presupposition of the search initiated by Thales. The primordial innocence--accepting things just as they appear to us—is thus forever lost. When Protagoras defends the thesis that *in fact* things are what they look like, i.e. that behind multiplicity there is no unity whatsoever, he cannot do so without contrastively stressing the expression 'in fact.'[1] For his thesis is not just a return to the innocent point of view. It can only be understood as a *philosophical* thesis, antithetical to the Ionic and Eleatic position, according to which there is no immediate correspondence between

J. J. E. Gracia et al. (eds.), Philosophical Analysis in Latin America, 285–312.

appearance and reality. Protagoras' thesis is not a return to common sense, but rather a use of common sense for philosophical purposes.

From this moment onwards, the issue of the relationship between common sense and philosophy becomes a philosophical problem, which cannot but haunt philosophers of all persuasions. Not all of them, to be sure, accept the symbiosis between common sense and philosophy, as proposed by Protagoras. Even those who defend the view that common sense has a privileged relationship with philosophy will be suspicious of the apparent simplicity of Protagoras' solution, namely the assignment of philosophical dignity to common sense's theses. They will envisage more complex forms of its philosophical use. Some, on the other hand, will try by all means to follow science's lead, and to set philosophy free from common sense. This should be achieved by transforming philosophy into a 'methodical' enterprise (Descartes, Bacon), or by grounding it in experience (Locke, Berkeley, Hume), or by making it 'critical' (Kant, Popper, Habermas), or by letting it discover 'the absolute' (Hegel), or else by turning philosophizing into something 'positive' (Comte), 'scientific' (Russell), 'rigorous' (Husserl), or 'formal' (Carnap). Yet, unlike science, philosophy's efforts to get rid of common sense did not succeed: after Bacon and Descartes, undefeated scepticism appealed again to common sense; after Hume, common sense was revived philosophically by Thomas Reid; after Kant and Hegel, it came back with Moore; after Comte, Russell and Carnap, it reappeared in ordinary language philosophy; after Husserl, existentialism brought it back; and nowadays it is still alive in the writings of those who criticize philosophy for distancing itself from the 'common sense view of the world,' thus becoming opaque and useless for the 'common man,' who happens to be not only the person most in need of it, but also the most authentic philosopher.[2]

Such a periodical reappearance of common sense at the very heart of philosophy both demonstrates the failure of those who tried to expel it once and for all, and requires us to examine more carefully its role in the philosophical enterprise. Perhaps it is a grave mistake to try to eliminate it entirely from philosophy, as science has purported to do. Maybe there is a deep connection between common sense and philosophy, such that any attempt to drive a wedge between them is doomed to failure?

The deprecating attitude of Bacon vis-à-vis the 'vulgar,' according to which common sense establishes artificial barriers between our understanding and the true nature of things (cf. "the primary notions of things which the mind readily and passively imbibes, stores up, and accumulates—and it is from them that all the rest flow—are false, confused, and overhastily abstracted from the facts"), or Kant's deep suspicion (common sense is "one of the subtlest inventions of modern times, which allows for the emptiest chatterer to face the deepest thinker and to win over the debate") lead to a deep estrangement between philosophy and common sense, without really addressing the problem of their relationship. For they merely presuppose, without argument, their incompatibility. If it were possible to prove such an incompatibility, as well as to show how it is possible to build an adequate philosophy without recourse to common sense, then the 'separatist' thesis would be fully justified. But I don't think this has yet been done. To be sure, many philosophers have developed systems which are very remote from common sense. Hegel's is a case in point. But is it not precisely its remoteness from common sense that constitutes, though not always explicitly, one of the main motives for strong opposition to the Hegelian system, by those who consider it speculative, dogmatic, and so on? On the other hand, although the internal criteria of validation of his philosophical theses are apparently independent of common sense, they are not necessarily incompatible with the latter. Furthermore, it is plain that such criteria are not able to validate themselves; for this purpose, some external factor must intervene—eventually, common sense itself.

Between total separation and full identification of common sense and philosophy, I think there is now room for exploring some alternative way. The problem is not that of exorcizing and purging philosophy from all traces of common sense. Nor is it that of teaching philosophers to abandon the distorted conceptions induced by pseudo-philosophy, and to open themselves again to the common sense view of the world--a task that requires an impossible return to primordial innocence. The problem is, rather, to redefine the relationship between common sense and philosophy: granted that they seem to be united in catholic marriage, how can they live together without making each partner feel constantly threatened by the 'imperialism' of the other? In other words, instead of declaring its absolute opposition to any interference of common sense in philosophical issues (or vice versa), philosophy must find the means to regulate common sense's role in philosophical argumentation, so that it will not endeavor, periodically, to replace philosophy itself. On the other hand, philosophy must also find the means to become relevant to the 'common man,' helping him, if not to solve, at least to conceptualize and understand the predicaments he faces in this chaotic modern world.

The way followed by science in its confrontation with common sense is an almost obligatory frame of reference for a discussion such as the present one, be it because of science's historical links with philosophy, or because of its dominant role in our days. _Grosso modo_, leaving aside for the moment the obvious impropriety of talking about science as a whole, one can say that science has succeeded in freeing itself from common sense. Galileo's role in this process was certainly more effective than that of Bacon or Descartes, and it deserves to be recalled. Galileo does not spare his criticism of current opinions, even when these are expressed by 'first rank men,' i.e. by the most experienced craftsmen in mechanical works: "Notwithstanding the fact that what the old man told us a little while ago is proverbial and commonly accepted, yet it seemed to me altogether false, like many another saying which is current among the ignorant."[3] And, if this is not clear enough: "The common opinion is here absolutely wrong. Indeed, it is so far wrong that precisely the opposite is true,..."[4] He doesn't say these things out of arrogance or _a priori_ contempt toward the 'vulgar,' but rather because he is confident that science has become able to replace the pseudo-certainties of common sense by well grounded knowledge. 'Common experience,' as Galileo puts it, is never conclusive; only when it is replaced by a much more rigorous form of observation, namely 'experiment,' can one reach truly reliable results. It is always useful to ponder again over one of the classical passages of the _Dialogues_, that stresses this and other relevant points:

Simplicio: Everyday experience shows that the propagation of light is instantaneous; for when we see a piece of artillery fired, at great distance, the flash reaches our eyes without lapse of time; but the sound reaches the ear only after a noticeable interval.

Sagredo: Well, Simplicio, the only thing I am able to infer from this familiar bit of experience is that sound, in reaching our ear, travels more slowly than light; it does not inform me whether the coming of the light is instantaneous or whether, although extremely rapid, it still occupies time. An observation of this kind tells us nothing more than one in which it is claimed that "As soon as the sun reaches the horizon its light reaches our eyes"; but who will assure me that these rays had not reached this limit earlier than they reached our vision?

Salviatti: The small conclusiveness of these and other similar observations once led me to devise a method by which one might accurately ascertain whether illumination, i.e., the propagation of light, is really instantaneous. The fact that the speed of sound is as high as it is, assures us that the motion of light cannot fail to be extraordinarily swift. The experiment which I devised was as follows:[5]

What follows is the description of a method that would allow for the measurement of the velocity of light, if one had sufficiently precise clocks. Experimentation, then, depending on technological progress, is what would enable the quantification of knowledge, raising it to a 'scientific' level. That big book of nature, always open under our eyes of common men, is nevertheless "written in mathematical language, whose characters are triangles, circles and other geometrical shapes; without such means it is impossible for us men to understand even a single word." Once the mathematical language and the experimental method, which alone allow one to extract from nature precise answers to precise questions, are mastered, we will be able to decipher the big book with such a certainty that we will have no qualms about abandoning the more primitive tools of common sense. Common observation, knowledge based on practical experience and tradition, non-mathematical language, and the other components of common sense, can only lead us up to a certain point, even when they help us to move towards truth. Full truth can only be reached by going beyond this point, by overcoming the intrinsic limitations of those tools:

> Some superficial observations have been made, as, for instance, that the free motion of a heavy falling body is continuously accelerated; but to just what extent this acceleration occurs has not yet been announced... It has been observed that missiles and projectiles describe a curved path of some sort; however, no one has pointed out the fact that this path is a parabola. But this and other facts, not few in number or less worth knowing, I have succeeded in proving; and what I consider more important, there have been opened up to this vast and most excellent science, of which my work is merely the beginning, ways and means by which other minds more acute than mine will explore its remote corners.[6]

Once the new methods are applied, and discoveries are made, the 'scientist' can boast:

> Please observe, gentlemen, how the facts which at first seem improbable, will, even on scant explanation, drop the cloak which has hidden them and stand forth in naked and simple beauty.[7]

Whereas Galileo adopts an 'intellectualist' position, which seems to posit a radical gap between common observation and systematic scientific observation,[8] other scientists claim that there is a continuity between common sense and science. Laplace is a case in point. To be sure, there is a great distance from the first sight our ancestors discovered in the skies to Laplace's days celestial mechanics, which gives us "a general view, today, of all the past and future states of the world's system."[9] Yet, according to Laplace, there is also, between these two extremes, a continuity of observations, of inquiries, of abandoned false hypotheses, which finally lead to the "complete explanation of all celestial phenomena, up to their minimal details."[10] When we watch the sky, we see celestial bodies that appear at dusk and disappear at dawn; we notice that some of them move, while others remain static, and so on. "Even at that moment, many interesting questions arise. What happens during the day to those stars we see at night? Where do those that begin to appear come from? Where do those that disappear go to? <u>The attentive examination of the phenomena yields simple answers to these</u>

questions":[11] the stars' light does not turn off, but is veiled by the sun's; the earth is not flat, but spherical, and the sky envelopes it from all sides; the earth turns around its axis; and so on. Careful observations allow for correction[12] of the illusions of common observation, as well as for their explanation. Observation also allows for the correction of the hypotheses created by the frail human mind in order to connect all these disparate facts. In this way, it finally leads to the discovery of "the true causes, or at least of the laws of the phenomena."[13] Along this lengthy route, one needs, besides systematic observation, to employ rigorous mathematical methods--notably the Calculus--"this wonderful tool, without which it would have been impossible to penetrate a mechanism as complex in its effects as it is simple in its causes."[14] True enough, only a super-human intelligence would be able to determine all the past and future states of the world's system, given the present states of all the particles of the universe. But our limited human intelligence can discover--with its mathematical tools--the laws of the system. It can also employ another mathematical tool, namely the calculus of probabilities, in order to apply that knowledge, even when lacking total information.

For Laplace, therefore, science is no doubt continuous with common sense. The latter's first observations and puzzles are the former's starting point, and the results of science can be applied to common experience so as to explain it. But this stage is reached only after a lengthy process, throughout which science piles up precise observations and develops very complex methods and theories, so that it can say categorically to common sense: "this is an illusion!"; "that is not as it looks like, but it is in fact so and so!"; and so on. Rather than a conciliation between common sense and science, what one finds is--even for Laplace--an asymmetrical relationship between them, whose ultimate result is the replacement of the former by the latter, insofar as our knowledge of reality is concerned.[15]

The result of this process of estrangement is the creation of a great gap between science and common sense, fully acknowledged today. According to Conant, for example, "even a highly educated and intelligent citizen without research experience will almost always fail to grasp the essentials in a discussion which takes place among scientists concerned with a projected inquiry." And this is so, not because he lacks knowledge of specific facts. This citizen suffers, in fact, from a "fundamental ignorance of what science can or cannot accomplish...; he has no 'feel' for the tactics and strategy of science."[16] The continuity between science and common sense, far from being a fact, has become something that must be restored, requiring a considerable amount of effort. In this vein, Bronowski speaks about the disappearance of a 'broad and general language,' shared by science, the arts, and other manifestations of our culture, and he considers it to be a fundamental task "to try to remake that one universal language which alone can unite art and science, and layman and scientist, in a common understanding."[17]

This restoration consists, partly, in pointing out the elements shared by science and common sense. For example, one may show that scientific experimentation has a lot in common with the trial and error methods employed, say, by a cook, by a craftsman, or by someone who wants to discover what key opens an old ark found in the cellar. One may recall that in certain fields of science--such as biological taxonomy--the line that separates common sense from science is very hard to draw. One may also point out the many commonsensical assumptions assimilated by science: the existence of objects located in a more or less tri-dimensional space; the permanence of such objects independently of the presence of an observer; the alleged uniformity of nature, which guarantees the reproducibility of phenomena in experiments performed under similar conditions; the reliability of induction; etc.[18] All these shared elements eventually justify the claim that there is a continuous gradient leading from common sense to science. But even those who defend this view will certainly acknowledge that the extremes of this continuous spectrum are quite different; that the difference in degree is so deep that it is equivalent to a qualitative difference; and that it is only by understanding this difference that one can understand what is science.[19]

It is not necessary to accept one or another conception of the nature of the 'method' of science in order to acknowledge the existence of such a difference. Nor is this necessary in order to acknowledge the widespread belief that, if there is some knowledge of any phenomenon legitimately called 'scientific,' then it is this knowledge rather than any other (e.g. the one deriving from common sense) which constitutes the knowledge of the phenomenon considered. Science cannot but be 'imperialist,' in this sense. When it provides knowledge, such knowledge is conceived as superior to that provided by any competitors (common sense, philosophy, witchcraft, religion). Any one of the other kinds of purported 'knowledge' is then downgraded to the level of 'illusion,' 'belief,' 'speculation,' etc., and can no longer compete with scientific knowledge.

It might be thought that such over-confidence of science comes from its (false) belief in its ability to provide certainty. But this is a mistake. Galileo, Newton, and Laplace eventually believed that their methods led indeed to certainties, if applied correctly. Yet, with the impact of Einstein's revolution in physics, and after Kuhn, Popper and Feyerabend have demolished the old doctrine of scientific method, most practitioners know that science does not produce certain and unchangeable truths. But none of them disputes the fact that science, due to the rigorous methods employed or to whatever makes it 'science,' is able to deliver the nearest surrogate to certainty, and that, in that, science is infinitely superior to common sense. Thus, the conciliation between them, whenever possible, is always one-sided: it consists in refining common sense, in educating it, so to speak, so as to let it be in a position to absorb the scientific truth; but it is never a matter of rejecting some scientific result because common sense is not prepared to accept it. The results of science remain as immune and indifferent to possible criticism from common sense as Galileo wanted them to be. In spite of all we know today about science's fallibility, who would dare to cast any doubt upon the theories accepted by physicists, biologists, or chemists merely because they contradict common sense?

Philosophy, unlike science, has not been able to produce methods which generate certainty or certainty-surrogates, nor to produce theories that can be 'applied' successfully. This has not prevented it from becoming esoteric and hermetic, i.e., far removed from common sense, at least superficially. Conant's dictum, quoted four paragraphs above, would also be true of the intelligent citizen who would try to understand a debate between two contemporary philosophers belonging to any school. This is due to the fact that philosophy has developed methods of analysis and criticism that enable one to cast doubt upon <u>any</u> philosophical theory, without ever leading to a clear victory of a theory over its competitors, or to a neat refutation of any of them. Furthermore, it has developed a rich 'technical' jargon for these purposes. The 'educated layman,' even if initiated into the mysteries of philosophical language, would hardly understand not only the tactics and strategy, but also the ultimate aim of an inquiry apparently destined to permanent inconclusiveness. Even the methods proposed by those philosophers who have made a similar diagnosis of philosophy's predicament have not enabled it to escape from its grip.

Given such a situation, it is inevitable for philosophy to seek 'external' help. Inevitable, of course, only for those who consider such a situation anomalous and want to find a way out of it. For those who accept it as philosophy's fate; for those who, like the late João Cruz Costa, claim that in philosophy there are no answers, but only questions, there is no need to look for redeeming help. But for those who seek, in philosophy, some answers--even though tentative--and have given up any hope that philosophy will be able to lift itself up from the mud by pulling its own bootstraps, it seems inescapable to appeal to external help. Wherefrom can such help come? For obvious reasons, neither from religion nor from art, since these disciplines find themselves in a situation very alike that of philosophy. Science would be another possible source. But, as we have seen, science is 'imperialistic' and would therefore try to swallow philosophy, as it has done with all those branches of knowledge which it has been able to cut off from the mother tree. Only common sense remains.

By appealing to common sense, philosophy often grants it the role of referee in philosophical conflicts: the conformity of a philosophical doctrine with common sense becomes the touchstone for its acceptability.[20] But what kind of judge can common sense be? Once doubt has crept into our minds, one can no longer presume the innocent and immediate certainty of one or another thesis of common sense. It doesn't help to shout ten or a hundred times "this is my hand, and it exists!" Nor does it help to invoke natural light's help, after having hypothesized the existence of an evil genius who can mislead us so completely that he may have also obfuscated our 'natural light.'

Wittgenstein has made this clear, by pointing out that there are answers which can bring to rest the doubt of 'reasonable men,' while they are useless for, say, an idealist philosopher. If I say "I know that here is a hand," a reasonable man can ask "How do you know it?"; to this, I can answer "For it's _my_ hand that I'm looking at." Under these conditions, says Wittgenstein, "a reasonable man will not doubt that I know.-- Nor will the idealist; rather he will say that he was not dealing with the practical doubt which is being dismissed; but there is a further doubt _behind_ that one.-- That this is an _illusion_ has to be shown in a different way."[21] That is to say, the mere appeal to the comfortable and familiar certainty of common sense does not even touch the problem posed by the idealist philosopher. Nor would a more sophisticated appeal to common sense, under the form of an analysis of the use of ordinary language, which might show, say, that sentences of the type 'There are physical objects' are meaningless, serve as a solution to the philosophical problem: "Is it an adequate answer to the scepticism of the idealist, or the assurances of the realist, to say that 'There are physical objects' _is_ nonsense? For them after all it is not nonsense."[22]

For the man who has never doubted the existence of his hand or of other physical objects, there is no need to point out that they do indeed exist. And against the idealist, who casts doubt upon such an existence, _in spite of appearance and current beliefs_, it is useless to show the hand, since he has presented arguments which he believes show the illusory character of the belief in the external existence of that very hand. Thus, to address common sense as a _source of alternative philosophical theses_ is of no help. For those who have lost their faith in common sense, it can no longer function as the supreme authority: innocence cannot be recovered by the emphatic reaffirmation of its value. The issue of the relationship between common sense and philosophy must be conceived in other terms.

In spite of the obvious uselessness of the appeal to theses of common sense in contraposition to philosophical theses, it should be recalled that this move has been often used in the history of philosophy--and not only by Moore. Hume, for example, when he discusses the 'instinctive' belief in the independent existence of the objects we perceive, comments:

> But this universal and primary opinion of all men is soon destroyed by the slightest philosophy, which teaches us, that nothing can ever be present to the mind but an image or perception, and that the senses are only the inlets, through which these images are conveyed, without being able to produce any immediate intercourse between the mind and the object.[23]

Well, it is precisely the result of Hume's 'slightest philosophy' that seems scandalous to Thomas Reid, who attributes to it the worst mistakes ever made in the theory of knowledge. A large list of philosophers, from Plato, through Aristotle, Malebranche, Descartes, Arnauld, Locke, and up to Hume, hold the perverse assumption "that we perceive not external objects immediately, and that the immediate objects of perception are only certain shadows of the external objects," their name--ideas, forms, species, phantasms, or impressions--notwithstanding.[24] Against such an impressive philosophical tradition, the _first_ objection raised by Reid is that

> it is directly contrary to the universal sense of men who have not been instructed in philosophy. When we see the sun or moon, we have no doubt that the very objects which we immediately see, are very far distant from us, and from one another. We have not the least doubt, that this is the sun and moon which God created some thousands of years ago, and which have continued to perform their revolutions in the heavens ever since. But how are we astonished when the philosopher informs us, that we are mistaken in all this; that the sun and the moon which we see, are not, as we imagine, many miles distant from us, and from each other, but that they are in our own mind; that they had no existence before we saw them, and will have none when we cease to perceive and to think of them; because the objects we perceive are only

> ideas in our own minds, which can have no existence a moment longer than we think of them.[25]

This first remark is in fact Reid's main argument: his opponents' _philosophical_ thesis must be false, since it contradicts common sense. To the philosophers who deny the apparent certainty of one of the central theses of common sense, Reid replies by asserting emphatically that thesis itself. No doubt it is hardly to be expected that a debate such as this should yield a rationally determinable victor.

The Reid/Hume case is not the only one. Even philosophers like Leibniz, who have developed metaphysical systems which are extremely distant from commonsense, and have employed in their philosophizing the technical apparatus of logic and mathematics, occcasionally appeal to common sense, in as simplistic a way as Reid. The quotation used as a motto in the present paper is a case in point: the evil genius hypothesis is a _philosophical_ hypothesis, whose purpose is to provide the most extreme possible formulation of scepticism. To mention against it the current notion of evidence, and "the experience and events of common life" is merely to forfeit the discussion of the philosophical problem of the grounds of our knowledge, by waving as a flag the fact that, in our practical endeavors, we do not actually _act_ as sceptical men. But the issue is not whether, in our capacity of common men, we trust our senses and other sources of knowledge _for practical purposes_, but rather whether we _have the right_ to trust them. The reassertion of the undoubted fact in question does not say anything about the latter, properly philosophical, issue.[26]

The relationship between common sense and philosophy must therefore be conceived in other terms. Neither opposition of theses, nor adoption of the theses of the former by the latter, but rather complementarity--not necessarily of _theses_--is what I suggest should exist between them. Before explaining this idea, however, I must argue against another possible solution, and clarify somewhat the meaning of 'common sense.'

The solution I don't accept is the one that posits a strict separation between common sense and philosophy. Defenders of this view may acknowledge that philosophy is in a desperate position, leading it to seek external help. But they deny that common sense is able, in principle, to provide such help. For, on their view, philosophy and common sense belong to 'different domains.' One way of performing the separation is to describe philosophy as an essentially 'theoretical' endeavor, and common sense as having the function of governing our 'practical' life. "Who conducts his life according to philosophy?", ask the proponents of this view. As a reply, one might refer them to Kant's arguments against the popular saying "This may be correct as far as theory goes but is worthless in practice."[27] Theory and practice are not opposed as two distinct domains; they rather complement each other. If the practice in question is everyday action, no ethical theory or deontic logic can be indifferent to it, since they must play a role as components of any explanation of human action. If the practice considered is the making of science or any other form of producing knowledge, can epistemology pretend to be entirely detached from it? Even metaphysics must take current 'practice' into account, e.g. perception: if it denies that things are as they appear to us, metaphysics has to explain (or to entrust someone with the task of explaining) why things seem to be what they are not. Practice, therefore, far from being a domain entirely detached from philosophy is, in fact, the test ground for any philosophical theory.[28]

Another way of expressing the same separatist idea would be to say, in the manner of Wittgenstein, that common sense and philosophy are different 'forms of life.' If so, they could not mix up, nor substitute for each other, nor even compete. Each of them would have its own 'theoretical' as well as 'practical' dimension. Such a position is no doubt tempting. Philosophers seem to have indeed a special form of acting, which manifests itself in all circumstances, and not only when they are thinking, or writing. Yet, is this any more than the so-called 'professional distortion,' to be found also in the physician, in the lawyer, in the cleaning woman, and in most other professions? Does really the philosopher--as such--make love, cook, and bathe differently from the common man? On the other hand, when the latter reflects about the meaning of life, God, or the limits of the physical universe, is it true that he does not reflect _philosophically_ just because philosophy is not his 'form of life'? But once one admits these points of contact, the thesis of a strict separation of the two forms of life can no longer be defended.

Let me now try to clarify the obscure notion of 'common sense' I have been using so far, in the footsteps of a very old tradition. I have talked about a primordial innocence,' about a belief in appearances which was inevitably lost once Thales began to see beyond them. The reader may have implicitly assumed that 'common sense' is, for me, nothing but this state of initial innocence or, rather, the set of naive beliefs that comprise it. No doubt common sense is always naive or innocent. But not because of the specific <u>content</u> of its beliefs; rather, because of the way in which such beliefs are accepted. In this sense, common sense is an <u>attitude</u>, and not a definite set of beliefs. Its innocence consists in its <u>a-critical acceptance</u> of the beliefs upon which its actions and judgments are based. Let me stress this point: beliefs that most typically belong to common sense (in a given culture and historical period) are those that manifest themselves, say, spontaneously, in the current actions of the vast majority of individuals (of that culture and historical period). They are not the result of reflection, nor are they beliefs put forth in the context of specific theoretical debates. Since they are not grounded on arguments that purport to justify them, they often appear to be based on 'intuition,' and offer tenacious resistance to argumentative criticism. If it happens that they are derived from some theory, or other form of systematic knowledge, this source has been usually forgotten by most people. For example, any housewife makes two holes in a juice can, before she pours it. If one asks her why, she will probably say that one of the holes is "to let the air in." The underlying belief, in this case, far from being derived from Torricelli's notion of atmospheric pressure, is presumably the Aristotelian principle according to which nature abhors vacuum: "the air must get in in order to fill the empty space left by the liquid that is poured"; but certainly the Aristotelian source of the housewife's practical belief has been forgotten.[29]

The set of beliefs generated and adopted by this a-critical attitude, manifested typically in action, and 'justified' in terms of intuition--and this is how I would describe 'common sense'[30]--is not <u>fixed</u>, but rather historically and culturally determined. With the evolution of science and of philosophy itself, new beliefs come to replace those that have been held by common sense for a long time, though--as shown by the above example--such changes may be very slow. From a dynamic point of view, the imprecise picture I have drawn at the beginning of this paper can be now easily mended. The fact of the matter is that the belief in appearances had already been abandoned by mythical thought, which looked for the deep unity of things in the realm of the supernatural. The 'innocence' of mythical common sense laid, therefore, not in believing in appearances, but rather in believing that behind them there was always an occult agent responsible for the observed events. What Thales did was no doubt a radical departure from <u>that</u> common sense, insofar as he substituted, for the mythical question "who," the entirely new question "what?" and demanded a 'rational' rather than supernatural answer.

Similarly, when Heracleitus proposes what Plato understood to be the most extreme form of dynamism, namely that things don't have an underlying unity, but are rather in constant flux, it was <u>another</u> common sense that he was facing: the common sense that had absorbed, however partially, the Eleatic and Ionic ideas of unity and identity. Similarly, common sense after Newton is not the same old Aristotelian common sense of the preceding centuries. To be sure, such changes are neither immediate nor complete. Only part of the new ideas of science, art, religion, or philosophy are slowly absorbed by common sense--and it would be interesting to find out which part and why. But the fact is that the contents of 'innocence' are constantly changing, without thereby making it less innocent. At any given point of history, it is within an innocently, i.e., a-critically and perhaps also unconsciously, accepted framework that human communities act and think. Once doubt is cast upon a part of such a framework, one can no longer defend it by appealing to its evidence, certainty, or intuitivity. On the other hand, such a framework is the historically given starting point for philosophical inquiry, even though such inquiry may finally lead to the overthrowing of the framework. Such an interaction can be better understood as a relationship of complementarity.

The model for such a relation comes from science itself. The relationship between a scientific theory and the observational data that verify, control, confirm, or refute it (let the reader choose the appropriate term, according to his or her methodological preferences) is always indirect. Data do not have the role of hypotheses or theories, nor do the latter have the role of the former. Data refute or confirm a theory not because they confront or support it directly, but only insofar as the predictions derived from the theory are opposed to/or in agreement with them. There is always a stage of 'derivation' mediating between theory and data. Similarly, there is always a process of 'elaboration' (which is systematic, according to some inductivists; and a-systematic according to anti-inductivists) which leads from the data to the theory; a theory is never reducible to the re-assertion of 'brute' data. Theoretical controversies are sometimes solved by an appeal to data, but not by simply abandoning a theory and replacing it by data. On the other hand, if a set of data is questioned, no purely theoretical argument will let one decide the issue: one will have to seek other data, connected with those questioned, e.g., by repeating the experiment, or by other means.

No doubt it is no easy matter to characterize with precision the relationship between data and theory in science, and many a philosophy of science has failed in this endeavor.[31] Among other things, philosophers of science of all persuasions have shown that the very notion of what is to be taken as 'data' is under the influence of the 'conceptual framework,' 'paradigm,' 'research program'--in short, of the theory to which the data are connected. Recall the case of the holes in the can. The Aristotelian principle of _horror vacui_, which 'explains' the need for the two holes, also explains how a suction pump works. Water goes up (when the piston is lifted) in order to fill the empty space resulting from the expulsion of the water in the tube. It so happens, however, that the water will not raise above 34 feet in the tube (at sea level). The above mentioned principle, which is purely qualitative, is of no use in explaining such a _quantitative_ limitation. Aristotelian physics would never consider such a phenomenon worth observing. But it is precisely by considering this and other _measurements_ as relevant 'data' that one can discover atmospheric pressure, and hence, an entirely different explanation of the familiar phenomena. Thus, the introduction of the notion of measurement induces a radical change in the notion of 'data.' Yet, this change turns out to be only the beginning of a continuous process of change. For there are measurements and measurements: what one theory takes to be a mere measurement error, within the limits of tolerance, can become a crucial piece of evidence for another. In spite of this fluid character of the 'data,' I believe the distinction between the two levels--theory and data--as well as the conception of their relationship being an indirect one, are essential to modern science, and therefore serve as a model for philosophy.

I propose, therefore, to conceive the relationship between philosophical theses and common sense beliefs (or intuitions) as similar to the relationship (in science) between theory and data. Common sense, according to this model, is no longer to be viewed as a reservoir of philosophical doctrines, but rather as the philosopher's laboratory. There, the philosopher will not look for ready-made theories, but only for the data, for the raw materials for building his theories, and mainly, for performing quality tests on them. Philosophical doctrines are, then, to be seen as the result of a process of elaboration of such observations. They can become very complex and as remote as necessary from their common sense sources. Philosophy may employ a very sophisticated logic, in the same way as science may employ highly refined mathematical tools. Ultimately, however, there must obtain some bridge between theory and data, in order to check the value of the theory. It is indispensable that philosophy, no matter how abstract it becomes, be able to return to the obscure concepts and intuitions of common sense which were its starting point, if only to correct them.

Philosophy does not have to satisfy all the demands of common sense, but neither can it simply abandon all of them, _en bloc_. Consider, for example, the well known 'paradoxes of confirmation.' They arise from the conflict between an _intuitive_ requirement (two logically equivalent propositions must be either confirmed or disconfirmed exactly by the same set of data) and the observation, equally _intuitive_, that the discovery of a red lamp does not increase in the least the 'degree of confirmation' of the proposition 'All ravens are black.' Any proposed solution of the _philosophical_ problem raised by these paradoxes, no matter how sophisticated the technical concepts of equivalence, confirmation, etc. employed, is always evaluated (at least in part) in terms of its capacity to explain those conflicting intuitions, or to justify, in some way, the rejection or modification of one of them.[32]

Where, then, are these data for philosophical work to be collected? Where is the metaphorical laboratory of the philosopher located? We might think--as Arne Naess did--that the philosopher should interrogate the 'men in the street' by means of questionnaires, thus yielding a collection of the views currently held by common sense. I don't think, however, that the relevant model to be followed is that of public opinion polls, but rather, that of contemporary linguistics. After Chomsky, linguistics has ceased to consider the tape recorder and the notebook the only legitimate source of empirical data. It has rehabilitated the linguistic 'intuitions' of each speaker and of the researcher himself as no less empirical and legitimate sources of data. Analogously, I think the philosopher must rehabilitate his confidence in the 'common man' lying inside himself, with whom he is directly acquainted. (If philosopher and common man lived in separate domains, according to entirely different forms of life, what I am now proposing would be impossible). No doubt, some of the techniques developed by phenomenology and analytic philosophy may be useful for the exploration of this implicit source of data. Yet, in the same way as the linguistic intuitions of a speaker are not linguistic theories, but only data which, when appropriately used, may lead to theories, so common man--even the one that lives inside each philosopher--is not the philosopher. Every speaker is able to pronounce intuitive judgments on the grammaticality of his language's sentences. The conformity with the _set_ of these judgments is the criterion for descriptive adequacy of a linguistic theory. But the theory cannot merely reproduce such judgments. Similarly, each common man pronounces, for instance, a set of judgments about that which he 'knows,' 'supposes,' 'believes,' 'thinks,' or 'ignores.' A philosophical analysis of the concept of knowledge and its cognates must take into account this set of judgments (as well as the behavior based on them), but cannot restrict itself merely to their reproduction and classification. It is the ability to _explain_ these intuitive judgments of common man that constitutes the criterion for the adequacy of the proposed analysis. Gettier's famous counter-examples are effective against the analysis of knowledge as justified true belief only because they show how this analysis leads one to count as knowledge that which intuitively is not knowledge. It is thus unable to explain our intuitions. The way from data to analysis and vice versa may (and perhaps should) be long and difficult, and the resulting analysis may be complex and sophisticated, but links with the data can never be totally severed.

One of the important features of the model here proposed is the idea that there must be a real and reciprocal interaction between common sense and philosophy. It is neither a relation of passivity nor of subordination. In particular, the idea is not to leave common sense's intuitions untouched, and to subordinate the validity of philosophical theses to their absolute conformity to them. Those intuitions are often vague, confused, and sometimes even contradictory. Particularly in borderline cases, not found in common experience, which demand some effort in order to be imagined, common sense simply may not have any intuition at all to offer to the philosopher. An example of such problematic cases is that of identifying individuals across possible worlds, which often occur in debates about the proper formalization of modal logics: if you had met an individual physically identical to Getulio Vargas, born in Rio Grande do Sul, the father of Ivete Vargas, dictator of Brazil, but who had not committed suicide in August 1954, would you consider him the same individual (though living in another possible world) as the historical Vargas we all know? And if that individual had committed suicide, but was not the father of Ivete Vargas, would you modify your reply to the former question? Presumably, the only answer of the layman to questions such as these would be a puzzled and eventually angry "What the hell do you want from me?" Philosophical theory, in such cases as this, must 'educate' or 'refine' common sense's intuitions, extrapolating to these problematic cases the theoretical devices that have proved satisfactory to deal with the non-problematic cases. Such a procedure is similar to that of the linguist, whose theory permits him or her to determine the grammaticality of very long or complex sentences, about which the speaker has no 'intuitive' opinion whatsoever. Along the same line of action lies the possibility of 'correcting' certain experimental data which, if they were accepted at their face value, would require the introduction of extreme complexities in the theories (e.g., in the mathematical functions) needed in order to explain the relevant set of data.

The possibility of 'correcting' or 're-educating' common sense's intuitions can go beyond the peripheral cases, and reach even the paradigmatic cases. In order to solve the paradoxes of confirmation, for instance, it seems to be necessary to modify, and eventually abandon, some of the most central intuitions we have about the concept of confirmation. Hence, there is nothing sacred about our intuitions, which would forbid giving them up. In this I entirely agree with Hintikka, who points out the fallibility of intuition.[33] This follows from the idea that common sense does not <u>compete</u> with philosophy in the role of producer of philosophical <u>theses</u>. Common sense's intuitions are not to be taken as undisputable propositions. But it is important to notice that a philosophical theory which decides to abandon an intuition cannot do it reasonably just by virtue of considerations pertaining to the internal economy of the theory. It must justify this step by showing what is gained in terms of <u>other</u> intuitions, and by showing what was wrong with the abandoned one. Hintikka himself, for one, justifies the rejection of a seemingly intuitive axiom of deontic logic by pointing out the fact that it has the 'nonsensical' (i.e., counter-intuitive) consequence that all the obligations are fulfilled.[34]

In its ability to face common sense by trying sometimes to modify its naive intuitions, philosophy resembles science. But the latter, as we have seen, not only replaces common sense beliefs by other better grounded beliefs, but it also dispenses entirely with common sense's services as a source of data or of control, functions now fulfilled by experimentation and systematic observation. Only at a very initial stage does science rely upon common sense. Beyond that stage, data have been accumulated, and new theories are based on preceding ones (even when they are in conflict with them) and on the accumulated data, no longer depending upon common sense. In philosophy, however, it is much harder to characterize 'progress.' There is no cumulative knowledge, though there is accumulation of theories and systems. But since it always deals with those problems which have not so far received a satisfactory 'scientific' treatment, philosophy must always return, ultimately, to its natural base, namely to common sense. This does not mean that a philosopher should not make use of the results of science. On the contrary, it would be absurd for someone to write nowadays about the problems of determinism and causality without taking into account what quantum mechanics has to suggest, or about the philosophy of language without inquiring into what has been achieved by linguistics. Relevant scientific information must be employed by the philosopher, together with common sense, as materials for his analysis. To be sure, the better and more intimate his familiarity with the science whose results he employs, the better are the prospects that the philosopher will not make gross mistakes when he employs them. Yet, to demand that the philosopher be actually initiated in the scientific _practice_ of the relevant disciplines,[35] is to demand too much. For science offers him systematized and conceptualized results, whereas common sense--in whose _practice_ I think the philosopher must participate--only offers 'intuitions' in their pre-analytic stage (even when they look as if they were full-fledged theories). The former can be absorbed by the philosopher without his having to go through the practice of their generation; the latter cannot.

The interrogation of common sense by the philosopher should not be direct, but rather indirect, according to the proposed model. In a sense, direct interrogation is self-defeating, since it alerts the subjects, thus leading them away from the state of innocence which guarantees, if not the objectivity, at least the philosophical neutrality of the data provided. If a psychologist, for instance, were to ask directly the subjects of a research on motivation "Are you motivated to perform x?", he would have to explain to them the theory of motivation he is trying to put to test in his experiment, a fact that would no doubt have some influence on the results. Instead, he looks for some indirect way of measuring the degree of motivation of the subjects. No doubt this presupposes a theory of motivation, which is actually employed in the interpretation of the measurements performed. But this is nothing but an instantiation of the general thesis that data are theory-laden, a kind of relativity from which there seems to be no escape in science. There is no reason why one should demand from philosophy to abide by standards of rigour and objectivity that science itself cannot achieve. The method of indirect interrogation contributes to reduce relativity to a minimum, though not to eliminate it entirely, by letting the theory interfere in the conception, performance, and interpretation of the experiment <u>by the researcher</u>, but not by the subject of the research.

A dissociation such as this is difficult to achieve in philosophy, especially when the researcher (philosopher) and the subject (common man) are the same individual. There are, however, forms of indirect inquiry, which have been employed by philosophers. For example, the method of 'semantic ascent' proposed by Quine,[36] or, more generally, the analysis of ordinary language. I do not deny that language is one of the most important sources of indirect data for philosophy. But I don't think it is the <u>only</u> legitimate source for such data, as those that employ the methods just mentioned presume. Any other means to extract from common sense relevant data for philosophical inquiry (e.g., phenomenology) must be employed, provided it is not conceived as a form of privileged access to <u>theses</u> which, having been sanctified by common sense, stand ready to replace philosophical reflection.

One of the objections that can be raised against the model here proposed is the following: what we want to know, when we engage in philosophical research, say, about the concept of knowledge, is what knowledge truly is, and not what common man (or the totality of men, for that matter) <u>believes</u> it to be. But the method proposed allows one to reach at best the latter result, never the former. This objection is probably based on an equivocation about the nature of the proposed model, presumably due to the analogy I have drawn with linguistics. The model does not propose to see common intuitions necessarily as expression of a <u>tacit</u> knowledge, basically correct though not explicit, which it is the task of philosophical research to make explicit. They are nothing but <u>data</u>--in some cases, the only available ones--upon which one must <u>build</u> the desired philosophical knowledge. Thus, I cannot see why common man's intuitions cannot be both, revealing of his <u>beliefs about knowledge</u>, and of what knowledge truly is. There are philosophers who assume that it is only within the framework of a 'folk psychology' that terms such as 'belief,' 'desire,' and 'knowledge' acquire their meaning, whence they conclude that the first task of a philosophy of mind is to make explicit this implicit psychology.[37] But this version of the relationship common sense/philosophy is, to my taste, too strong, and it is not at all required by the proposed model. In this, the model differs not only from the position of the philosophers just mentioned, but also from some formulations of Chomsky's views, where he takes the purpose of linguistic research as coinciding with the explicitation of the tacit, internalized knowledge mastered by every common speaker of a language.

When we go on, from the philosophy of mind to other areas of philosophy, such as metaphysics, the preceding objection seems to gain force. What we are interested in discovering is what being is, and the intuitions of John, Anthony or Marcelo-the-common-man about being, seem to be, even indirectly, totally irrelevant for metaphysical inquiry. To this, I can only reply that I suspect that such a being which is supposed to be entirely disconnected from any set of human intuitions and human behavior, is so close to the Kantian Dinge an sich that the inquiry into its nature becomes an impossible mission and, therefore, irrelevant as a philosophical enterprise.

I am perfectly aware that the model I have here proposed for the relationship between philosophy and common sense is full of imperfections. Nowadays, when our innocence no longer accepts faith in absolute certainties, except perhaps for the certainty that all of today's certainties will be overthrown tomorrow, we must try, against all odds, to build uncertain models which will guide our action and our research, under the prevailing conditions of uncertainty. Let this uncertain model be attacked, by the no less uncertain critique which will, for sure, shake its foundations.

NOTES

* This is a modified version of my paper "Senso comum e filosofia," Manuscrito 3, 1 (1979).

1. In this case, as in Austin's well-known examples, the expression 'in fact' is not the one that wears the pants. It merely indicates a contrast with a view that one wishes to refute. From a linguistic point of view,

'in fact' always bears contrastive stress, a fact that assigns to utterances containing this and similar expressions certain interesting semantico-pragmatic properties (cf. F. Dretske, "Contrastive statements," The Philosophical Review 81 (1972), 411-437).

2. A position of this kind is defended, for example, by Oswaldo Porchat Pereira in "O conflito das filosofias," "Prefácio a uma filosofia," and "A filosofia e a visão comum do mundo," in Bento Prado Jr., Oswaldo Porchat Pereira, and Tércio Sampaio Ferraz, A Filosofia e a Visão Comum do Mundo (São Paulo: Brasiliense, 1981).
3. Galileo Galilei, Discorsi e Demonstrazioni Matematiche intorno a Due Nuove Scienze, in Opere, ed. P. Pagnini, vol. 4, (Firenze: Salani Editore, 1964), pp. 119-120 (quotations are from the Henry Crew and Alfonso di Salvio translation, New York, Dover, 1914, p. 1).
4. Ibid., p. 120, (p. 2).
5. Ibid., p. 178, (pp. 42-3).
6. Ibid., p. 317 (pp. 153-4).
7. Ibid., p. 125 (p. 4).
8. However, this position can be justified if one recalls the historical context of Galileo's fight against the predominant Aristotelianism.
9. P. S. Laplace, Exposition du Système du Monde, 6ème éd. (Paris: Bachelier, 1835), p. 1. Translations are mine.
10. Ibidem.
11. Ibid., p. 3 (underlining added).
12. "C'est une illusion que les premiers observateurs ne tarderent pas à rectifier." Laplace, op.cit., p. 4.
13. P. S. Laplace, Traité de Mécanique Céleste, in Oeuvres Complètes de Laplace, vol. 3 (Paris: Gauthier Villars, 1878-1914), p. xi.
14. Ibid., p. xii.
15. I have discussed Laplace in virtue of Joseph Agassi's comments on the first version of this paper. Agassi pointed out that Laplace considered any opposition between common sense and science as being problematic, and therefore he tried to show how scientific theses, apparently opposed to common sense, were in fact anchored in it. Yet, I have not been able to find, in a brief inspection, examples of this procedure in Laplace. At any rate, the question is *how* did Laplace perform such a conciliation: by modifying common sense or by modifying science? It seems to me that the answer to this question is clear.

16. James B. Conant, _Science and Common Sense_ (New Haven: Yale Unversity Press, 1951), pp. 3-4.
17. J. Bronowski, _The Common Sense of Science_ (Harmondsworth: Penguin Books, 1960), p. 17.
18. Conant, _op. cit._, pp. 51, 222, 33-4.
19. _Ibid._, pp. 49, 237.
20. In some cases, the contempt for common sense is so strong that the relationship is the inverse one: if a doctrine is in conformity with common sense, then it is unacceptable, and vice versa, namely _credo quia absurdum est_.
21. L. Wittgenstein, _Uber Gewissheit - On Certainty_, ed. by G. E. M. Anscombe and G. H. von Wright (Oxford: Blackwell 1969), par. 19. See also par. 24.
22. _Ibid._, par. 37.
23. David Hume, _Enquiries Concerning the Human Understanding and Concerning the Principles of Morals_, ed. by L. A. Selby-Bigge (Oxford: Clarendon Press, 1902), p. 152.
24. Thomas Reid, _Essays on the Intellectual Powers of Man_, ed. by Baruch A. Brody (Cambridge, Mass.: M.I.T. Press, 1969), p. 124.
25. _Ibid._, p. 212-3.
26. To be sure, Leibniz argues in quite different ways against Descartes and against scepticism. On the former, see Y. Belaval, _Leibniz Critique de Descartes_, (Paris: Gallimard, 1960). On the latter, see M. Dascal, "La razón y los misterios de la fe," _Revista Latinoamericana de Filosofía_ 1 (1975), 193-226. It seems to me that the well-known ordinary-language-philosophy criticism of Cartesian hyperbolic doubt (cf. O. K. Bouwsma, "Descartes' Evil Genius," _The Philosophical Review_ 43 (1949), 141-151) suffers from the same flaws as the Leibnizian criticism here discussed: if ordinary language is not well suited for the formulation of a philosophical problem, does this make the problem vanish?
27. I. Kant, _Werke_, vol. 6 (ed. E. Cassirer, Berlin, 1922 ff.), 359 ff. Though Kant confines himself here to a discussion of the relationship between ethical theory and practice, he employs in his argument analogies with other fields, e.g. the relationship between the engineer's practice and theoretical mechanics.
28. The complex and interesting issue of the relationship between theory and practice certainly deserves a more detailed discussion, which I cannot undertake here. For

a historical account, see N. Lobkowicz, Theory and Practice: History of a Concept from Aristotle to Marx (Notre Dame: University of Notre Dame Press, 1967), especially chapter 9, on Kant.

29. This example is Conant's (op. cit., p. 65). His conclusion is that "the common-sense view of the world even today is more Aristotelian than we sometimes like to think."

30. In this sense, common sense has many of the characteristic features of what Popper calls a 'tradition.' The difference lies in the former's essentially a-critical character, whereas Popper believes in the possibility of establishing a 'critical tradition.' Perhaps the critical attitude will, at some time in the future, become a part of Western common sense. If this is the case, 'common sense' in the sense I have given to this term will cease to exist. Unfortunately, I see no signs that this is about to happen.

31. Some of these difficulties are discussed in my "Empirical Significance and Relevance," Philosophia 1 (1971), 81-106.

32. For an analysis of the various solutions proposed for the paradoxes, see Menachem Fish's M.A. thesis "The Paradoxes of Confirmation and their Solutions" (Tel-Aviv University, 1981). The reader will have noticed that I am assuming that common sense has 'intuitive' notions of confirmation and equivalence. To be sure, these notions may have come from conceptions of science (Bacon's) and of logic (Aristotle's?) which have become, in the course of history, part of the set of beliefs of common sense. This comes as no surprise in view of common sense's dynamic character.

33. J. Hintikka, "Intuitions and Philosophical Method," Revue Internationale de Philosophie 135 (1981), 79.

34. Ibid., p. 82. He goes on to explain the confusion that has led to the adoption of the apparently intuitive axiom. Later on (p. 83), Hintikka defends a certain version of his game-theoretical semantics, provided it is combined with certain "extremely natural assumptions," i.e., extremely intuitive ones.

35. G. Mounin, for example, suggests that "it is impossible to philosophize about the experimental sciences without being a good experimenter," and claims that "only the practice of linguistic analysis...really permits one to

reflect philosophically about language" (<u>Linguistique et Philosophie</u>, Paris, Presses Universitaires de France, 1975, p. 9).

36. W. V. O. Quine, <u>Word and Object</u> (Cambridge, Mass.: M.I.T. Press, 1960), pp. 271-6. On the difficulties that arise when this method is employed carelessly, see my "Levels of Meaning and Moral Discourse," in A. Kasher (ed.), <u>Language in Focus: Foundations, Methods and Systems</u> (Dordrecht: D. Reidel, 1976), pp. 587-625.
37. For example, S. Stich, "On the Ascription of Content," in A. Woodfield (ed.), <u>Thought and Object</u> (Oxford: Oxford University Press, 1981).

Zeljko Loparić

DECIDABILITY AND COGNITIVE SIGNIFICANCE IN CARNAP*

INTRODUCTION

We find in Carnap first a strict criterion and later different liberalized criteria for cognitive significance. The strict, constructivist criterion amounts to a set of general, direct and deterministic decision procedures for certain (infinite) classes of empirical questions. We call these procedures general because they (supposedly) give answers to all the members of the class of questions for which they are proposed, direct because they apply to questions themselves or to equivalent questions, deterministic because the questions answered are what-questions or yes-or-no-questions and not how-probable-questions. The strict criterion received alternatively semantic (phenomenalistic and physicalistic) and syntactic (linguistic, formal) formulations. These alternatives reveal Wittgensteinian (and later Tarskian) and Hilbertian influences, respectively. But the different formulations notwithstanding, Carnap's strict criterion never ceased to be a set of procedures of decision. It can safely be interpreted as an attempt to realize in the domain of empirical questions the old rationalist idea of a universal effective method for solving problems.

The liberalized criteria, on the other hand, consist of partial or indirect or probabilistic decision procedures, equally for empirical questions. Conceptually and historically, the origins of the liberalized criteria are the negative solutions for decision problems for certain formal questions. It is only secondarily that the liberalization is due to the impossibility of finding strict procedures for some important classes of questions in empirical science, such as questions about the applicability of dispositional and theoretical terms.

Our interpretation is at variance with most current interpretations. The main difference is that none of them,

J. J. E. Gracia et al. (eds.), Philosophical Analysis in Latin America, 313–337.

to our knowledge, reconstructs the systematic connection that exists between the significance criterion and effective procedures. As a consequence, most of them are not consistent with the results of our interpretation.

1. FORMULATION OF THE STRICT CRITERION

Carnap offered no single complete formulation of the strict criterion. Moreover, in no earlier writing did he give either the precise characterization of the descriptive language to which the criterion applies nor the complete list of linguistic entities to be classified as significant or non-significant. The difficulty of the task of reconstructing the criterion itself is exacerbated by the fact that Carnap oscillated with regard to its precise formulation. In his first writings on the subject in 1928, he adopts semantic formulations, although in the early thirties he banishes semantics from logic and epistemology as beset with 'insoluble difficulties and contradictions' (Carnap [6], p. 454), and in the late thirties, under the influence of Tarski, he switches back to it.

The reconstruction of the descriptive language that satisfies the strict criterion of cognitive significance becomes much easier if we take into account some later texts, in particular Carnap [8], [10] and [11]. The first of these texts contains a precise characterization of what Carnap considers to be a constructivist (or intuitionist or finitist) formal language L_I. The purely logical part of L_I is an attempt at a formalization of the intuitionist requirements for a language adequate for constructing the intuitionist elementary arithmetic. And the purely descriptive part of L_I is intended to be used in the formal reconstruction of empirically significant descriptive sentences. By relying on L_I we avoid unnecessary complications in early descriptive languages, such as those of the language of the AUFBAU based on the Russellian theory of types, without jeopardizing the exact comprehension of the significance criterion.

The logical part of L_I contains classical sentential connectives, variables for natural numbers, constants for number-theoretic properties, relations and functions, identity sign and a number of restricted operators, including limited sentential operators (quantifiers). The axioms of L_I are given by axiom schemata. The purely logical axioms consist of the axioms for the sentential calculus, the restricted quantifiers and the identity. The arithmetic axioms characterize the K-operator and the successor relation. Besides explicit definitions and definitions in use, recursive definitions are also admitted (Carnap [8], part I). The descriptive part of L_I obeys the same logical axioms as the logical part and satisfies, in addition, some other requirements, identical to the requirements for the observation language L_I stated in Carnap [10]. The variables of L_I take their values in (not necessarily infinite) domains of concrete empirical entities ('things' or 'total experiences') and every value of a variable is designated by an expression in L_I . All primitive descriptive constants in L_I are observational and all other constants reducible to the primitive ones either by explicit definitions or by definitions in use (Carnap [10], p. 41). The descriptive part of L_I so characterized is the language to which the strict significance criterion applies.

The guideline for reconstructing the significance criterion for such a language is given by some general semantic and syntactic properties of the logical part of language L_I. In a late text Carnap mentions the following properties of this language: "That language form provides a way to formulate unrestricted universal propositions about numbers, viz. with the help of open sentential formulas that are admitted as sentences. However, that language form cannot provide the formulation of unrestricted existential propositions... Each closed logical sentence S_i of that language is decidable, i.e., exactly one of S_i and $\neg S_i$ is provable and there is a (decision) procedure for discovering the proof. Each closed logical numerical expression A_i is computable, i.e., there is a procedure for discovering the numerical expression A_j in normal form ('0', "0", etc.) such that $A_i = A_j$ is provable" (Carnap [11], pp. 163-164). In order to be significant the descriptive part of L_I will be submitted to the same restrictions of its expression power and to the analogous requirements of constructivity of its sentences, predicates and functions. The Carnapian strict significance criterion is nothing other than a set of methods which ensure that the descriptive part of L_I satisfies the constructivist requirements. It is clear, however, that these methods cannot be the same as the decision procedures for the logical part of L_I. The decision problem for the question of significance of descriptive sentences, e.g., cannot, in general, be solved by formal proof procedures. The descriptive part of L_I is not axiomatized and, moreover, it is certainly not finitely axiomatizable. As an example of Carnap's views on this question, let us consider his analysis of the problem of deciding the truth values of the sentences regarding the relations of situation and the qualitative relations between given space-time points. This problem cannot be solved, observes Carnap, by a (formal) proof in L_I . "A relation of situation in the simplest case will be expressed by means of an analytic (or contradictory) sentence (e.g., 'Positions 7 and 6 are neighbouring positions'). On the other hand, a qualitative relation in the simplest case, will be expressed by means of a synthetic descriptive sentence (e.g., 'Position 7 and position 6 have the same color'). The former sentence is determined by a logical operation, namely, a proof; the latter, on the other hand, can only be decided on the basis of empirical observations, that is to say, by derivation from observation-sentence" (Carnap [8], p. 45). We see in this example some of the peculiarities of the decision problem for descriptive sentences. The positive solution of the problem implies the sharp division between logically provable and non-logically provable sentences (the famous analytic-synthetic distinction) and is conditional on the existence of general decision procedures either by observation (the semantical, empirical, formulation) or by derivation (the syntactic, linguistic, formulation).

General solutions for the decision problems relative to different classes of expressions of the descriptive part of L_I are given by the Carnapian strict criterion of cognitive significance. To put it another way: the criterion in question amounts to a set of general, direct and deterministic decision procedures for certain (infinite) classes of empirical questions formulable in L_I. As we already said above, we call these procedures general, because they (are supposed to) give answers for all the members of the class of questions for which they are proposed, direct, because they apply to questions themselves or to equivalent questions, deterministic because the questions answered are yes-or-no-questions or what-questions and not how-probable-questions. Yet, as we also already noted, in no place did the strict criterion receive a complete and precise formulation. Moreover, it was the origin of many hesitations in the questions of detail. What we want to do now is to pick up what seem to be the most important statements of (parts of) this criterion, without limiting ourselves to the formulations relative to L_I. In different texts the criterion is applied to individual, function, predicate and relation constants, sentences and problems of descriptive languages (L_D) and even to the descriptive languages themselves. In most cases the criterion appears in both semantic and syntactic formulations.

1.1 <u>Syntactic criterion for individual constants</u>

A descriptive individual constant is significant if it is primitive in L_D or defined in it by a definition chain in which no unrestricted operators occur.

This definition is a generalization for descriptive language L_D of a definition given by Carnap in [8], p. 45, for L_I. Our procedure, employed also below, is justified by the fact that Carnap thinks it to be possible to give a purely syntactic characterization of the difference between descriptive and logical expressions and that he considers all empirically significant languages to be equivalent (Carnap [6], p. 462).[1] Following the same source, we shall call a descriptive expression 'definite' if it is either primitive in a L_D or defined in it by a definition chain in which no unrestricted operators occur. In a later text, Carnap declared that his term 'definite' is synonymous to the term 'effective' in modern computability theory (Carnap [9], p. 193).

1.2 <u>Semantic formulation</u>

A descriptive individual constant is significant if it designates a concrete, observable entity that is or can be generated by constructive procedures from basic relation(s) among total experiences.

In the AUFBAU these constructive procedures are formulated in the so called "language of fictitious constructive operations." The operations considered by Carnap are supposed to be able to simulate the behavior of human subjects, at least in certain cases of data processing, for instance, in the (hypothetical) case of the construction of quality spaces _out_ of total experiences. Methodologically, the operation rules for the construction of objects (concepts) are considered necessary in order to assure univocity, non-emptiness and extensional character of terms defined in the formal language of the AUFBAU. This requirement shows that in 1928 Carnap still did not have a clear idea of the syntax of a formal language that would satisfy the requirements of constructivism.

2.1 _Syntactic criterion for function constants_

A descriptive function constant is significant if it is definite, i.e., computable (Carnap [8], p. 45, and [9], p. 193).

In the AUFBAU the descriptive functors are defined by finite lists of significant descriptive individual constants.

2.2 _Semantic formulation_

A descriptive function constant is significant if it designates a set of pairs of total experiences (or things).

This definition is a consequence of the relation theory used in the AUFBAU.

3.1 _Syntactic criteria for property and relation constants_

a. A descriptive property or relation constant is significant if it is definite (Carnap [8], p. 45).

b. A descriptive property or relation constant occurring in a given atomic sentence of L_D is significant if the protocol sentences of L_D (cf. definition 4) _from which_ the given atomic sentence is derivable have been stipulated (Carnap [5], p. 224).

3.2 _Semantic formulation_

A descriptive property or relation constant is significant if the property or relation it designates is decidable (Carnap [8], p. 161 and [9], p. 193).

4. Semantic criteria for protocol sentences

a. Significant protocol sentences constitute a subset of the sentences of the language of the physical system. A sentence belongs to the language of the physical system if it correlates a certain numerical value (or interval or probability distribution of numerical values) of a state variable to each member of a set of space-time points. Protocol sentences in such a language differ from other sentences in two aspects. First, they do not refer to single space-time points but the space-time regions. Second, they do not correlate specific values of state variables to these regions, but rather large classes of unknown values (Carnap [6], p. 458).

b. A sentence is a significant protocol sentence if it is about states of affairs that obtain among total experiences that constitute the fundamental relation(s) of the AUFBAU (Cf. Carnap [3], 119 and 180).[2]

5.1 Syntactic criteria for sentences

a. A sentence in L_D is significant if it is definite, i.e., if all constants in it are definite and all variables are bounded by restricted quantifiers (Carnap [8], pp. 45-46).

b. A sentence in L_D is significant if it is derivable from a set of protocol sentences in L_D (Carnap [5], p. 224).

c. A sentence in L_D is significant if sentences of the protocol language in L_D are derivable from it and not necessarily vice-versa (Carnap [6], p. 440).

d. A sentence in L_D is significant if protocol sentences of the language of the physical system are derivable from it and not necessarily vice-versa (Carnap [6], p. 463).

5.2 Semantic formulation

A sentence is significant if it expresses a conceivable state of affairs whose existence or non-existence is decidable by a set of operations based on precisely characterized possible experiences (Carnap [4], pp. 325-326).

We see that all significant sentences are verifiable or falsifiable by possible, i.e., theoretically conceivable experiences, and that, conversely, all sentences so decidable are significant. An important consequence of this definition is that we may know that a sentence is significant without knowing whether it is true or false (Carnap [4], p. 324).

6. Criterion of significance for problems

A problem is significant if it is either a yes-or-no problem concerning a significant sentence or a what-problem concerning a significant functor (Carnap [3], 180).

It is clear that all Carnapian significance problems are effectively soluble either by purely formal or by empirical procedures. Let us emphasize that, according to Carnap, not all important human problems are significant in his sense. That means that Carnap is very far from identifying the objectives of his constructivist theory of science with the ambitions or the classical rationalism. On the contrary, he is explicit about the limitations of the scientific rationalism: "The proud thesis that no question is in principle unsolvable for science agrees very well with the humble insight that, even after all questions have been answered, the problem which life poses for us has not yet been solved. The task of cognition is a definite, well-circumscribed, important task in life, and it can certainly be demanded that mankind should shape that aspect of life which can be shaped with the aid of knowledge by a determined application of this knowledge, that is, by using the methods of science. Even if modern movements frequently underestimate the importance of science for life, we do not wish to fall into the opposite error. Rather, we wish to admit clearly to ourselves, who are engaged in scientific work, that the mastery of life requires an effort to all our various powers; we should be wary of the shortsighted belief that the demands of life can all be met with the power of conceptual thinking alone." These words, put in the close paragraph of the AUFBAU, should serve as a serious caveat to all those who want to challenge Carnap as a defender of a totalitarian scientific reason.

7. Criterion of significance for languages

a. A descriptive language is significant if it is definite, i.e., if all expressions in it are definite (Carnap [8], p. 46).

b. A descriptive language is significant if all closed expressions in it are definite (*ibid.*, 15).

2. ORIGINS AND NATURE OF THE STRICT CRITERION

Our review of the formulations of the strict, constructivist, Carnapian significance criterion for descriptive languages L_D makes it clear that this criterion is identical with a set of purely syntactic (formal) and empirical decision procedures and that the decision problems presumably soluble by these procedures have close parallels in the logical part of L_I which supposedly formalizes the elementary arithmetic of the intuitionists. In particular, the decision problems for the question of verifiability of all observation sentences and of computability of all empirical functions correspond to the decision problems for provability of all logical sentences and computability of all numerical expressions in L_I, respectively. Now we want to make it clear that this parallel is not a mere accident but a technical counterpart of some purely philosophical views of the foundations of mathematics and empirical sciences, common to Carnap and to other students of the problem of foundations.

The constructive properties of L_I, such as decidability of formulas, computability of numerical expressions, solubility of all number-theoretic problems formulable in it and so on, were intended by Carnap to be a formalization of the vague intuitive and philosophical requirements on the foundations of elementary arithmetic laid down by the Intuitionists (Carnap [8], p. 46). In the intuitionist writings, only constructive expressions were considered to be *justifiable* in a purely rational way, and that was meant to imply that non-constructive expressions were *meaningless*. Accordingly, a language that would fulfill the intuitionist requirements for rational justifiability of its expressions would also fulfill the intuitionist criterion of significance. Coming back to Carnap, it is clear that the constructiveness assures the rational justifiability in L_I. On the other hand, Carnap himself made it clear that L_I was also significant in the intuitionist sense: "The language form under discussion here [L_I] agrees with certain philosophical views sometimes called 'finitism' or 'constructivism.' According to these views it is the case, e.g., that unrestricted existential quantifiers with respect to infinite domains give rise to meaningless sentences, and that predicates and functors are meaningful only if there is a fixed procedure by which their applicability in any concrete case can be decided" (Carnap [11], p. 164). It is now easy to see that Carnap's conditions on the significance for L_D given by his strict criterion are closely parallel to the formalized intuitionist significance criteria for mathematical languages. And we already know that the basic logic for L_D is the same as the logic of the purely logical part of L_I. This shows that the concepts of justification and of significance associated with the construction of L_D are closely related to earlier intuitionist concepts.

It has often been said (by Hempel, for instance), that the Carnapian significance criterion reflected basic principles of contemporary empiricism. We see now that this thesis is not to be taken in a restrictive sense. Indeed, the Carnapian criterion also reflects the basic principles of contemporary constructivism. Moreover, there are some other important parallels between the Carnapian significance criterion and influential contemporary views on the foundations of mathematics. The insistence on the question of problem solving was a basic characteristic of various programs in foundational research in continental Europe. Since 1900, Hilbert repeatedly made public his convictions that every mathematical problem was soluble, although not by a specifiable general method. In 1925, for instance, he wrote: "As an example of the way in which fundamental questions can be treated I would like to choose the thesis that every mathematical problem can be solved. We are all convinced of that. After all, one of the things that attracts us most when we apply ourselves to a mathematical problem is precisely that within us we always hear the call: here is the problem, search for the solution; you can find it by pure thought for in mathematics there is no *ignorabimus*. Now, to be sure, my proof theory cannot specify a general method for solving every mathematical problem; that does not exist. But the demonstration that the assumption of the solvability of every mathematical system is consistent falls within the scope of our theory" (Hilbert [14], p. 384). In his Tractatus (1921) Wittgenstein expresses similar optimism as to the problem solving capacity of human cognitive apparatus: "If a question can be framed at all, it is also possible to answer it" (Tractatus, 6.5). Wittgenstein also offered what he believed to be final solutions to many kinds of fundamental problems such as the problem of formulating a constructive criterion of validity of all mathematical and logical sentences. However, his algorithms for validity apply neither to all classical mathematics (Wittgenstein was an intuitionist) nor to the full classical logic. In 1928, Hilbert formulated what can perhaps be considered the most ambitious objective for his proof theory: to solve the problem of validity in the full classical predicate calculus by finding a decision procedure for valid formulas. Hilbert declared this problem to be the central problem of mathematical logic (Hilbert and Ackermann [15], 11).

It seems safe to say that Hilbert's and Wittgenstein's optimism is dominant among the contemporary students of foundational problems, despite some discordant voices, such as Brouwer's. In [2] Brouwer declared, indeed, that Hilbert's "axiom of solubility" of every mathematical problem and its equivalent, the principle of the excluded middle, were both false. As for Carnap, he sided resolutely with the optimists. In the AUFBAU, proposing his first solutions to the decision problem for empirical questions, he wholeheartedly subscribed to the "proud thesis of the omnipotence of rational science" expressed in the Tractatus (Carnap [3], 183, my underlining).

Not only is Carnap's insistence on problem solving related to contemporary studies in the foundations of mathematics but obviously so are his specific solutions. We explicated above Carnap's debt to Intuitionism in this respect. We want now to recall that the whole Carnapian program of a syntax of the language of science and, in particular, his syntactic formulations of the significance criterion stand under direct influence of Hilbert's proof theory. The fact is well known and scarcely deserves to be documented from texts. We mention only Feigl's report of a discussion with Carnap, at about 1928: "Quick with labels for new conceptions (rather than carrying them out in exact detail) I told Carnap that the syntax (he then still called it 'semantics,' a term which he applied--following Tarski after 1934 or 1935--in the later well established sense) he was thinking of as formulated in a metalanguage, amounted to a 'Hilbertization' of *Principia mathematica*. He accepted, smilingly, my designation as essentially correct" (Feigl [12], p. XIV).[3]

As for the semantic formulations of the criterion, they stood under the direct influence of Wittgenstein's algorithm of validity.

Thus, both from a historical and a conceptual point of view, the strict significance criterion for descriptive languages has only superficially to do with Humean empiricism. It is as such more appropriately traced back to the Leibnizian *calculus ratiotinator* or to Cartesian general methods for solving geometrical problems. A direct evidence supporting this conclusion is Carnap's own approximation between his attempt to apply the modern theory of relations to the task of analyzing empirical reality (a task which included the formulation of the significance criterion) and some Leibniz's projects: "The fundamental concepts of the theory of relations are found as far back as Leibniz's ideas of a *mathesis universalis* and an *ars combinatoria*. The application of the theory of relations to the formulation of a constructional system is closely related to Leibniz's idea of a *characteristica universalis* and of a *sciencia generalis* (Carnap [3], 3). There seems to be no doubt, indeed, that what Carnap actually tried to do by formulating his strict significance criterion was to realize in the domain of empirical questions the old rationalist ideal of solving problems by mere calculation.

In support of this conclusion let us examine just one difficulty in any interpretation of the strict significance criterion framed exclusively in terms of traditional empiricism. The expression 'conceivable experience' used in semantic formulations is highly theoretical. Indeed, it is analytic that 'conceivable experience' is not a definite descriptive property constant. Again, it is analytically not a significant disposition term. In 1956 Carnap is quite clear about this: that conceivable experience has been always understood by him and by Reichenbach as an experience compatible with the fundamental laws of physics and not dependent upon our empirical knowledge of observers' abilities (Carnap [10], pp. 53-4). Accordingly, Carnap's criterion presupposes a highly theoretical account of possible experiences (never undertaken by anybody, by the way) and not mere introspection or, still less, experimental analysis of immediate sense data. We also note that the recognition of the impossibility of relying exclusively on introspection and constructive method, i.e., on the traditional empiricism plus modern deductive logic, is implicit already in the AUFBAU, where Carnap uses abstract geometry and physical analogies in constructing properties of phenomenal objects.

By these remarks I do not want to imply that Carnap did not say that his significance criterion was an empiricist one. He certainly did. What I do want to say is that he could as well have called his criterion a rationalist or a formalist criterion of empirical significance and that _we_ can safely do so. The strict criterion and other of Carnap's contributions to contemporary philosophy of science are not to be credited to either of the two sides in the _old_ epistemological and theoretical conflict between rationalism and empiricism but rather considered as an effort to produce a synthesis of sound elements belonging to both of these traditional doctrines (Cf. Carnap [3], p. VI).

We have seen that Carnap offered both semantic and syntactic formulations of the strict significance criterion. He never tried, however, to show that these alternative formulations characterized _equivalent_ descriptive language as significant. In fact, this is not the case, as is easily seen by comparing, for instance, the operations involved in the decision procedures in the two kinds of formulations. But Carnap did not always consider them even as _admissible alternatives_. In his first writings on the subject in 1928, he proposed only semantic formulations and contrasted the genetic method employed in defining concepts (objects) with non-constructive definitions by formal axiom systems like the one used by Hilbert in _Foundations of Geometry_ (1899). The semantic genetic method characterizes (defines) only single definite objects (concepts), as illustrated by Dedekind's cuts, by inductive definition of natural numbers and, of course, by Carnap's own constructive definitions of empirical (phenomenal) objects in the AUFBAU. In contradistinction, the formal, syntactic method of implicit definitions by a (syntactically consistent) set of uninterpreted axioms, such as Hilbertian ones, laid down at the outset, defines only classes of objects. But at that time, Carnap considered classes as 'improper' or 'quasi-objects,' in close agreement with Russell's 'no class theory.' And whereas the genetic method of introducing objects guaranteed the definiteness of all objects introduced, the axiomatic method did not in general assure the univocity (categoricity) of the object systems (models). After 1928, remaining as we have seen, under direct influence of Hilbert's formal proof theory, Carnap became reconciled with the formal axiomatic method and switched to syntactic formulations of his strict criterion. He went even so far as to identify logic with formal syntax (metamathematics): "We have already seen that this formal method of general syntax can also represent concepts which are sometimes regarded as not formal and designed as <u>concepts of meaning</u> (or concepts of a logic of meaning), such as, for instance, consequence relation, content, relations of contents, and so on. Finally, we have established the fact that even the questions that refer to the interpretation of a language, can be handled within the domain of a formal syntax. Accordingly, we must acknowledge that all questions of logic (taking this word in a very wide sense, but excluding all empirical and therewith all psychological reference) belong to syntax" (Carnap [8], p. 233). As we shall show below, Carnap not only believed that semantics was dispensable in methodological and logical questions but also that it led to "unsoluble difficulties and contradictions." Even after Gödel's theorems of incompleteness and of improvability of the consistency of the elementary number theory by finitary syntactic methods, Carnap continued to attempt syntactic formulations of this significance criterion, presented hereafter in liberalized forms and characterizing non-constructive languages. After the Paris Conference on the Philosophy of Science in 1935, under the strong influence of Tarski, Carnap returns to semantic formulations and treats his problem of meaning as "belonging to the field which Tarski calls semantics," a theory of "the relations between expressions of a language and things, properties, facts, etc., described in the language" (Carnap [7], p. 73). The languages here considered were not exclusively constructive languages. The Carnapian significance criteria written in a Tarskian key were already liberalized.

This outline of the changes in the formulations of the Carnapian significance criterion shows clearly that these changes are related, in an essential way, to some general philosophical questions, and to developments in basic fields of logic and mathematics, and not to peculiarities and eventual advances of philosophical empiricism or of empirical psychology.

Let us now take a closer look at two "insoluble difficulties" in the use of semantics in the foundations of logic and methodology, and consequently in the formulation of the significance criterion. Both difficulties concern the empirical semantics developed according to the model of the _Tractatus_.

We have seen that, according to the semantic formulation of the strict significance criterion, a sentence of L_D is significant if it expresses a state of affairs whose existence or non-existence are decidable by conceivable experiences. According to the phenomenalist view of the AUFBAU, an empirically decidable state of affairs is a structure (configuration) of immediately given total experiences (vertical slices of the solipsistic time-flow) or, more precisely and formally, list of lists of numbers taken from an enumeration of such data. Now, the immediately given total experiences are generally not common to different human subjects. Accordingly, the semantic formulation of the strict criterion is essentially solipsistic: it does not guarantee inter-subjective public significance for descriptive sentences (Carnap [6], p. 454). Under Neurath's influence (Neurath [17]), Carnap made the proposal that even specifying the significance of a given sentence, we should stop speaking about "sense data" and the comparison between the sentence and the "reality" and instead limit ourselves to the analysis of syntactic derivability relations between the sentence and the class of protocol sentences, whose vocabulary, formation and transformation rules would have to be syntactically characterized.

A second difficulty with empirical semantics is that it does not allow for a general definition of logical consequence. In order to be able to show why Carnap thinks so, we must recall Wittgenstein's definition of logical consequence. Let us call the n-tuples of truth-values of atomic sentences occurring in a truth-functionally composed molecular sentence which make that sentence true, the "truth-grounds" of that sentence (Tractatus 5.101). The concept of logical consequence is now defined as follows: "If all truth-grounds that are common to a number of propositions are, at the same time truth-grounds of a certain proposition, then we say that the truth of that proposition follows from the truth of the others" (Tractatus 5.11). Wittgenstein himself adds the following comments on this definition. If q follows from p, then the truth-grounds of q are contained in the truth-grounds of p. Also, the sense of q is contained in the sense of p. But the sense of a sentence is the state of affairs it describes and affirms or negates (Tractatus, 4.064, 4.2). Thus, if q follows p, the state of affairs described by q is contained in the state of affairs described by p. Carnap now observes that in such semantics, no universal logical consequence concept can be defined, and in particular, not the one needed for logically connecting sentences describing phenomenal and physical states of affairs. He argues in the following way. The sentences of the first kind mentioned describe structures of solipsistic total experiences, and those of the second kind configurations of physical objects, such as electrons, electromagnetic fields, etc. But the states of affairs described in the first case and in the second case belong to completely separate material spheres of objects. Accordingly, they cannot be contained in one another. Now, suppose that the logical implication relation exists between a sentence of one of the two kinds and a sentence of the other kind. Then by the Wittgensteinian semantics, the state of affairs described by the implied sentence is a part of the state of affairs described by the implying sentence. But this is impossible. Consequently, empirical semantics based on the semantics of the Tractatus is inadequate as a foundation of the logic of the descriptive languages which contain phenomenalistic and physical constants (Carnap [6], p. 453). Carnap's way out was the general syntax.

This analysis shows that in the early thirties Carnap was still far from the idea of the interpretation of descriptive terms over abstract domains which themselves admitted empirical interpretations, i.e., from Tarskian or formal semantics. It furnishes also an interesting consequence for the interpretation of Carnap's contemporary ideas about the structure of descriptive theories. As far as the semantic formulation of the strict significance criterion of the AUFBAU is concerned, no physical sentence, in particular, no theoretical sentence of a physical theory can be reduced to phenomenalistic sentences. More drastically, to require it would be semantically absurd. This result is contrary to many influential reconstructions of the early Carnapian ideas on the subject. Our contention is that the reductionist tendencies of early Carnap were greatly exaggerated by later commentators. But to show that in detail is beyond the scope of the present paper.

We want, finally, to point out that the strict criterion is closely related to the Carnapian thesis of extensionality. This thesis says that in every statement about a concept, this concept may be taken extensionally, i.e., it may be represented by its class or relation extension (Carnap [3], 43). As applied to the constants of L_D the strict significance criterion is nothing else than a method of decision for the extensions of the concepts designated by the constants. This remark is particularly important for appreciating some standard objections to the criterion. Popper, for instance, considers it to be an obvious absurdity to use verifiability as a meaning criterion for sentences, because, he argues, it is necessary to "understand" a sentence in order to be able to judge whether or not it could be verified (Popper [18], p. 63). It is easily seen that this objection, in order to apply to Carnap at all, must be so interpreted as to imply either that the Carnapian extensionality thesis is outright nonsense or that it is absurd to require that the extensions be decidable. But such claims do not seem to be very obvious, indeed. In a similar vein, Marhenke [16] argues that Carnap's significance criterion for sentences is false because one cannot know that something is evidence for or against a sentence unless one knows what that sentence means. But Marhenke fails to note that his objection presupposes the thesis that (at least) empirical sentences are intensional. Accordingly, he spends no effort in establishing this intensionality thesis.

3. LIBERALIZATION OF THE STRICT CRITERION

After rejecting semantics from the foundations of logic and epistemology, Carnap was left with syntactic formulations of the strict criterion only. In these formulations he used syntactic terms like 'provability,' 'derivability,' 'computability,' and 'definability.' The use of the criterion was justified by the belief that the decision problems for applicability of those terms in exhaustively characterizing large classes of descriptive and logical languages were or could be given positive solutions. The early Carnap was, in particular, optimistic, as we noted above, about the possibility of giving a complete and possibly also purely constructive criterion of validity of the sentences and of significance of the expressions of *classical* logic and mathematics. Soon, after Carnap's adoption of this view (around 1928), Gödel's limitation theorems dealt it a terrific blow. Carnap's monumental *Logical Analysis of Language*, completed in 1933, is a long meditation on the possible way out. The solution found was the liberalization of constructive requirements.

We are not going to discuss the substitute for the constructive criterion of validity. We only mention that Carnap explicitly gave up the search for a constructive criterion of validity of the whole of mathematics as well as the Hilbertian program of constructing a complete formal criterion of validity, which, although not itself definite (effective) would still be based on definite rules (Carnap [8], pp. 99-100). As for the problem of cognitive significance, the strict significance criterion was substituted by a liberalized one which admitted the use of indefinite syntax languages (metalanguages) and of indefinite logical and descriptive object-languages. The thesis of the equivalence of the significant descriptive languages was abandoned. The principle of tolerance was announced: Everyone is at liberty to build up his own language; all that is required in doing so is to state clearly the syntax of the proposed language.

The use of indefinite syntax languages, conditioned only by the informal principle of non-contradiction, is now justified exclusively by methodological considerations and no longer by stronger epistemological views such as those of constructivism. No question of justification or of correctness is raised. The terms 'analytic' and 'contradictory,' for instance, have been proved indefinite in most relevant formal systems (Carnap [8], p. 165). Carnap uses them in his syntax because they are useful "from the standpoint of certain general considerations" (*ibid*., p. 175). One important theoretical advantage of these terms consists in the fact that "with their help the complete division of sentences without descriptive constants into analytic and contradictory is possible," whereas the corresponding classification of the same class of sentences into provable and refutable ['provable' and 'refutable' being also indefinite] is incomplete as a consequence of Gödel's theorem (*ibid*., p. 173). Another methodological virtue of these terms is that they are "more closely connected with material interpretation of language." (*ibid*., p. 175).

In an analogous fashion, Carnap now admits *indefinite* mathematical terms. From a technical point of view this amounts to the admission of definite-chains where unlimited quantifiers occur. A definition of the form

$$P_1(x) = (\exists y)\ (Q(x,y))$$

where 'Q' is a definite number-theoretic relation constant and 'P_1' a new constant, are now tolerated, although the question whether, for instance, '$P_1(5)$' is true or not is not answerable by a definite method. In the standard intuitionist objection that '$P_1(5)$' is a meaningless formula, Carnap now replies: "... it is true that we know of no method of searching for the answer, but we do know what form the discovery of the answer would be--that is to say, we know under what conditions we would say that the answer had been found. This would be the case, for example, if we discovered a proof of which the last sentence was '$P_1(5)$'; and the question whether a given series of sentences is a proof of this kind or not is a definite question. Thus there exists *the possibility of the discovery of an answer*, and there appears to be no cogent reason for rejecting the question" (*ibid*., p. 161). Carnap justifies the use of impredicative constants in just the same way. He sums up his position on admissibility of meaningfulness of mathematical constants as follows: "In general, since there are sentences with unrestricted operators which are demonstrable, there is always the possibility of coming to a decision as to whether or not a certain indefinite or impredicative term is applicable in a particular individual case, even though we may not always have a method at hand for arriving at this decision. Hence such terms are justified even from the standpoint which makes the admissibility of any term dependent on the possibility of a decision in every individual case. (Incidentally, in my opinion, this condition is too narrow, and is not convincingly established)" (ibid., p. 164). The parenthetical remark at the end of this text expressed exactly the central aspect according to which the strict criterion for mathematical object-languages had been liberalized: it is no longer required that it be general.

In keeping with this liberalization, Carnap also gave up the requirement of generality of the significance criterion for descriptive languages. Accordingly, and this is the essential point, the use of a descriptive expression need no longer be, in all classes, justified as correct by an algorithm. Questions that have earlier been formulated as decision problems now become questions of choosing the form of language, i.e., questions 'of the establishment of rules of syntax and the investigation of the consequences of these' (Ibid., p. 164).

This is, then, some of the available evidence for the main thesis of this section: from both conceptual and historical points of view the first Carnapian liberalizing departure from the strict significance criterion is intimately related with general developments in the theory of formal systems, and in particular with Gödel's limitation theorems. On the other hand, there does not seem to exist any evidence that the liberalization was mainly motivated by specific difficulties in the application of the strict criterion. In particular, there is no evidence that it had to do with the problems in the logic of disposition terms (Hempel [13], p. 3). Even after completing the liberalization by introducing the principle of tolerance in constructing languages, Carnap continued to treat disposition terms like 'breakable' as being on the same logical footing as terms for kinds or things ('horse'), for kind of substances ('iron'), and even for directly perceptible qualities ('warm,' 'soft,' 'sweet'). All these kinds of terms were considered to express properties of domains of things. No peculiarity of the disposition terms was mentioned (Carnap [8], p. 150).

Among the rules of syntax of liberalized descriptive languages, physical rules or P-rules are also admitted. Carnap distinguished several kinds of such rules. Some of these rules are theoretical postulates, such as the Maxwell's equations. The other kind is constituted by reduction sentences. Examples of these will be given below. Still another kind contains the correspondence rules, such as the rule that says that the temperature of a gas is the mean kinetic energy of its molecules. In accordance with his general non-constructive position after 1933, Carnap considers the acceptance of P-rules, not as a logico-philosophic problem but, to a large extent, a matter of convention and, hence, a question of expedience (Carnap [8], p. 180). All P-rules are used as non-logical rules for derivations and as means of introducing new terms. But the reduction sentences have an additional special usage: they serve as partial empirical decision procedures, based on empirical laws, for the concepts they introduce. The first treatment of reduction sentences is given in Carnap [7]. We shall now examine briefly how they work.

Let L_D be a first-order descriptive language, with unrestricted quantifiers, whose variables take their values in the domain of space-time points. Let Q_j, $j \neq 3$, be a sequence of descriptive predicate constants significant in language L_D, and 'Q_3' a new constant whose intended interpretation is a disposition property. It can readily be shown that we cannot introduce 'Q_3' by explicit definitions. We can nevertheless introduce it into L_D by admitting the following P-rules, called reduction sentences, as valid:

$$R_1 \quad Q_1 \rightarrow (Q_2 \rightarrow Q_3)$$

$$R_2 \quad Q_4 \rightarrow (Q_5 \rightarrow \neg Q_3)$$

In these formulas, 'Q_1' and 'Q_4' describe empirical conditions (and, in particular, test conditions) realized at a space-time point, 'Q_2' and 'Q_5' specify observable events (and, in particular, positive test results) at the same point, and 'Q_3' is the new constant. It is supposed that as a matter of empirical fact Q_1 is regularly followed by Q_2, and Q_4 by Q_5, equally without exception. It is clear that R_1 and R_2 constitute an <u>empirical</u> decision procedure for the problem of applicability of 'Q_3' (under the condition, of course, that we know the empirical procedures for the application of 'Q_1', 'Q_2', 'Q_4' and 'Q_5'). But this procedure is <u>not general</u>. For the class of points that satisfy the condition

$$(1) \qquad \sim((Q_1 \wedge Q_2) \wedge (Q_4 \vee Q_5))$$

the applicability of 'Q_3' is not decidable by means of R_1 and R_2. For these regions Q_3 is not determined, and hence, "without meaning." (Carnap [7], p. 60). Now if the condition (1) were P-valid, 'Q_3' would be without any meaning at all in L_D.

Carnap also tried to formulate procedures for diminishing the indeterminacy of the application of newly introduced predicates by admitting as valid in L_D whole sequences $R^i_{1,2}$ of pairs of reduction sentences of the form

$$R^i_1 \qquad Q^i_1 \rightarrow \ (Q^i_2 \rightarrow \quad Q_3)$$

$$R^i_2 \qquad Q^i_4 \rightarrow \ (Q^i_5 \rightarrow \quad \neg Q_3)$$

where the sentences R^i_1 specify classes of space-time points $Q^i_{1,2}$ to which 'Q_3' applies and the sentences R^i_2 the classes of space-time points $Q^i_{4,5}$ to which it does not apply. Thus, a sequence $R^i_{1,2}$ of pairs of reduction sentences appears to constitute a decision procedure for the applicability of 'Q_3' stronger than the one based on any of its elements. In 1956, however, Carnap recognized that the use of a sequence $R^i_{1,2}$ for introducing a new property constant into L_D would violate the requirement of extensionality for L_D. As a matter of fact, the new constant introduced would remain intensional until an effective rule were laid down for the construction of the $R^i_{1,2}$. But, on the other hand, it is impossible to give such a rule, for to give it would be equivalent to specifying the content and the order of discovery of all unknown empirical laws. Having recognized that an $R^i_{1,2}$ cannot possibly determine the exact extension of 'Q_3' and continuing to adopt a strictly extensional view on the meaning of descriptive constants, Carnap proposed that only single pairs of reduction sentences be used as (partial) definitions and that, in order to avoid ambiguity, we would call concepts so introduced by different names, as has already been required by Bridgeman (Carnap [10], p. 64). This correction of an error in Carnap [7], shows in a particularly clear way what the reduction pairs really are and what they are intended to be: decision procedures for the extensions of the constants they introduce.

In conclusion we can say the following: First, the strict criterion of cognitive significance was proposed by Carnap as a solution of decision problems for questions concerning the use of different kinds of descriptive expressions. In offering his solution, Carnap imitated solutions of decision problems for questions in mathematical language. By its very nature, the strict criterion belongs to the history of problem solving by effective methods. Second, the liberalization of the strict criterion was in the first place a consequence of internal limitations of effective methods discovered in the meantime. Third, the problems and methods peculiar to traditional empiricism had little to do with the origins, the nature and the development of the Carnapian significance criterion.

NOTES

* Published by the Centro de Lógica, Epistemologia e História da Ciência of the Universidade Estadual de Campinas.

1. Some authors, like Neurath, believed that there was only one significant descriptive language.
2. The expression 'state of affairs' is used here and below in the sense of the *Tractatus*.
3. The main targets of the Hilbertization of the *Principia mathematica* were, of course, the axioms of infinity and reducibility.

BIBLIOGRAPHY

[1] A. J. Ayer, *Logical Positivism*. New York: The Free Press, 1959.
[2] L. E. J. Brouwer, "Intuitionistische Mengenlehre," *Jahresbericht der Deutschen Mathematiker-Vereinigung* 28 (1920), 203-208.
[3] R. Carnap, *Der Logische AUFBAU der Welt*. Berlin: Weltkreis-Verlag, 1928. (Translated into English as

The Logical Structure of the World, by R. A. George. Berkeley: Univ. of California Press, 1967).

[4] R. Carnap, *Scheinprobleme der Philosophie*. Berlin, 1928. (Translated into English as *Pseudoproblem in Philosophy*, by R. A. George. Berkeley: Univ. of California Press, 1967).

[5] R. Carnap, "Überwindung der Metaphysik durch die logische Analyse der Sprache," *Erkenntnis* 2 (1931-2), 219-241. Translated in Ayer [1], pp. 60-81.

[6] R. Carnap, "Die Physikalische Sprache als Universalsprache der Wissenschaft," *Erkenntnis*, 2 (1931-2), 432-465.

[7] R. Carnap, "Testability and Meaning," in *Readings in the Philosophy of Science*, H. Feigl and M. Brodbeck (eds.). New York: Appleton-Century-Crofts, 1953, pp. 47-92. (Originally appeared in *Philosophy of Science*, 3 (1936), 420-468 and 4 (1937), 1-40).

[8] R. Carnap, *The Logical Syntax of Language*. London: Routledge and Kegan Paul, 1937.

[9] R. Carnap, *Logical Foundations of Probability*. Chicago: University of Chicago Press, 1950.

[10] R. Carnap, "The Methodological Character of Theoretical Concepts," *Minnesota Studies in the Philosophy of Science* 1 (1956), 38-76.

[11] R. Carnap, *Introduction to Symbolic Logic and Its Applications*. New York: Dover, 1958.

[12] H. Feigl *et. al.*, "Homage to Rudolf Carnap," in *Boston Studies in the Philosophy of Science*, vol. 8, R. S. Cohen and M. W. Wartofsky (eds.). Dordrecht: Reidel, 1971, pp. XI-XV.

[13] C. G. Hempel, "Rudolf Carnap, Logical Empiricist," in *Rudolf Carnap, Logical Empiricist*, J. Hintikka (ed.). Dordrecht: Reidel, 1975, pp. 1-13.

[14] D. Hilbert, "On the Infinite," in *From Frege to Gödel*, J. van Heijenoort (ed.). Cambridge, Mass.: Harvard Univ. Press, 1967, pp. 367-392. (Originally appeared in *Mathematische Annalen* 9 (1925), 161-190).

[15] D. Hilbert and W. Ackermann, *Grundzüge der Theoretischen Logik*. Berlin: Springer, 1928.

[16] P. Marhenke, "The Criterion of Significance," in *Semantics and the Philosophy of Language*, L. Linski (ed.). Urbana: University of Illinois Press, 1952, pp. 139-159.

[17] O. Neurath, "Soziologie im Physikalismus," *Erkenntnis* 2 (1931-2). (Translated in [1], pp. 282-317.)

[18] K. R. Popper, "Autobiography," in _Philosophy of Karl Popper_, vol. 1, P. A. Schilp (ed.). La Salle, Illinois: Open Court, 1974, pp. 1-181.
[19] L. Wittgenstein, _Tractatus Logico-Philosophicus_, in Annalen der Natur-und Kulturphilosophie, vol. 14 (1921). (English translation by Pears and McGuiness. London: Routledge and Kegan Paul, 1961).

João Paulo Monteiro

NATURAL CONJECTURES*

I. How do common sense and theoretical knowledge relate to each other? Hume seems to suggest a smooth and swift continuity between both: "Philosophical decisions are nothing but the reflections of common life, methodized and corrected."[1] Philosophical inquiry is characterized by the search for evermore general principles, but this transition has already begun at the level of common sense: "From our earliest infancy we make continual advances in forming more general principles of conduct and reasoning;... the larger experience we acquire, and the stronger reason we are endowed with, we always render our principles the more general and comprehensive; and... what we call *philosophy* is nothing but a more regular and methodical operation of the same kind."[2]

There is no breach, therefore, separating philosophical from common knowledge. The difference is only of degree, as is explained by Quine in a passage that is a perfect echo of Hume: "Science... differs from common sense only in degree of methodological sophistication."[3] And certainly for Hume it was a question of contrasting common sense with *science* in general, not only with what we today call philosophy, since the latter was for him an aspect of the first, in its character as a "science of man." As Chomsky has emphasized, it would be a mistake to suppose that our present distinctions between philosophical and scientific concerns would have any meaning for Hume.[4] Hume is concerned with the relations between common knowledge and what we may call *theoretical* knowledge (scientific or philosophical, covering both what in his day was called the natural *philosophy* of Newton, which today we call science, and what Hume considered his own *science* of human nature, which we today call philosophy). Hume's thesis of a fundamental continuity concerns the relations between the theoretical domain as a whole and the territory of common knowledge.

J. J. E. Gracia et al. (eds.), Philosophical Analysis in Latin America, 339–364.

But how are we to understand that from such a perfect continuity the most acute contradiction is born? On the one hand, according to Hume, philosophy destroys the most deeply rooted, and the most natural beliefs, of common sense: "This universal and primary opinion of all men is soon destroyed by the slightest philosophy, which teaches us, that nothing can ever be present to the mind but an image or perception, and that the senses are only the inlets, through which these images are conveyed, without being able to produce any immediate intercourse between the mind and the object."[5] On the other hand, for Hume philosophical reflections can lead to an epistemological despair from which the only possible escape is abandoning ourselves to our natural tendencies. Each movement of that reflection threatens to make us fall into skepticism: "The understanding, when it acts alone, and according to its most general principles, entirely subverts itself, and leaves not the lowest degree of evidence in any proposition, either in philosophy or common life."[6] It is only nature acting within us, not philosophical reason, that can give us back our confidence in knowledge: "Most fortunately it happens, that since reason is incapable of dispelling these clouds, nature herself suffices to that purpose, and cures me of this philosophical melancholy and delirium, either by relaxing this bent of mind, or by some avocation, and lively impression of my senses, which obliterate all these chimeras."[7] That is, before, we had philosophy concluding against our natural tendencies, but now it is our natural tendencies that bring us to reject the conclusions of our most "refined and metaphysical" reflections.[8] This is a conflict that in no way seems to support the thesis of continuity between philosophy and common sense.

Hume could argue that the breach with common sense is the work of preceding philosophies, from Pyrrhonian skepticism to the modern theory of representation (Cartesian or Berkeleyan), but that his own philosophy escapes such a destiny, in that it is a theory of human nature. The nature of our species would be only "one nature" among the various that make up Nature in general, and the science of man would be a theoretical endeavor within the framework of which common knowledge, defined as natural, would appear at the same time explained and rehabilitated. Corrected and methodized, without a doubt, as Hume's first formula indicates; but corrected solely in its most naive aspects as in the belief in a direct apprehension of things, and methodized only in the search for a greater generality of the principles to be discovered, as the second of these formulae indicates. Theoretical knowledge would simply be an enlargement of common sense, not its global refutation. In Hume's philosophy, there is a conciliation, or rather a reconciliation, of philosophy with common sense so that they find themselves, ultimately, "in the same boat."

However, Hume's philosophy, at least in one of its aspects, while it might perhaps be in the same boat as common sense, certainly does not share the same deck of the same ship with it. As I tried to show elsewhere,[9] there is a point after which Hume's philosophy does not limit itself to the attempt to make the principles encountered by common sense more general. After that point, it is no longer concerned with proceeding to inductive generalizations; it is concerned with constructing hypotheses about certain mechanisms that lie beyond the limits of the observable. Inductive generalization, typical of common sense and of those sectors of scientific knowledge that we can qualify as empirical and not theoretical, does not go beyond those limits: the causes attributed to the phenomena are always causes as observable as the effects which are explained and predicted by them. The "general causes" discovered by the science of man, on the contrary, are unobservable principles constructed through theoretical artifices which are irreducible to the original instruments of common sense.

Thus one of Hume's chief discoveries, that of custom or habit as the general cause and principle of inductive inferences, is not reached merely through one more inductive step, distinguished from the preceding ones only by its superior generality. It is not a matter of finding, starting with observable resemblance and repetition, a cause that is also observable, and simply capable of accounting for a greater number of phenomena than those previously examined. On the contrary, the Humean contribution consists in <u>attributing</u> to the cognitive phenomena of induction an unobservable cause, a device that was <u>invented</u>, or constructed by the philosopher himself, and whose attribution to the phenomena is justified by its capacity to explain them better than any other known alternative. In a similar way, as I tried to show elsewhere,[10] the Newtonian discovery of the force of gravity, as the cause of a whole series of visible phenomena, cannot be understood, within the Humean epistemological framework, as an inductive step more general than the previous ones, but only as the true invention of a mechanism in itself unobservable, and whose supposition is equally justified by its capacity to account for the phenomena, or to "save the appearances." Both the science of man and the science of nature, therefore, are characterized at least sometimes by a radical distance in relation to the usual limits of common knowledge, which are the limits of the observable. The Humean epistemological perspective thus seems to force us, against the explicit statements of the philosopher himself, to conform to the inexistence of a true continuity between the work of common sense and the construction of theoretical knowledge. The leap beyond the frontiers of the observable, characteristic of the latter, seems to qualify it as a radically distinct endeavor from common knowledge, rendering any thesis of a continuity between them simply untenable.

II. But would it not be possible to sustain that common sense too goes beyond the observable? I am not thinking of the possibility of invoking the "unobservables" created by that vulgar form of imagination whose products for centuries have been considered deserving of the generic name of "superstition." It is not customary to include in common sense this type of product of fantasy; on the contrary, the most usual attitude is the condemnation of superstition in the name of common sense itself. He who is the possessor of "a solid common sense" does not believe in elves. As to religion, certainly it also speaks of the invisible; but it is not usually conceived as a product of common sense, but turned accessible through a different path, which is that of faith. The historic alternatives to this answer, like that of natural theology from the XVIIth to the XIXth century, do not have recourse to common sense either, but to its opposite, to an argumentation that presumes itself to be scientific, as in the case of the famous Design Argument. The problem of a possible transcendence of the observable, as we shall see shortly, is connected to another type of argument, extraneous to any reference to superstition or religion.

It may, for instance, be argued that common sense dispenses with the direct observability of its objects. This is the "problem of Berkeley": why do we commonly believe that physical objects exist even when they are not objects present in our perception? The answer to that argument (leaving aside here the Berkeleyan question in itself) is that the definition of common sense objects as observables in no way implies their constant _direct_ observability, or rather, that they are posited as always _observed_. An _observable_ object is, by definition, an element of a class of objects that are obviously all observable, but in which only _some_ elements are, or have already been, observed. All future events are evidently _unobserved_, but not all of them are _unobservable_. The heat that, in the classical Humean example, we are certain will accompany all future fires, is certainly unobserved, but in no way is it "unobservable." It is in this sense that inductive reasoning always goes beyond the observed, inasmuch as it allows the prediction of future events, or the retrodiction of past events; but it never goes beyond the observable, for it does not posit objects or events that do not belong to _classes_ defined as observable. Now, the same applies, even more so, to every object presently unobserved but defined as the _same_ observable object previously given in our experience. Thus common sense does not transcend the observable, although it is not bound to what has already been, or can be immediately observed.

On the other hand, one might bring up the case of other people's mental events and processes (the problem of "other minds"), which are never directly observed by us and at the same time are trivial objects of common sense. It is "common sense" to believe that other people have mental processes just like us: shouldn't we conclude that, at least in this case, common sense goes beyond the observable? Our own mental events, at least in Humean terms, are directly accessible to our inspection; and it is a mere question of linguistic convention to define them as observable objects, understanding observability as the simple belonging to the realm of perception, "external" or "internal." But, in this sense, the mental processes of others are not objects of our observation. I "observe" my thoughts, but not those of others. A provisional answer might be the following: the supposition that other people have mental processes like ours involves only an operation of identification by resemblance. And for this very reason it is a _common_ supposition, not a _theoretical_ invention, like that of atoms or that of genes; only the latter involve an operation different from resemblance, which is analogy. It is true that Wisdom, for example, says that we know when someone is angry using "the analogy of our own feelings;"[11] but here the term "analogy" is equivalent to "reasoning by resemblance"--it does not have the strong meaning that it acquires when it refers to the analogies constructed by theoretical science. The difference between resemblance and analogy shall be discussed below. But before that it is worthwhile to insist a little more on the problem of other minds, using an argument from Hume himself.

In the Enquiry, Hume says that we naturally know only the superficial (that is, observable) qualities of objects, for "nature has kept us at a great distance from all her secrets": the most fundamental "powers and principles" of nature are unobservable, "hidden" by her from our eyes. But, although we do not know these unobservable "natural powers," we still spontaneously believe, in all cases, that objects with the same sensible qualities, and that present themselves as similar to our observation, "have the same secret powers;" and we always expect from them the same effects already experienced in the past.[12] Now, the mental processes of others, although at first they can be put on the same level as the "secret powers" of observable objects, leave this level from the moment we remember that we have access to our own mental processes and we proceed to the operation above indicated by Hume for the same objects. In Hume's example,[13] we conclude that one piece of bread, being similar to another piece of bread, has the same "secret powers" as the other. In the case of the minds of others, given that it is clearly observable that other people are similar to us, it is also natural to attribute to them mental processes similar to our own; and the latter are not "secret," for they are objects of our internal experience. That which is mysteriously "within" a piece of bread, its unobservable powers, to which it owes its nutritive effects, can only be the object of theoretical conjectures; but that which is "within" the mind of a being similar to us can be more easily inferred, that is, simply by induction. We simply conclude that beings similar to us have minds similar to ours, with mental processes similar to our own. I do not presume that this answer resolves any of the philosophical problems concerning this knowledge of "internal and invisible states"[14]--only that it is a Humean answer to this question, in terms of common sense. Our common belief in the existence of the mental processes of others is not an example of any authentic transcendence of the observable by common sense. That transcendence would only occur if we tried to know the productive mechanisms of these processes--which would be theoretical knowledge, not common, and which would present similar problems, whether it dealt with others or with ourselves.

Suppose that suddenly, by a strike of magic, our apparatus of vision is transformed into the "microscopic eyes" imagined by Locke (adapting Pope), which would allow us to see directly "things several millions of times less than the smallest object of our sight," coming to discover "the texture and motion of the minute parts of corporeal things."[15] Suppose still that we can apply this marvelous "faculty" to only one singular object of those which make up a certain class, for example to only one of all the pieces of bread in the universe. In that moment, looking at the inside of a piece of bread we may perhaps directly discover a minute mechanism; and maybe we would attribute to that mechanism the nutritive capacity of that piece of bread. This "secret power" ceases then to be mysterious for us, although it has only become visible in that single piece of bread. And from now on, if before we attributed to two pieces of bread, due to the resemblance between them, the same "secret powers," without doubt we now begin attributing to <u>all</u> pieces of bread the same cause of the nutritive capacity which they all have in common--although we have observed this mechanism in only one member of that class. Let us compare this cognitive situation with that of our knowledge of other minds. Is it not evident that we find ourselves, in relation to these, in a situation analogous to that of our privileged "Lockean" observer? As he, we have had occasion to look "inside" one specimen of a class of beings: human beings, in this case. And as with that observer, it was enough to see once the normally unobservable mechanism of a piece of bread in order to conclude that all pieces are governed by the same mechanism, so it is enough for each of us to observe our own mental states, and the relations between them and our behaviour, to conclude that in all beings of the same species similar mental states must occur. There is only one important difference between the two cases, yet it favors our comparison, for it shows that the situation of the knower of the mental states and processes of others is even more priviledged than the situation of the observer in our "Lockean" miracle. For this observer had to cope with the "ultimate" mechanism, which is the "true" and deepest cause of the nutritive capacity of bread, while we simply confront an _effect_ of our cognitive mechanisms, which presents its own problems but certainly is not as "deep" an object as its own unobservable cause. Knowledge of other minds does not involve, therefore, an _inference_ of a different nature from that which is typical of other aspects of common sense: it is simply an inductive inference. Its peculiarity does not occur on the inferential level, only on the observational level; it simply happens that knowledge of minds, like that of the nutritive mechanism of bread in our imaginary example, becomes possible merely through observation of a single member of the class or species of beings in question. Actually, the very formulation of the problem of _other_ minds masks one of its fundamental aspects: that it is basically about knowing _minds_ in general; and that this knowledge becomes possible for us, contrary to what occurs in other cases, by means of an observation which is restricted to a single member of the class that is its object. Within common sense, this is a special case, but in no way is it so special that it allows us to conclude that, in it, common sense goes beyond the observable. _Minds_ do not constitute a class of unobservables; only the "secret principles" that govern them are unobservable in this case. As to _other minds_, one cannot assert without qualification that they constitute a class of unobservables; they are _part_ of a class of objects, and not an entire class, and the class to which they belong is not, as such, a class of unobservables.

Common sense, then, is entirely restricted to the limits of the observable. The leap beyond these limits, by means of theoretical hypotheses positing unobservables, is the mark of scientific activity alone. Common knowledge has its inferential activity limited to the territory of induction, in which only the observed is transcended, never the observable. But there is a more fundamental realm than that of inductive inference, and in a certain sense preliminary to it, that is, belief in the existence of observable realities external to the knowing subject--a realm which is for Hume, although probably not for ancient philosophy, the classical place for skeptical doubts about the legitimacy of common sense. Now, Hume has very little to say, from the point of view that matters to us here, concerning this more fundamental level without which the process of inductive inference would not even have the possibility of beginning. In the Enquiry, what he tells us is simply that this realm is established by the action of an instinct: "It seems evident, that men are carried, by a natural instinct or prepossession, to repose faith in their senses; and that, without any reasoning, or even almost before the use of reason, we always suppose an external universe, which depends not on our perception, but would exist, though we and every sensible creature were absent or annihilated."[16] This explanation is identical to that of induction, which had already been attributed in the same work to "some instinct or mechanical tendency."[17] And it only seems to confirm the impossibility of the thesis of continuity: it is spontaneously, not by conjectural constructions, that all that territory normally attributed to common sense is constituted.

We do not find in Hume's works as systematic an usage of the expression "common sense" as that which later became characteristic of Scottish philosophy in the XVIIIth century. But Hume indicates as the "right path" for philosophy a constant appeal to "common sense and the natural sentiments" of man, as well as the representation of common sense "in more beautiful and more engaging colors."[18] And in his moral philosophy he points at the intrinsic rationality of common sense, when he says that only those who admit the reality of moral distinctions are "on the side of common sense and reason."[19] Common sense is for him a kind of "faculty" or "sense" common to all mankind, which is a condition of possibility of all knowledge. But it is defined as a kind of natural instinct, which, although not always contradicted by theoretical knowledge, seems different enough from the latter (understood as a construction of hypotheses or conjectures) to forbid statements like those he himself presented to us, long before Quine, concerning the existence of an essential continuity between common sense and scientific or philosophical knowledge. Thus he seems to have left the problem of the relations between common sense and philosophy without a solution. But did his immediate posterity find a better solution? Did Thomas Reid, head of the Scottish philosophy of common sense, who dedicated himself more than Hume to these problems reach a satisfactory conception of the relation between philosophy and common sense?

III. For Thomas Reid, the common man makes no hypotheses.[20] The common man is guided by the principles of common sense. His thought stays always faithfully linked to the observable, and is faithful also to its more pure and primitive origins. Quine would perhaps say that the Reidean common man remains faithful to the heritage of pre-historic man:

> The unrefined and sluggish mind
> Of Homo javanensis
> Could only treat of things concrete
> And present to the senses.[21]

Reid's philosophy reserved for itself the task of rigorously determining the principles of common sense--also without making hypotheses. Hypotheses, according to him, may suggest experiments or inspire investigations, but only "just induction" can determine our discoveries.[22] The philosophy of common sense triumphs because it closely follows the epistemic attitude of common man: always sticking to the facts, refusing with total intransigence the uncertainty of conjectures.[23]

The principles of common sense, which Reid also calls "first principles," are self-evident, not derived from any reasoning or inference.[24] They are of two kinds: those that allow us to reach "necessary and immutable truths, whose contrary is impossible;" and those from which are derived truths that are "contingent and mutable, depending upon some effect of will and power, which had a beginning, and may have an end."[25] The first principles of necessary truths are the axioms of logic and mathematics, the rules of grammar, the axioms of taste, and moral and metaphysical principles.[26] The first principles of contingent truths, on the other hand, are for instance that of the existence of the things of which we are conscious, or that of personal identity, or that of the non-fallacious character of the natural faculties by which we distinguish truth from falsehood.[27] But the two species of principles apparently constitute a capital of common sense, forming a single block where they are all immediately evident, intuitive, irresistible; they are expressed in propositions "which are no sooner understood than they are believed."[28]

However, the interpretation proposed by Paul Vernier in a recent paper on Reid separates from the others the _contingent_ first principles, denying that the philosopher attributed to these a character of absolute certainty. These, not the necessary ones, are those that Reid seeks to defend from the threat of skepticism.[29] Knowledge of contingent truths, like that of necessary truths, needs without doubt to be founded on non-inferential propositions, otherwise it would fall in an infinite regress in the task of justifying those truths; contingent principles are also not deduced from others, and Vernier does not question their intuitive and irresistible character in Reid's philosophy. But the Reidean answer to skepticism involves something less than the intention of attributing to these principles the character of an absolute knowledge of objective truth. The Reidean theory of justification differs from the Cartesian in its recognition of our fallibility when affirming the foundations of our knowledge. The non-acceptance of this fallibility, as a matter of fact, led the Cartesian project to failure, bringing many philosophers to conclude that there is no escape from skepticism.

Reid's thesis is weaker, and for this same reason it is less vulnerable. It states that contingent first principles are justified simply by the absence of any decisive reason to doubt them--not because their evidence involves the absolute certainty of their truth. This justification is obtained by the confrontation between skeptical arguments and those principles, and the discovery that the reasons presented by the skeptic in support of his doubts are not sufficient to justify them. And Vernier concludes that, according to Reid, we cannot presume to be certain of the truth of contingent principles "beyond any conceivable doubt." We can only assert that they lie beyond any _reasonable_ doubt, for skeptical arguments are incapable of founding this kind of doubt.[30]

The validation of contingent first principles, from this point of view, depends on the refutation of the skeptical arguments which are opposed to them, not on the demonstration that skeptical doubts are inconceivable. It is enough to show that the skeptical argumentation is incapable of proving the unreasonableness of the refusal of its doubts. Thus the essential process of validation consists of a struggle where the principles of common sense and the skeptical principles confront each other, and where the latter are finally defeated. Neither common sense nor skepticism is able to impose to the audience the absolute certainty of the validity of its principles. But skepticism is defeated in the very territory of its presumption, that of demonstrating the unreasonableness of its rivals. Common sense is triumphant precisely because it renounces the presumption of absolute certainty, and on the same battlefield where Cartesianism had failed owing to this presumption. Common sense is victorious because, instead of insisting on its absolute self-foundation, it renders the attack on the adversary, and the weakening of his position, an integral part of the process of its own validation.

This means that, to an extent, the principles of common sense have a _conjectural_ character. They have it inasmuch as their justification depends on the refutation of the conjectures which oppose them. No doubt that it still is legitimate, in another sense, to affirm that the common man makes no hypotheses but this is more properly in a psychological than in an epistemological sense. It is in the sense that one does not deliberately proceed to the construction of explanatory philosophical systems, like the Cartesian or the peripatetic that Reid opposes to common sense. The common man is bounded instead to believe spontaneously that the things he perceives are external objects, whereas those philosophers invented the hypothesis that the things that they perceive are but ideas in their minds.[31] The common belief in the existence of external things is not obtained through the deliberate construction of a hypothesis: it simply appears, irresistibly, in the mind of the common man. From the point of view of its psychological genesis it cannot be mistaken for a hypothesis. But from the point of view of its validation, by the philosophy of common sense, it appears as a conjecture: a proposition whose superior rationality becomes manifest through the refutation of its contrary, for it is not a possible object of any kind of direct proof. Epistemologically, this belief is a hypothesis.

This is an interpretation which Thomas Reid would probably have rejected with an indignation comparable to that of his general and omnipresent rejection of hypotheses in science. But it is equally clear that in his texts the term "hypothesis" has a much more restricted meaning than it may have today: deliberate intention is part, at least, of its connotation in Reid's texts, and today it is common to employ it as a synonym of "conjecture" or "theory." And in this sense, today, inasmuch as a conjecture or theory may be deliberate, it can also be spontaneous and unconscious. As we read in Quine's Word and Object: "Theory may be deliberate, as in a chapter on chemistry, or it may be second nature, as in the immemorial doctrine of ordinary enduring middle-sized physical objects." It is this "doctrine" that Reid labels common sense, as is also true of Quine in numerous passages.[32] In Quine we have the explicit recognition of the theoretical character of common sense. In Reid, if Vernier's interpretation is correct, we have only implicitly, but in a discernible form, the admission of the theoretical character of at least one face of common sense--that where the principles called "contingent" are formulated.

IV. The idea of a conjectural common sense is one of the most central in Quine's philosophy. The objects of common sense--physical observable bodies--are entities which it would be illegimate to consider less conjectural than the entities designated by theoretical terms that are proposed to us by the sciences, like, for example, molecules: "What are given in sensation are variformed and varicolored visual patches, varitextured and varitemperatured textual feels, and an assortment of tones, tastes, smells, and other odds and ends."[33] Quine takes the parallel even farther: "If we have evidence for the existence of the bodies of common sense, we have it only in the way in which we may be said to have evidence for the existence of molecules. The positing of either sort of body is good science insofar merely as it helps us formulate our laws--laws whose ultimate evidence lies in the sense data of the past, and whose ultimate vindication lies in anticipation of sense data of the future."[34] And he concludes with a reiteration of his thesis of a continuity between common sense and science, and of a differentation between them thinkable only in terms of degree: "The positing of molecules differs from the positing of the bodies mainly in degree of sophistication. In whatever sense the molecules in my desk are unreal and a figment of the imagination of the scientist, in that sense the desk itself is unreal and a figment of the imagination of the race."[35] Common sense tells us that each and every perceptible physical object exists in reality as an object distinct and separate from our perceptions. Quine tells us that each of these is, in a relevant way, the equivalent of a theoretical conjecture: in the same way as molecules, the bodies of common sense are simply presupposed in the course of the organization of our responses to the stimulation to which the world subjects us; also, for the existence of bodies "man has no evidence... beyond the fact that their assumption helps him organize experience."[36] Common knowledge thus appears as a system of conjectures, or hypotheses, derived from a natural, pre-scientific, cognitive capacity, thought of as a mechanism underlying that system. This allows us to go back to the problem of continuity between common sense and theoretical reason, yet now on a different level: this continuity is recovered, not in the traditional manner, in terms of a prolongation of a "positive" common sense by a science which is also "positive" (observational, or inductive), a prolongation rendered dubious by the conjectural character of scientific and philosophical knowledge, but now taking theoretical knowledge as the prolongation of a common sense already having that conjectural character. Rather than an inductive science prolonging a common sense in which nothing is hypothetical, we have a conjectural common knowledge in tranquil continuity with a theoretical knowledge characterized too as a system of hypotheses.

Each science, and each philosophy, is a coherent system of conjectures, and in the Quinean perspective common knowledge similarly appears as a system of conjectures. But one fundamental difference should at least be this: if scientific systems are the result of the common effort of mankind, if scientific conjectures, as well as philosophical ones, are derived from the active work of a deliberately organized imagination, common knowledge in turn is the spontaneous product of the underlying mechanism of common sense, and therefore we can only admit common knowledge to be a system of conjectures, admitting also that it is a system of <u>natural</u> conjectures. The expression "imagination of the race," chosen by Quine to indicate the productive subject of the fictions that are the objects of common sense, physical observable objects, thus appears as strongly tainted with metaphor, and not simply derived from a legitimate analogy between the production of the theoretical concepts of science and the genesis of the concepts of visible things. In what way could common sense be taken as an "imagination" characteristic of the human species?

Less equivocal, perhaps, than considering common sense an imagination would be to define it as the "functioning" of a system of natural and innate devices, of a biological mechanism produced by the evolution of the species. This is the direction indicated by Popper in Objective Knowledge, when he proposes the discussion of the problem of common sense from the point of view of an evolutionary epistemology, adopting the perspective of scientific realism.[37] From this perspective it is perfectly clear, says Popper, "that if our actions and reactions were badly adjusted to our environment, we should not survive. Since 'belief' is closely connected with expectation and with readiness to act, we can say that many of our more practical beliefs are likely to be true, as long as we survive. They become the more dogmatic part of common sense which, although not reliable by any means, true, or certain, is always a good starting point."[38]

The thesis that common sense depends on a system of natural mechanisms is part of the Popperian theory of "expectations." Its central nucleus maintains that knowledge is fundamentally "dispositional and expectational": "It consists of dispositions of organisms, and these dispositions are the most important aspect of the organization of an organism."[39] The innate dispositions are those that Popper defines as expectations, in each of which one could speak of a pre-existent knowledge, previous to any kind of experience or observation. The existence of these innate dispositions is explained in Darwinian terms--which justifies the qualification of this Popperian theory as part of an evolutionary epistemology: all animal species, including the human species, live in a natural environment, in which certain traits of the organism and of the behavior characteristic of each species are necessary for survival. Now, besides organic and behavioral traits, certain kinds of "spontaneous knowledge," genetically transmitted from generation to generation, which Popper calls, in a very broad sense, "anticipatory theories," are also necessary for survival. An example: "The eye of a cat reacts in distinct ways to a number of typical situations for which there are mechanisms prepared and built into its structure: these correspond to the biologically most important situations among which it has to distinguish. Thus the disposition to distinguish among these situations is built into the sense organ, and with it the _theory that these, and only these, are the relevant situations for whose distinction the eye is to be used_."[40] In this way, Popper maintains, all the sense organs, in the different animal species and in the human species, are _theory impregnated_.

Belief in the existence of the objects of common sense, or "perceptual faith," to borrow an expression from Merleau-Ponty, could be the product of one, or of a set of these innate dispositions. Already in Hume the concept of a disposition of this type is generalized, as in Popper, for non-human animal species. In the same text in which he attributes observational belief to a natural instinct, Hume adds: "Even the animal creations are governed by a like opinion, and preserve this belief of external objects, in all their thoughts, designs, and actions."[41] That is, what Hume calls "the primary instincts of nature," or "the infallible and irresistible instinct of nature," or "the instincts or prepossessions of nature,"[42] constitute a mechanism which is common to mankind and the other animal species. Thus common sense, as perceptual faith, would not only be a biological common sense, but it would also be "common" to many other living species besides the human. Its productive mechanism would be universal and _a priori_, common to all men and previous to any sense experience—more than this, it would be a condition of possibility of all experience—and besides its "commonness," its universality, as a condition of experience, would go far beyond the limits of the human species. From this perspective, common sense acquires the right of citizenship, and if with Popper and Quine we accept its "theoretical" character, it also acquires the right to be considered the true precursor of scientific and philosophical knowledge. But this is obtained only at the cost of raising difficulties for the establishment of the concept of a _specifically human_ common sense. We would have a conjectural common sense, prolonged without interruption by theoretical reason, but it would be a mechanism much more general than our species—we would be forced to concede the right of citizenship to an _animal common sense_. Cases like that of personal identity could be considered as specific to man (although this question is not simple); but the largest part of common sense would be seen as _shared_ by man with numerous other animal species.

V. But why should we admit that common sense, and the perceptual faith which is its central nucleus, is derived from a genetically transmitted biological mechanism? For the same reasons, I think, that lead Quine to an identical supposition concerning the patterns of similarity underlying human and animal inductive operations. This is not simply concluding their innate character from their "irresistibility." In the case of the patterns of similarity, or "quality spacings," the fact that these patterns are distinct among themselves leads Quine to conclude that they cannot be produced by learning, some of them being innate. And part of the problem of induction consists in explaining why those innate mechanisms accord so well with natural phenomena, allowing for prediction, that is, for predominantly correct inductions.

Quine's explanation is biological, evolutionary (although he does not call his epistemology evolutionary, contrary to Popper),[43] and is based on the Darwinian concept of natural selection: if patterns of similarity are genetically transmitted dispositions, in this case the patterns underlying the statistically more correct inductions must have predominated in the evolutionary process through the action of natural selection. Animals whose patterns do not match their environment are eliminated by natural selection, since it is obvious that only species gifted with "adequate" patterns could survive--which is an explanation as legitimate as the Darwinian explanations of the anatomical traits of organisms.[44] Or as Quine says in The Roots of Reference: "Our innate standards of perceptual similarity show a gratifying tendency to run with the grain of nature. This concurrence is accountable, surely, to natural selection. Since good prediction has survival value, natural selection will have fostered perceptual similarity standards in us and in other animals that tend accordingly."[45] Or still in his good-natured statement of "Natural Kinds": "Creatures inveterately wrong in their inductions have a pathetic but praiseworthy tendency to die before reproducing their kind."[46]

That this explanation of the existence of innate mechanisms underlying induction, as well as of the success of induction in general, is essentially Humean, is what I have tried to show in other studies.[47] For Hume, it was the

Wisdom of Nature that gave us an instinct or tendency capable of producing agreement and harmony between the course of our ideas and the course of nature.[48] Once the Leibnizian explanation by pre-established harmony is rejected, the best remaining explanation of that inductive instinct, or of that inductive device, is an explanation based on the selective mechanisms of nature.

The same explanation that revealed itself as effective in accounting for Quinean similarity patterns ought to work in the case of perceptual faith. The human species and other animal species have a genetically transmitted innate mechanism, that irresistibly leads them to conceive and accept the real existence of material objects or common sense objects, because this mechanism is indispensable to the survival of all those species. In a parody of Quine, we certainly could say, concerning radical skeptical doubts, that the species of skeptics, if they were coherent, would reveal a pathetic but praiseworthy tendency to die before having time to reproduce themselves. Fortunately, as Hume has pointed out, the skeptic cannot be coherent, being forced to live with the same "carelessness and inattention" as the rest of mortals in relation to his doubts.[49] It is quite clear that this carelessness and this inattention are for the skeptic an indispensable condition of survival.

VI. Explaining perceptual faith and common sense by biological mechanisms, explainable in their turn by evolution and natural selection, could lead us into a vicious circle if we had the presumption of justification manifested, for instance, by those who look for a justification of induction. Concerning induction, Quine himself declares the illegitimacy of a justification of this kind; for according to him the law of natural selection, in its turn, is based on induction--natural selection only gives "a reason why induction works, granted that it does."[50] That is, for Quine this is an explanation, not a justification, and the validity of the argument depends on this restriction. Two things could be discussed here. In one direction, we could discuss whether even the simple aim of explaining induction would not be vitiated by recourse to a concept that is based in turn on inductions--which would destroy the Quinean argument. And in the opposite direction, we could discuss whether Quine's restriction, forbidding any justifying aims, would not pay undue hommage to the inductivistic conception of science, notedly Darwinian science--and whether the refusal of this conception of science could not tend to raise the Quinean argument to the dignity of an element in an authentic justification of induction, that is, of a solution of that which, since Kant, is usually called "the Problem of Hume." But these questions should be set aside here and left as the suggestion of a parallel between the problem of induction and that of common sense, both conceived in biological and evolutionary terms.

In the case of induction, suspicion of invalidity of the evolutionary argument is derived from the justificatory pretension: it is not legitimate to attempt to justify an operation which is secretly assumed in the premises of the argument itself. But explaining common sense by a system of innate mechanisms does not involve any pretension of justification. Neither are we trying to propose here any kind of "solution" to the problem of common sense. We are only trying to understand what it can be, and what its _origin_ can be; and how our perceptual faith is to be explained when we know that solipsism is logically irrefutable and we do not possess any true justification for the belief in the existence of common sense objects. Thus the problem of common sense runs parallel to the problem of induction. Abandoning any hopes of justification, we can defend, as a program of epistemological investigation, the search for an evolutionary explanation of the various facets of what we usually call common sense, beginning with perceptual faith in general. For this program--from an interdisciplinary perspective tending to combine contributions from psychology, neurology, the theory of evolution, physiology, genetics, ethology, epistemology, etc., defined as an endeavour placed in the framework of an evolutionary epistemology--the hypothesis of the existence of biological mechanisms capable of producing, in the human or animal knowing subject, natural conjectures positing the existence of the various objects of common knowledge, might be an acceptable starting point.

Examples of this program are the physiological investigations of von Holst, referred to by Lorenz in _Behind the Mirror_, which have led to the discovery of the biological mechanisms on which the perception of the tridimensional space of Euclidean geometry as an evident reality is dependent (tactile corpuscles, the labyrinth of the internal ear, the nervous system, etc.). To this Lorenz adds: "It seems to me totally aberrant to pretend that all these organs and their functions, which in all evidence have developed for the preservation of the species, adapting themselves to the data of reality, have nothing to do with our *a priori* apprehension of space." The theory of relativity affirms the existence of at least a fourth dimension of space, but that structure of reality does not impede that what we can perceive be a simplified version of space, the "Euclidean" version, which is given to us by the organization of sense organs and the nervous system of our species.[51] Similar investigations have led to the discovery of true "mechanisms of time," biological devices which fulfill a role described by Lorenz as that of "internal clocks," which "determine the course of time as we perceive it on the phenomenological level."[52] Now, the existence of tridimensional space, the existence of a uniform irreversible time, and the objective reality of both independently of the knowing subject, are some of the most deeply rooted (or "dogmatic," as Popper says) of common sense beliefs. If these depend on known biological mechanisms, nothing is more plausible than to conjecture that the other common sense beliefs also depend on mechanisms of an identical nature.

VI. Common sense beliefs certainly are not instances of an apprehension of reality in itself--if it is at all meaningful to evoke the possibility of such an apprehension. They are conjectures about that reality, which we naturally and spontaneously frame, led by biological, or neuro-physiological mechanisms, which in turn are the objects of conjectures proposed by science and epistemology. In this sense, theoretical knowledge prolongs common knowledge, without any essential break, given that the production of conjectures is no novelty introduced by science, but an immemorial cognitive procedure, whose origin is lost in the long duration of the evolution of the species. Thus the only fundamental difference between natural conjectures, and the conjectures usually called theoretical, seems to be the spontaneity of the first. After the reception of certain visual and tactile stimuli, my belief that I have before me a real table, for example, would be a conjecture whose only "confirmation" would be derived from the efficient role of the concept of that table as existent, and possessing certain properties, in the organization of my experience, analogously to what occurs with the confirmation of theoretical concepts referring to unobservable entities. All epistemological privilege being denied to observability, the vast field of knowledge appears as a surface devoid of any discontinuities, extending from animal expectations to the most sophisticated scientific theories. Common sense and science, or epistemology, would have nothing fundamental separating them, as Hume and Quine have asserted, but a certain degree of methodological sophistication.

There are, however, some reasons to doubt that this dethronement of observability could be quietly achieved. For after all the term "conjecture" is a term that acquired its strong sense (i.e., "theory," or "theoretical hypothesis") within a framework where the concept of observation already had a central position. Even admitting all that has been said about the conjectural character of common observation and the beliefs connected to it, there is still room for questioning the meaning of the term 'conjecture' when applied to those beliefs. That is to say, there are still very good reasons to ask whether it would not be legitimate to speak at least of two different types of conjectures--different in an important way, as occurs with the spontaneity of natural conjectures, contrasting with the methodological sophistication and the deliberateness of the theoretical conjectures of science. Is it not true that a hypothesis, employing theoretical terms referring to unobservables, is constructed in part through the artifice of <u>analogy</u>? This is what Quine himself admits when he reminds us that molecules are conceived by analogy, for instance, with "a swarm of dust motes in a shaft of sunlight,"[53] that is, with objects observable in common experience. Scientific conjectures, and perhaps also epistemological conjectures concerning the natural devices underlying common knowledge, are constructed by analogy with observable objects. What about those natural conjectures that posit the existence of these same observable objects? Even if we accept, in some sense of the term, their character of hypotheses, it does not seem possible to say anything whatsoever concerning their construction by analogy, contrary to what is allowed to us in the case of theoretical hypotheses. A common sense object, after all, would be posited by analogy with what? Before we have this kind of object, we have only sense data--and I think there is no way to make these have, in the construction of concepts of common objects, a similar role to that fulfilled by the latter in the construction of theoretical concepts.

It is also Quine who reminds us that sense data, in their turn, are also conjectures—they are conjectures of psychological theory, conceived as hypothetical components of subjective experience.[54] This means that, after all, our "data," like theoretical concepts in physics or biology, are conjectural constructions produced by science. As we have seen above, Thomas Reid said that philosophers had _invented_ the hypothesis of "data"--which in their time they called "ideas" or "perceptions." Psychology did nothing but develop this conjecture. It certainly would be senseless to try to speak of a construction of common sense objects by analogy _with_ sense data. Of this construction we know nothing, whereas we know something, although admittedly little, about the construction of theoretical concepts in science. Evolutionary epistemology may persuade us that the most plausible conjecture is that of the existence of natural innate mechanisms, producers of beliefs which in their turn have a conjectural character. But it certainly is incapable of elaborating any theory concerning a hypothetical analogical construction of common sense objects. This, no doubt, would be a "mere hypothesis," in the pejorative sense so common in Newton and Hume, that is, a supposition devoid of any kind of foundation or ground.

Hence, enough reasons come together perhaps to reassert, against Hume and Quine, the reality of the rupture between common sense and theoretical knowledge, a break that occurs in the very spot where, in science, philosophy or epistemology, the intervention of the cognitive artifice of analogy takes place. This doubtlessly is the same spot where the use of theoretical terms properly so called begins, with the invention of unobservable entities designed to explain observable events which otherwise would remain unintelligible. Having recourse to the Humean vocabulary, we might say that common knowledge works only with the resources of _resemblance_, and that analogy is not reducible to simply resemblance. For Humean or observational induction, resemblance is certainly sufficient, together with contiguity in space and time, as a basis to proceed to its operations, never going beyond the limits of the observable. But theory construction through the artifice of analogy is placed on a different level, distinct from that of Humean induction, although before Hume it had been possible for Newton, for instance, to give the same name of "induction" to both these operations--which was the root of several ambiguities. These ambiguities have come to an end with the analyses of Moulines, who made clear that the Newtonian concept of induction also covers inference from a domain to another domain <u>distinct</u> from the first, but in some sense taken as analogous to it: "Newtonian induction includes analogical reasoning, besides 'normal' induction."[55] "Normal" induction is restricted to the observable. The shift to another level, the "non-normal" level, characterizes and distinguishes scientific and philosophical invention, in contrast to what has been called by Quine the "racial imagination" of common sense.

Thus, between common sense and theoretical knowledge, both conceived as essentially conjectural, there is a certain closeness, derived from the hypothetical element common to both. If in earlier times it was possible to believe in the continuity between common sense conceived as the simple and direct apprehension of the facts of nature, and theoretical knowledge consisting of the same type of operation on a higher level of generality and methodological perfection, now the <u>simultaneous</u> crumbling of both these myths might preserve that continuity, for both these conjectural domains would not prolong each other less naturally than the two "factual" domains presupposed by the epistemology of other eras. But this double shift of epistemological paradigm gives rise to new problems, especially related to the role of analogy in the construction of, if I dare say, "properly so called" conjectures. Between theoretical conjectures and the natural conjectures of common sense there is still a fracture, insofar as the first call human imagination to a role that has no equivalent on the level of common sense. Common knowledge is constituted by <u>natural</u> hypotheses or conjectures, derived from innate biological mechanisms, but such hypotheses ought to be carefully distinguished from <u>theoretical</u> conjectures produced by scientific work. Between science and common science there is thus an unbreachable abyss, leaving no place for the Humean myth, prolonged by Quine, of an unbroken continuity between those two levels of human knowledge.[56]

NOTES

* Translated by Barbara Ann Norton and the author. Original article in Manuscrito 3 (1979), 57-80.

1. Hume, An Enquiry Concerning Human Understanding XII, iii, in Enquiries, ed. L. A. Selby-Bigge and Rev. P. H. Nidditch (Oxford: Clarendon Press, 1975).
2. Hume, Dialogues Concerning Natural Religion, I, ed. N. C. Smith (New York: Bobbs-Merrill, 1947), p. 162.
3. Quine, Ontological Relativity and Other Essays (New York and London: Columbia Univ. Press, 1969), p. 129.
4. Chomsky, Reflections on Language (New York: Random House, 1975), p. 224.
5. Hume, An Enquiry..., XII, 1, p. 152.
6. Hume, A Treatise of Human Nature, I, iv, 7, ed. L. A. Selby-Bigge (Oxford: Clarendon Press, 1888), pp. 267-8.
7. Ibid., p. 269.
8. Ibid., p. 268.
9. Monteiro, "Indução e Hipótese na Filosofia de Hume," Manuscrito I (1978) and "Hume's Conception of Science," Journal of the History of Philosophy 19 (1979).
10. Monteiro, "Hume e a gravidade newtoniana," Ciência e Filosofia 1 (1979).
11. Wisdom, Other Minds (Los Angeles: University of California Press, 1969), p. 230.
12. Hume, An Enquiry..., IV, ii, pp. 32-3; cf. ibid, pp. 36-8.
13. Ibid., p. 33.
14. Wisdom, Other Minds, p. 3.
15. Locke, An Essay Concerning Human Understanding, II, xxiii, 12, ed. A. C. Frazer, vol. I, (New York: Dover Publications, 1959), p. 403.
16. Hume, An Enquiry..., XII, i, p. 151.
17. Ibid. V, ii, p. 55.
18. Ibid, I, p. 7.
19. Hume, An Enquiry Concerning the Principles of Morals; I, in Enquiries, p. 170.
20. Reid, Essays on the Intellectual Powers of Man, II, viii, (Cambridge, Mass., and London: MIT Press, 1969), p. 152.
21. Quine, From a Logical Point of View, (Cambridge, Mass.: Harvard University Press, 1964), p. 77.
22. Reid, Essays..., II, iii, p. 91.

23. _Ibid._, VI, v, p. 628.
24. _Ibid._, IV, iv, p. 593.
25. _Ibid._, VI, v, pp. 614-5.
26. _Ibid._, VI, vi, pp. 644, 646, 649, 650.
27. _Ibid._, VI, v, pp. 617, 625, 630.
28. _Ibid._, VI, iv. p. 593.
29. Vernier, "Thomas Reid on the Foundations of Knowledge and his Answer to Skepticism," in _Thomas Reid: Critical Interpretations_, ed. Stephen Barker and Tom Beauchamp (Philadelphia: Philosophical Monographs, 1976), p. 15.
30. _Ibid._, pp. 19-21.
31. _Ibid._, pp. 19-21.
32. Quine, _Word and Object_ (Cambridge, Mass.: MIT Press, 1960), pp. 11 and 3.
33. Quine, _The Ways of Paradox_ (New York: Random House, 1966), p. 237.
34. _Ibid._
35. _Ibid._
36. _Ibid._, p. 238.
37. Popper, _Objective Knowledge_ (Oxford: Clarendon Press, 1973), pp. 68-70.
38. _Ibid._, p. 69.
39. _Ibid._, p. 71.
40. _Ibid._, p. 72.
41. Hume, _An Enquiry_..., XII, i, p. 151.
42. _Ibid._, pp. 152-3.
43. _Ontological Relativity_, however, is included in the bibliography presented by Donald Campbell in his "Evolutionary Epistemology," p. 463.
44. Quine, _Ontological Relativity_, pp. 123-6.
45. Quine, _The Roots of Reference_ (LaSalle, Il.: Open Court, 1974), p. 19.
46. Quine, _Ontological Relativity_..., p. 126.
47. Monteiro, "Indução e Seleção Natural," _Revista Latinoamericana de Filosofía_ 2 (1976), and its enlarged version, "Hume, Induction, and Natural Selection," in _McGill Hume Studies_ (San Diego, Calif.: Austin Hill Press, 1979).
48. Hume, _An Enquiry_, V, ii, p. 55.
49. Hume, _A Treatise_, I, iv, 2, p. 218.
50. Quine, _The Roots of Reference_, p. 20.
51. Lorenz, _L'Envers du Miroir_ (Paris: Flammarion, 1975), p. 17.
52. _Ibid._, p. 18.
53. Quine, _The Ways of Paradox_, p. 235.

54. *Ibid.*, p. 239.
55. Moulines, "Los fundamentos metodológicos de la filosofía natural de Isaac Newton," *Diánoia* 22 (1976), 30.
56. I wish to thank Eduardo Rabossi and Ezequiel de Olaso for their useful critical comments on an earlier version of this paper.

Jorge J. E. Gracia

IV. PHILOSOPHICAL ANALYSIS IN OTHER LATIN AMERICAN COUNTRIES

Outside of Argentina, Mexico and Brazil there are so far no strong nuclei of analysis in Latin America. There are some important analytic philosophers, but they are for the most part isolated from each other. This situation may be changing, however, for in countries such as Peru and Venezuela, some efforts are being undertaken to develop an interest in analysis and to widen the number of those who actively participate and publish within this philosophical tradition.

The Latin American countries with which we are concerned fall into three categories: the Andean countries, Central America, and the West Indies. I shall take each of these geographical areas in turn and provide a general but brief description of the state of analysis in each.

1. ANDEAN COUNTRIES

Of the Andean countries the only ones that merit special mention are Chile, Colombia, Peru and Venezuela. In Bolivia and Ecuador there is no evidence that philosophical analysis has had any impact.

1.1 Chile

Chile is one of the few countries in Latin America where there has been from the nineteen-forties onwards, interrupted only by the political upheaval of the last decade, a continued and growing interest in mathematical logic. This is perhaps a result of the lasting influence of positivism, which was still an important philosophical force in the forties in spite of the work of Molina, Millas, Schwarztmann and others. One has to consider only that Luis Lagarrigue, one of its most important exponents, died in 1949 and that two years earlier his Comtian book, La religión de la humanidad, was still being re-edited.

J. J. E. Gracia et al. (eds.), Philosophical Analysis in Latin America, 365–380.

This interest in mathematical logic, however, is not by itself an indication of any impact that analysis may have had in Chile. It only signals some acquaintance by Chilean philosophers of the logical work of some philosophers who were members of the analytic movement, since it is these who have contributed most effectively to the development of this field. But other areas of philosophy, such as ethics, epistemology and even the philosophy of science, where analysis has been strongest elsewhere, seem to have been generally dominated by other philosophical traditions. In particular the influence of Bergson and post-Hegelian philosophy, including phenomenology and existentialism, make a good showing. This can be easily confirmed by looking at the issues of the Revista de Filosofía, founded in 1949 as the official organ of the Sociedad Chilena de Filosofía. Indeed, very often one finds that the same authors who write on logic and discuss and/or adopt the views of logicians such as Carnap, Tarski and others, who are members of the analytic movement, will be very critical of the ideas of these and other thinkers of an analytic perspective in other areas. The most obvious case of this ambivalent attitude is Juan Rivano (b. 1926). Rivano is the author of the official textbook on logic (Curso de lógica antiqua y moderna, 1964) used in the Universidad de Chile, has reviewed a well known text on the same subject prepared by José Ferrater Mora and Hughes Leblanc entitled Lógica matemática (1955) and has translated Ayer's Elminination of Metaphysics (1957). In addition he has written a number of articles on topics which are of particular interest to logicians and analysts, such as the principle of identity (1957), sentences and propositions (1958), etc. and has taught regularly logic courses in Santiago (1960-75) and Concepcion (1957-60). All this points to an interest and acquaintance with contemporary logic and to some extent with the analytic tradition. But, when it comes to his own ideas about the philosophy of science, metaphysics and other philosophical fields, he feels more comfortable within the neo-Hegelian tradition. In 1960, for example, he translated Bradley's Appearance and Reality into Spanish. This neo-Hegelianism comes through quite strongly in some of his more recent works, such as Introducción al pensamiento dialéctico (1972).

The major exception to this attitude, which favors mathematical logic but rejects philosophical analysis, is

Gerold Stahl (b. 1926), who, although primarily a logician, has also written on other areas of philosophy from an analytic point of view. His work is contemporary with that of Rivano. He published a textbook of logic, Introducción a la lógica simbólica, in 1956 and other logical books, such as Enfoque moderno de la lógica clásica in 1960 and Elementos de la metalógica y meta-matemática in 1964. In the interim period between these books, he wrote several articles on other themes: the application of logic to physics (1956), semantic paradoxes (1958), causality (1962), the identity of indiscernibles (1964), and truth (1966). Stahl is without a doubt one of the most prolific logicians and exponents of philosophical analysis in Chile.

The first philosophers who drew attention to mathematical logic in Chile were García Bacca and Ferrater Mora, who had emigrated to Chile from Spain as a result of Franco's victory (the latter after spending a year in Cuba), and Granell, also of Spanish origin. But their efforts were limited. García Bacca, as mentioned in the section on Venezuela, has an interest in logic and the philosophy of science, but is not a logician and certainly not an analyst. Moreover, he did not stay in Chile for very long. And Granell seems to have done little to maintain an interest in logic and/or analysis and, like García Bacca, moved to Venezuela where he has worked ever since. The case of Ferrater Mora is somewhat different, however. He stayed in Chile from 1941 to 1947 and, as his Diccionario de Filosofía (first edition, Mexico 1942), indicates, was well acquainted with the analytic movement and developments in mathematical logic, even though, like Granell and García Bacca, he cannot be considered an analytic philosopher. The seed had been planted, however, and others took care that it germinate. Carlos Grandjot, for example, spoke in the Sociedad Chilena de Filosofía about mathematical logic and its advantages over classical logic in 1950, and when the Asociación Chilena de Lógica y Filosofía de las Ciencias was founded in 1956, he presented a paper on "La lógica y el numeral." Around the same time, Carlos Videla offered a course on logical theory at the Universidad de Chile--in addition to the courses taught by Rivano, to which we have already referred, and those taught regularly by Stahl. On the other hand, throughout the fifties the Revista de Filosofía published some extensive reviews of works on logic or about the analytic movement, such as Bochenski's Formale Logik,

Kraft's The Vienna Circle, and Ayer's Language, Truth and Logic, and in 1960, on Bunge's Causality.

In 1956 two important events took place. First, the Asociación Chilena de Lógica y Filosofía de las Ciencias was founded, as mentioned earlier. Second, the Inter-American Congress of Philosophy took place in Chile. These events constitute a clear recognition of increased philosophical activity in Chile and of its importance in the country. But, more than this, it allowed Chilean philosophers to establish contact with important logicians and members of the analytic movement elsewhere, such as Mario Bunge.

In the sixties, others became interested in logic and/or analysis. Roberto Torretti (b. 1930, see section on the West Indies), displayed his increased concern with these when he reviewed Quine's From a Logical Point of View and Ayer's Logical Positivism in the same year (1962). His main work at this period, however, deals still with Kant. There are also efforts on the part of others. Nathan Stemmer published two articles on Russell's theory of definition and verifiability in metaphysical systems in 1962 and 1964 respectively. And Augusto Pescador, of Spanish origin, wrote an article on "Logic and Language" in 1965, even though he did not adopt an analytic perspective. All of these articles appeared in the Revista de Filosofía.

Moreover, in 1970 a seminar and colloquium on mathematical logic was organized in Santiago. This was the first of its kind that takes place in Latin America. And in 1972 appeared a translation of Frege's Logic and Semantics, prepared by Alfonso Gómez Lobo. Unfortunately, owing to the events that took place in the early seventies and as a result of the political situation that continues to this day, the period of philosophical normalcy in Chile, as Romero would have called it, came to an abrupt end. The last number, before a prolonged interruption, of the Revista de Filosofía, the main philosophical organ in the country, appeared in 1970. It was not until 1973 that a successful effort was undertaken to continue the philosophical tradition established by this journal. The result was the journal Teoría, subsequently entitled Escritos de Teoría, edited by Humberto Giannini with the help of Renato Cristi and Carlos Ruiz. This journal is not aligned with any philosophical tradition in particular. In the analytic and logical areas it has published translations of articles by John Searle ("Los nombres propios," 1974) and Tarski ("Verdad y prueba," 1975). From this and some of the theses recently written in Chile, it is clear that in spite of the political turmoil that has agitated the country in the last decade, the interest in philosophy in general and in logic in particular has been dampened but not destroyed. Further evidence of this fact was Tarski's visit in 1974-75, who was invited to Chile by R. Chuaqui. Professor Chuaqui is himself a logician who has substantial publications in that area. It is too early to know, however, whether the interest in logic and analysis will continue or whether it will succumb in the face of the strong social and political concerns that agitate Chilean intellectuals and the disruptive influence of the government. But more recently there have been indications that the state of normalcy is returning and that the interest in logic and analysis has continued. The Revista de Filosofía, which reappeared in 1977, has published an article by Jorge Estrella entitled, "Cosmovisión del positivismo" (Vol. XV, n. 1). The same author, who is professor of Philosophy of Science in the Facultad de Filosofía y Letras of the Universidad de Chile, brought out a second article, "Qué explican las explicaciones" (Vol. XVI, 1978). Earlier, Augusto Pescador wrote a couple of articles of logical interest in Cuadernos de Filosofía, a journal published by the Universidad de Concepción ("El problema de la modalidad y los fundamentos críticos de la lógica," 1974-75; "Categorías e implicaciones modales," 1976).

1.2 Colombia

The Instituto de Filosofía y Letras de la Universidad Nacional de Colombia in Bogota was created in 1945 as part of the Facultad de Derecho and became the Facultad de Filosofía y Letras in 1952. Since 1965 it is a department of a larger unit, the Facultad de Ciencias Humanas. In 1948 appeared the short lived Revista Colombiana de Filosofía, official publication of the also defunct Sociedad Colombiana de Filosofía. In 1951 was founded a more enduring journal, Ideas y Valores, published by the Department of Philosophy of the Universidad Nacional. In 1954 it was discontinued until 1962, when its publication was resumed. It is presently edited by Rubén Sierra Mejía. All of this points to a late start of philosophical activity in Colombia and to its limited scope. Moreover the restrictions placed on freedom of expression by successive governments, and the emphasis on a conservative traditionalism identified with the Spanish Catholic culture, have made it difficult for philosophy to develop freely in Colombia.

In this rather weak environment one could not expect great inroads from analysis. In general the strongest philosophical schools are continental (existentialism, phenomenology, Marxism and neo-scholasticism). The only indication in the fifties of an interest in analysis was a review of Hans Reichenbach's _La filosofía científica_, published in 1953. In the sixties there were only a handful of articles dealing with ideas related to the analytic movement. Alfredo Trendall's "Estado actual de la lógica," published in _Ideas y Valores_ (17, 1963), as its title indicates, is a general exposition of the present state of symbolic logic. The theories of Gödel, Ackermann, Carnap, Kneale and Hilbert on the nature of logic are briefly examined in it. A similar didactic exposition was the intent of Carlos Patiño Russelli's "Los 'juegos del lenguage' de Wittgenstein," (_Eco_ 43, 1963). This expository attitude extended to the seventies, although the caliber of the material improves markedly. This can be seen particularly in the works of R. Sierra Mejía, who has written a couple of articles on Russell (for ex., "Lógica y filosofía del lenguaje en Bertrand Russell," _Cuadernos de Filosofía y Letras_ II-4, 1979) and a general article on the nature of philosophical analysis ("La filosofía analítica," in _Ensayos filosóficos_, 1978) in addition to having published some translations of analytic texts in _Cuadernos de Filosofía_, vol. 4, 1981.

Critical attacks on the movement, sometimes quite misguided, are also to be found, although these are more frequently directed against the classical position adopted by early logical positivists than against the philosophical currents which developed after the later Wittgenstein. These attacks come frequently from phenomenologists, but particularly from Marxists. An example of the first is Carlos B. Gutiérrez's article, "El solipsismo de Wittgenstein," (_Ideas y Valores_ I, 1964, and II, 1965) and of the second, Guillermo Hoyos' "Crítica al positivismo desde la racionalidad dialéctica," (_Razón y Fábula_ 40-41, 1976).

From all this we can infer that thus far the impact of philosophical analysis in Colombia has been very limited and that the reaction to it has been primarily negative.

1.3 Peru

The state of philosophy in Peru is peculiar. On the one hand, Peru has a long established and distinguished philosophical tradition. Scholastic philosophy was taught in the country throughout the colonial period in the Universidad Nacional Mayor de San Marcos, founded in the sixteenth century. And Peru has had in this century three of the most important philosophers that Latin America has produced: Alejandro Deústua (1849-1945), Francisco Miró Quesada (b. 1918) and Augusto Salazar Bondy (1925-1972). Moreover, philosophy has played more than an academic role in Peru through the efforts of the mentioned authors and an active group of Marxists such as Mariátegui. But, on the other hand, Peru has failed to form a nucleus of philosophical activity strong enough to initiate and maintain the publication of a journal dedicated exclusively to philosophy. This has prevented the concentration of the philosophical efforts in the country, since most Peruvian philosophers have had to scatter their publications throughout a wide variety of journals, most of them non-philosophical, and some foreign. It has also stood in the way of international recognition.

Analysis was introduced in Peru by Francisco Miró Quesada. His interest in symbolic logic goes back to 1946, when he published the first text on that subject in Spanish America. In the whole of Latin America it was preceded only by Ferreira da Silva's Elementos de lógica matemática, which appeared in 1940, and by the publication of Quine's O sentido da nova lógica in São Paulo, in 1944. Miró's earlier work had a phenomenological flavor clearly evident in El sentido del movimiento fenomenológico (1941), but after 1950 most of his articles and books lean toward philosophical analysis. He published Problemas fundamentales de la lógica jurídica in 1956, another textbook of logic in 1958, the important work Apuntes para una teoría de la razón in 1963, and the first volume of a treatise entitled Filosofía de las matemáticas in 1980. The last book is an introduction to logical theory. This work is in addition to numerous articles on logical theory, epistemology, ontology, and philosophy of science, and political theory.

Miró describes his change of philosophical orientation thus: "In the beginning we believed the fundamental problem had been resolved, for we had the impression of having explained the various types and levels of explanatory structures of the sciences. We believed that, having determined the explanatory types and having found the unity which underlied their specific variations, the rest was a matter of detail. The explained derivative processes seemed to us clear and their validity definitively established. But in order to reach the theoretical level that would make possible such explanation it was necessary to use the instrument of logical conceptualization. And knowledge of modern logic led us to the belief that fundamental concepts, such as 'deductive process,' 'logical derivation,' 'analytic truth,' etc., were far from clear." (Cited by F. Larroyo in <u>Historia de las doctrinas filosóficas de Latinoamérica</u>. Mexico: Parma, 1968, p. 226).

Augusto Salazar Bondy shared some of Miró Quesada's preoccupations. Trained in the continental tradition, he began to take an interest in analytic authors in the early sixties. In 1962 he published a work entitled <u>Tendencias contemporáneas de la filosofía moral británica</u>. His most analytic work, however, appeared only in the early seventies, in <u>Para una filosofía del valor</u>, and in several articles on the meaning of 'good' and other ethical issues, published in <u>Crítica</u>. His early death prevented him from developing his ideas further in this direction.

Present analytic activity in Peru is concentrated around a program on the philosophy of science organized in the mid-seventies by Miró Quesada, Alberto Cordero (b. 1948) and others in the Universidad Cayetano Heredia, and more recently in the Instituto de Investigaciones Filosóficas, whose main purpose is the study of the philosophy of science. A new generation of analysts seems to be gathering around these organizations. Alberto Cordero, who studied physics and philosophy with Rom Harré in Oxford and philosophy in Cambridge, is very active, even though his publication record is small at present. He has written among other things an article on "El lenguaje del mundo," published in a book (Las ciencias naturales y la concepción del mundo de hoy, ed. I. Prado, Lima, 1979) and several articles which have appeared in Jornadas Científicas de la Universidad Peruana Cayetano Heredia, such as "Carácter empírico de los predicados científicos y significación local" (Dec, 1979) and others. Also in that group are Jorge Secada, who has studied in Cambridge. Juan Bautista Ferro and David Sobrevilla, although not analysts, are interested in the movement and have participated in discussions of philosophical analysis.

Among linguists who are interested in and have contributed somewhat to philosophical analysis is Ernesto Zierer. He teaches at the Universidad de Trujillo and is editor of some interdisciplinary works to which philosophers such as Miró Quesada have contributed.

Although the predominant philosophy in Peru is still continental (phenomenology, existentialism, Marxism and neo-scholasticism), the situation of philosophical analysis does not seem at all bad and the future may be bright.

1.4 Venezuela

Properly speaking, contemporary philosophical activity did not begin in Venezuela until the decade of the 1940's, when the Escuela de Filosofía of the Universidad Central de Venezuela was founded by a group of enterprising philosophers encouraged and aided by Risieri Frondizi, who had been invited to join the Faculty at the time. The other philosophers, all of Spanish origin, were García Bacca, Eugenio Imaz and Domingo Casanova. There was, however, no evidence of any impact from the analytic movement until the late 60's. True, García Bacca, a Spanish emigré who left Europe due to the victory of Franco's forces in Spain, settled in Venezuela and maintained an interest in the philosophy of science. Early in his career he published two textbooks of logic entitled Introducción a la lógica matemática (1934) and Introdución a la lógica moderna (1935), and in 1942 two books, Tipos históricos de filosofar físico and Filosofía de las ciencias, which can be classified as in the area of philosophy of science. Both of these texts, as most of his writings, were traditional in style and took little account of developments in the analytic movement. His book Nueve filósofos contemporáneos y sus temas (1947), for example, does not include any analytic philosophers even though at the time analysis was well established as a powerful philosophical force in the United States and Britain. In recent years he has become an outspoken critic of analysis.

The journal Episteme (not to be confused with the journal published in Argentina with the same name), founded by García Bacca, also maintains an interest in epistemological and logical problems and in issues related to the philosophy of science. But again, the styles of philosophical writing and the authors discussed were seldom analytic until recently, when more attention is being paid to this philosophical movement.

This situation begins to change in the late sixties, when Juan Nuño (b. 1927), another philosopher of Spanish origin but educated in Venezuela, where he studied with Frondizi and García Bacca, became interested in contemporary semantic theory and logic. Nuño's training had been classical and continental; Plato and Heidegger seem to have had substantial influence in his thought. Indeed, as late as the early sixties Nuño was still writing on Sartre and existentialism, but in 1965, after having worked with Bochenski, he changes direction with his book, Sentido de la filosofía contemporánea. There he compares logical empiricism and Marxism as two philosophical movements which aim to transform philosophy: Marxism practically, logical empiricism theoretically. For Nuño, the theoretical transformation of philosophy proposed by logical empiricism is nothing less than an intellectual revolution which seeks to lead philosophy through a more rigorous and scientific course. This change of direction in Nuño began to bear fruits later in several articles, which he published in Diánoia and Crítica, as well as in other works on Tarski, Carnap, Black, Quine, and Wittgenstein, where he criticizes continental philosophy.

In addition to Nuño, there is a small but active group of philosophers who practice philosophical analysis in Venezuela. These are associated mainly with the Instituto de Filosofía of the Universidad Central. Most of them have been educated in Oxford, Cambridge, London and several universities from the United States. Among them are Rafael Burgos, author of several articles, and Pedro Lluberes, who has written an important book, Ciencia y escepticismo. Two others, Roberto J. Vernengo, originally from Argentina, and Ulises Moulines emigrated to Mexico. The first published a translation of Max Black's The Labyrinth of Language in Caracas, in 1969. And Moulines published La estructura del mundo sensible in Barcelona, in 1973. Ernesto Battistella, of Argentinian origin, belongs to an older generation, but works also within philosophical analysis. The efforts of these philosophers have already had an impact on the philosophical atmosphere in Venezuela. In 1977 and 1979, for example, Luis Castro taught courses on analytic philosophy of History and Philosophy of Law in the Universidad Central. And in 1978 the Instituto de Filosofía of the same university organized a series of conferences on Bertrand Russell in which Nuño, Battistella, and others participated. More important still, in 1977 the Facultad de Humanidades y Educacion of the Universidad Central established a program of graduate studies in philosophy, one section of which is dedicated to logic and methodology. And in 1979, in addition to the foundation of the Asociación Venezolana de Epistemología, a new journal of philosophy, Semestre de Filosofía, of a somewhat analytic orientation, appeared. Moreover, Episteme has become more analytic in orientation with the passage of time and Nuño's progressive influence.

In spite of these vigorous inroads on the part of philosophical analysis, as a whole philosophy in Venezuela can still be characterized as "continental" for there are strong existentialist, phenomenological, Marxist and scholastic groups. This is evident from the list of courses offered both in the Universidad Central and the Universidad Simón Bolivar, two of the most important intellectual centers in the country. It can also be seen in the Revista Venezolana de Filosofía (founded in 1974), where there has been little notice taken of analysis, except for a recent issue (n. 4) in which an analytic article appears (Battistella's "Interpretación del atomismo lógico del Tractatus mediante la lógica modal de von Wright").

We may conclude, then, from all of this that the analytic movement has had some impact in Venezuela, but that most of what it has produced to the present can be still classified as "assimilation." The number of philosophers interested in analysis are relatively few and their work is not extensive. Yet, their vigorous efforts may indicate a change of course in philosophy and a much brighter future for analysis in that country.

2. CENTRAL AMERICA

Except for Costa Rica, philosophical activity in Central America has been very limited and began quite late. The Universidad de San Carlos in Guatemala, founded in the 17th century, had a long scholastic tradition, but the serious study of modern philosophy in that country does not begin until the Facultad de Humanidades was founded in 1945, and then only by a small group of faculty--three as late as 1968. In Nicaragua there was no Facultad de Humanidades until 1964 and in El Salvador a Center for the study of philosophy was founded only in 1961. In spite of the late beginning, some progress has been made recently, however. The Catedra de Filosofía in Honduras maintains a faculty exchange program with the University of Kansas. The First Ibero-American Symposium of Philosophy, organized by the Sociedad Iberoamericana de Filosofía, took place in Guatemala in 1960 and there are regularly held colloquia for professors of philosophy in that country. Paradoxically Guatemala is the birthplace of one of the perhaps two or three major figures in the analytic trend in Latin America. Héctor-Neri Castañeda (b. 1924) began his career as a teacher of Spanish and mathematics (1945-48) and became a professor of philosophy in the Universidad de San Carlos in 1954, after a five year stay in the U.S.A. and a year in Oxford. His works, "Nota sobre el concepto de relación en Platón," "La lógica general de las normas y la ética" and _Naturaleza de las cuestiones filosóficas_, appeared in the mid-fifties. Since 1956, however, Castañeda has resided in the U.S.A. (see appendix below). Nevertheless, he maintains an interest in the development of philosophy in Latin America and continues writing in Spanish. His presence is strongly felt in the published literature as well as in congresses, conferences and other professional meetings which take place in the area.

In addition to Castañeda, two other non-Costa Rican Central American philosophers have published material related to the analytic movement. In El Salvador, Luis Melgar Brizuela has published an article on semiology and its relation to epistemology and, in Panama, José de J.

Martínez published a didactic article on Gödel, in 1971.

2.1 Costa Rica

The center of philosophical activity in Central America is without a doubt, Costa Rica. In its Universidad de Sto. Tomás scholastic philosophy was taught until 1888, the year when the University was reorganized into professional schools. The present Universidad de Costa Rica was founded in 1941 and within it was created the Facultad de Filosofía y Letras. But it was not until 1956, when the Facultad was reorganized into departments, one of which is the present Department of Philosophy, that the study of philosophy was more consistently pursued. In 1957 the Department began to publish its _Revista de Filosofía de Costa Rica_, which to this day continues to be the most important philosophical organ in Central America and one of the well established journals of philosophy in Latin America. In addition to these developments, the Asociación de Filosofía de Costa Rica, founded in 1958, conducts a series of lectures and seminars and helps to organize local and international events such as the Segundo Congreso Interamericano de Filosofía and the Primer Congreso Centroamericano de Filosofía which took place in Costa Rica in 1959 and 1971 respectively.

The orientation of the Department is clearly continental. There is a strong emphasis on French and German authors, and the influence of Ortega y Gasset and Unamuno is quite evident. In spite of this, analysis has made some inroads in the last two decades. In the fifties there was still little evidence that points to any interest in philosophical analysis. But in 1960 an article by Claudio Gutiérrez on "El consentimiento civil a la luz de la lógica moderna," in which symbolic logic is used to clarify some problems in the philosophy of law, appeared in the _Revista de Filosofía_. This was the first of a series of articles by Gutiérrez that maintained the presence of philosophical analysis in Costa Rica. In the Segundo Congreso Interamericano Extraordinario de Filosofía that took place in Costa Rica in 1961 he read a paper entitled "Sistemática de enunciados indiferentes." The only other analytic papers in the Congress were from Castañeda, who had been especially invited, Luis González and Luis Camacho. From then onwards the influence of Gutiérrez has been felt not only in publications but also in the University curriculum. A course on Symbolic Logic which he taught in 1962 used works by Leblanc, Quine, Tarski, Suppes, Copi and others. And in a course on Methodology he used Tarski and Wittgenstein. The rest of the faculty, however, seemed indifferent to analysis at this time. In Olarte's course on the philosophy of language, for example, the only text (out of ten) of an analytic bend used was Ayer's Language, Truth and Logic. This was in 1962. And in the same year, a course on contemporary philosophy included no analytic author at all. The climate was already changing, however, for two years later Christian Rodríguez gave a lecture on "El auge de la filosofía analítica" and in 1965 the curriculum included such subjects as logic, epistemology, philosophy of science and, what is still more significant, "analytic philosophy."

Claudio Gutiérrez went to Chicago in 1965, where he received a Ph.D. After he returned to Costa Rica he continued publishing actively. Among his publications are a textbook of logic, Elementos de Lógica (1968), his doctoral dissertation, Epistemology and Economics, published in the Revista de Filosofía in 1969, and various articles on systems of natural deduction and paradigms (for example, in Crítica, 1968), and in the early seventies on cybernetics. His philosophical activity decreased somewhat, however, when he was made President of the University. During this period he has published three articles among which are "Knots and Blanks: The Pragmatic Foundation of Principles" (Theory and Decisión 6, 1975) and "Ambigüedad y communicación" (Revista Latinoamericana de Filosofía III, 3, 1977). It was at this time that Luis Camacho's work began to appear: First a paper in a colloquium of philosophy in 1974, then reviews of a volume of the Boston Studies in the Philosophy of Science and of Sellars' Science, Perception and Reality in 1975. Moreover, the Revista de Filosofía has also published analytic material by Castañeda and Zurcher in recent years.

From all this it is clear that philosophical analysis is represented in Costa Rica and that its influence will probably continue to grow. Its future in the rest of Central America, on the other hand, is not so clear, particularly in light of the recent political turmoil.

3. WEST INDIES

The countries of the West Indies that are culturally a part of Latin America are Cuba, the Dominican Republic and Puerto Rico. The last of these is not politically independent, but has maintained its Hispanic heritage nonetheless.

3.1 Cuba

In general philosophical terms Cuba is the most prominent of these countries, having a long and distinguished philosophical tradition that goes back to Félix Varela (1788-1853). It was in Cuba, moreover, that positivism found one of its two (the other being Ingenieros in Argentina) most important exponents, Enrique José Varona (1849-1933). Unfortunately, the lasting influence of Varona, an early and strong interest in existentialism, and the advent of the Cuban Revolution and its imposition of Marxist ideology prevented analysis from taking hold in Cuba. In the various issues of the Revista Cubana de Filosofía, founded in 1949, one finds nothing to indicate an interest or even firsthand acquaintance with the analytic movement. Only one article, published in 1950 by Mario O. González, entitled "La crisis de los fundamentos de la matemática," concerns itself with issues associated with the analytic movement. And, although the author cites and briefly discusses some of the pertinent views of Russell, Gödel, Lukasiewicz and other philosophers associated with analysis, the style of the article is not analytic. The two Cuban authors included in the Spanish version of this collection, Sosa and Gracia, work in the United States and therefore will be discussed elsewhere.

3.2 Puerto Rico

The situation in Puerto Rico and the Dominican Republic is somewhat differrent from that of Cuba. Neither of those two countries has a distinguished philosophical tradition. Eugenio María de Hostos (d. 1903), in Puerto Rico, was an important figure in the development of positivism, but cannot be classified as a philosopher strictly speaking. In both countries philosophy as an independent discipline developed late. The University of Puerto Rico, founded in 1903, did not have a department of philosophy until the forties. Risieri Frondizi visited the island in 1950 and helped to develop further the curriculum. And it was not until 1964 that Diálogos, a journal dedicated entirely to philosophy, is published in the country. The Dominican Republic had its first journal of philosophy, entitled Revista Dominicana de Filosofía, in 1956, thanks to the initiative of Waldo Ross. But in neither case did these publications indicate an interest in analysis. This situation changed, however, in Puerto Rico owing largely to the influence of Roberto Torretti, who came to the island from Chile in 1970 and has become one of the most important philosophers working there. It should be noted, however, that Jorge Enjuto (n. 1922), of Spanish origin and with studies in France, Puerto Rico and Kansas, had helped to promote the study of North-American thought in the department of philosophy, although his interests were primarily in the speculative metaphysics of A. N. Whitehead. In the case of Torretti, his background was Kantian, but his concern with the philosophy of science led him to study logic, the history of mathematics and the founding fathers of the analytic movement. This shift in interest began to show in his publications in the early sixties, when he was still in Chile (see section on Chile). The selection we include from him in the Spanish version of this Anthology is from the late sixties. By that time, Torretti, helped by other members of the Department of Philosophy in Rio Piedras, such as Delacre, Tollinchi and others, had undertaken a massive effort to absorb and digest contemporary analytic literature. This is clearly seen in the analytic articles and numerous reviews on analytic material that begin to appear in Diálogos, and in the philosophers invited to teach and/or give lectures in Puerto Rico. This effort can only be rivaled in Latin America by Crítica from Mexico, by Manuscrito from Brazil, and perhaps by the Revista Latinoamericana de Filosofía from Argentina. Both Diálogos and Crítica accept articles in English, an indication of the desire of their editors to maintain active communication with developments in Anglo-American philosophy.

Pedro Lluberes

POPPER'S SOLUTION TO THE PROBLEM OF INDUCTION

> "Hume's philosophy... represents the bankruptcy of eighteenth-century reasonableness... It is therefore important to discover whether there is an answer to Hume within a philosophy that is wholly or mainly empirical. If not, there is no intellectual difference between sanity and insanity. The lunatic who believes that he is a poached egg is to be condemned solely on the ground that he is in a minority..."
>
> B. Russell

The observation made sometime ago by C. D. Broad identifying the lack of an adequate solution to the problem of induction as the great scandal in modern philosophy is well known among specialists. It seems somewhat peculiar, however, that for more than two centuries the attempts made by the best minds in the West--the Homeric Kantian attempt included--to try to solve this puzzle officially bestowed by Hume have failed.

I say this because the difficulties that arise, together with the "solutions" so far proposed, lead one to suspect that it is a problem without solution. Thus, every attempt to solve the traditional problem of induction under the shade of deductive inferences inevitably invites the "squaring of-the-circle" type of ghost. It is perhaps to Hume himself, more than anyone else, that we owe such motivations, when for example he invokes a fundamental principle which would allow inferences from the observed to the unobserved, namely, the principle that "the future resembles the past."

But, of course, recourse to such an expedient disguises the original question which was not concerned in the first place with deductions. Moreover, even conceding the putative relevance of the principle, the problem of

J. J. E. Gracia et al. (eds.), Philosophical Analysis in Latin America, 381–396.

justification would not have been overcome but would have only been transferred to the principle itself, which would remain in need of making explicit its own credentials.

There are at least two interconnected contributions made by Popper in this area: on the one hand, his criticisms of the traditional formulation of the problem (see, Objective Knowledge, ch. 1, Oxford University Press, 1972; hereafter OK), and, on the other, his own characterization of it. Popper is quite explicit in emphasizing the importance of his own formulation since this is what--in his opinion--leads to the solution of the problem. Let us, then, first of all, briefly survey Popper's approach to the subject; afterwards we may go on to discuss the type of solution he puts forward.

With regards to Hume's discussion, one ought, according to Popper, to distinguish between two different aspects: one logical, the other psychological. With respect to the latter, Hume, as is well known, appealed to habit and custom as explanatory tools to deal with our faith or beliefs in the repeatability of certain experiences, an approach with which Popper is in considerable disagreement (see, P. A. Schilpp, ed., The Philosophy of Karl Popper, LaSalle, Illinois, Open Court, 1974; vol. II, pp. 1019 ff.). This is not, however, an area to be explored here. The present paper will focus on the logical aspect of the problem. Popper distinguishes between two different questions (OK, 4):

> L1: "Can the claim that an explanatory universal theory is true be justified by empirical reasons?"
>
> L2: "Can the claim that an explanatory universal theory is true or that it is false be justified by empirical reasons?"

Popper, in agreement with Hume, answers L1 negatively. The difference comes out in his affirmative answer to L2: sometimes, according to Popper, one is able to justify the claim that an explanatory universal theory is false. Let us note, in passing, that Popper's point is logical, that is, it deals with the possibility of providing a formal demonstration of the falsity of specific laws or scientific theories. He himself has identified his own position as a "deductivism" (The Logic of Scientific Discovery. London: Hutchinson, 1959; p. 30; hereafter LSD); though his approach shows some basic differences with respect to traditional deductivist types of solutions to the problem of induction.

In theory, Popper's solution to account for the formal demonstration of the falsity of a law or theory seems quite simple, namely: the application of the modus tollens of classical bivalent logic. In practice, however, the scheme involves a series of refinements and complexities, surely the source of frequent disagreement among interpreters. This situation is well attested by the fact that, for example, in the Schilpp volume he accuses well known philosophers--his own collaborators included--of misinterpreting even the very statement of the falsificationist criterion. The rationale of the approach is, however, as I have just said, quite straightforward: from a logical point of view the finding of a counter-example to a generalization, such as "All swans are white," entails the falsity of the law "All swans are white" (Schilpp II, 1020).

But, putting aside the question of how the formal scheme should be set up for simple cases like the above generalization, what I am interested in, however, is in more complex cases, such as the ones found within the field of science itself. For in a very important sense, cases of the type "All ravens are black" could be extremely misleading.

The nature of the problems to be found once one moves to scientific fields proper are not difficult to pinpoint, since it is not isolated generalizations that are to be subjected to falsificationist tests, but rather groups of statements which involve at least auxiliary hypotheses (Schilpp II, 998) as well as something rather vaguely characterized by Popper as "background knowledge." The latter is the type of knowledge partially translatable into statements tacitly or explicitly accepted with the theory or hypotheses subjected to tests (Schilpp II, 1080; Conjectures and Refutations. London: Routledge and Kegan Paul, 1969; p. 240, hereafter CR). If one introduces the concept of necessary and sufficient conditions the situation could be described as follows:

> A theory is falsifiable if and only if for the theory T, in conjunction with auxiliary hypotheses Hl, H2,... Hn, and background knowledge B the class of potential falsifiers is not empty (see, LSD, 76, CR, 238-40; Schilpp II, 998).

Now, the nature of the difficulties associated with such an approach is well known: given the fact that we are dealing here with a conjunction '$(T \wedge H1 \wedge H2 \wedge \ldots Hn \wedge B)$' rather than with an isolated statement, one would need additional criteria in order to determine which one among the conjuncts, in every specific test whose outcome would involve the negation of the conjunction, has in fact been falsified.

What then are these criteria? In a note in LSD (n. 2, p. 76) Popper refers to the "scientific instinct of the investigator." More recently he again takes note of the problem and says that in such cases one has "to guess" (Schilpp II, 998). It seems clear, however, that on Popper's own constraints about "deductivism," recourse to expedients of this nature do not take us very far.

There might, however, be a different line of approach via Popper's talk about "methodological rules." The trouble, however, in spite of the prominent role they are supposed to play within Popper's methodology, is that whenever one would like to hear about a systematic account of these rules, the results are clearly unsatisfactory, a situation of which Popper is not unaware (see OK, 13). I have tried to gather a series of Popper's references and descriptions of methodological rules with the aim of finding some organization among them, and the most that can be said so far, is that important differences in kind can be detected.

Thus, and just as a preliminary approach to the subject, one might distinguish between at least three different categories or types of methodological rules:

> (i) Those methodological prescriptions having to do in a rather general way with the way that work in the field of science ought to be carried out. Examples of these rules would be: "try out, and aim at, bold theories with great informative

content; and then let these theories compete, by discussing them critically and by testing them severely" (Schilpp II, 1022). Or again: "try to construct alternative theories—alternatives even to those theories which appear to you inescapable" (OK, 266); "The game of science is in principle without end" (LSD, 53).

(ii) Those statements having to do with the way the very process of falsification should be set up and developed. When Popper says, for example, that "only those (auxiliary hypotheses) are acceptable whose introduction does not diminish the degree of falsifiability or testability of the system in question, but on the contrary, increases it" (LSD, 82-3), we are within the area of the tests themselves. From a formal viewpoint, indications of the way _modus tollens_ should be applied, and presumably how to identify what has really been affected by any test, belong to this category of methodological rules.

(iii) Finally, one would have to distinguish the two previous types of rules from those related to the problem of _acceptability_ of theories already submitted to falsificationist tests. Thus, criteria such as, for example, having "the greatest informative content and the greatest explanatory power" (OK, 15), or again the capacity to predict phenomena which have not been observed (CR, 241), as marks of the better among competing theories, are to be included among the methodological devices of the third category.

It is important, I believe, to carry on such distinctions among methodological rules if only for the following reason: that criteria of evaluation of the rules may radically vary according to which category of rule we are dealing with. Thus, rules belonging to the first type ought to be primarily evaluated or criticized in terms of concrete results to which their observance in the scientific work may lead, in the sense that the appeal to historical examples is not necessarily so important for the respective rule. In other words, if historical analyses would show that, say, Popperean prescriptions of the first type have not been followed at all in real scientific work, there remains for the defender the possibility of maintaining that in _future_ cases their observance may lead to fruitful results for scientists. So, in such cases the relevance of those rules might be an open question.

A different picture emerges with rules of the second type, since Popper explicitly and emphatically maintains that the _actual_ procedure of science is through conjectures and attempted refutations (see for ex., CR, 53). Since, according to Popper, the putative inductivist's methods are no more than a myth or an optical illusion prevalent among practitioners or philosophers of science who also attempt to describe scientific procedures, he is seriously committed to bringing the historical dimension into the analysis and evaluation of rules of this type.

As for the third kind of rules, their analysis and evaluation would have to be linked to the development of a _logic of preference_ among theories, a subject to which Popper has devoted only rather preliminary attention. Moreover, such logic seems to be naturally associated with questions about _rationality_ (something on which I shall make some brief observations toward the end of this paper). It must be said that the development of criteria for guiding choices among competing theories seems in this context a particularly pressing matter in the light of the _backward_ direction of Popperean tests: "Corroboration (or degree of corroborations) is thus an evaluating _report of past performance_, _but it says nothing whatever about future performance_, or about the reliability of a theory" (OK, 18). That is to say, even if a satisfactory solution could be found for solving the difficulties that hover around the applications of the falsificationist tests themselves, there remains the no less formidable problem of structuring reasonable preference canons given that the tests are no more than mere "reports of past performances."

In what follows I am going to summarize three types of criticisms prevalent in the literature which bear particularly upon the second type of rules just described. I should add that, to my knowledge, the first of the criticisms discussed below has not been as sufficiently articulated and developed as the other two. As for the second and third types of criticisms, it seems worthwhile to include them here, since we have Popper's attempted answers in the Schilpp volume.

As for the first type of criticism, it can be gathered from Popper's own recognition of the problem (Schilpp II, 982), but I do not know whether he is implicitly referring to some specific person or persons who might have developed the objection at length. The problem, in Popper's own words, is the following: "A biologist offers the conjecture that all swans are white. When black swans are discovered in Australia, he says that it is not refuted. He insists that black swans are a new kind of bird since it is _part of the defining property_ of a swan that it is white."

Now, if it is true that some generalizations of the "All S is P" type could be made to evade falsifications by just postulating 'P' as a defining property of the objects denoted by 'S,' then one could distinguish _prima facie_ between at least two categories of universal hypotheses: on the one hand, those in which the appeal to the defining property syndrome might look reasonable, and on the other, those in which this is definitely not the case (e.g., a generalization like "All Japanese are short").

Moreover, one could also identify a third type of generalization not uncommon in science, and whose putative way of functioning before a falsificationist test would differ from the second above mentioned, namely, hypotheses such as "All metals under the action of heat expand" or "All solids under the action of heat melt." Generalizations of this type could be protected from attempts at falsification by appealing to rather peculiar strategies since, as it has been pointed by G. Maxwell (Schilpp I, 295), we are not in a position to test, say, a solid object to an infinite range of temperatures (just in case that, for example, we would find an object with all the characteristics of a solid, but which would not melt under all temperatures).

The fundamental point in terms of a logic of falsificationism is, of course, that no such diversity of generalizations is taken care of within classical bivalent logic, which is Popper's basic formal tool for implementing his program (OK, 305). Thus, in order to provide an acceptable answer to the sort of problem delineated above, one would have to show either, (a) that remaining inside classical bivalent logic it is possible to deal in a satisfactory manner with the diversities just described, or (b) to provide a logic which would take care of such variations among universal hypotheses, so as to ensure an adequate application of the falsificationist tests.

It seems to me that this sort of problem belongs, together with the difficulties already indicated of determining what exactly has been falsified within a complex premise submitted to Popperean tests, to a purely _formal_ level and already establishes, at such a level, formidable obstacles in the way of a satisfactory solution for the application of the falsificationist scheme.

The second criticism I will be mentioning here is perhaps the best known in the literature. The interest of repeating it lies, as already suggested, in the fact that there is a subsequent reply by Popper himself, and this surely may bring new considerations into the debate.

In its bare bones the objection displays the following structure: examination of several relevant examples in the history of science shows procedures which as a rule diverge from Popperean's criteria. Thus, in the face of putative counter-examples a standard attitude has been _not_ to reject the theories or generalizations tested, but rather to work out adjustments and changes so as to assimilate the presumptive refutation, therefore making the theories less and less vulnerable to potential falsifiers. A standard example has been the survival of Newtonian Celestial Mechanics in the light of a persistent potential falsifier, namely, the perihelion of Mercury.

In fairness to Popper it must be said that he became aware of the problem quite early; much earlier in fact than his critics. Thus, already in _The Logic of Scientific Discovery_ one reads for example that "no conclusive disproof of a theory can ever be produced; for it is always possible to say that the experimental results are not reliable" (p. 50). But, then, a rather curious situation seems to take shape, since Popper's more recent rebuttal to criticisms of this sort (see for example his replies to Lakatos and Putnam in the Schilpp volume II, 998 and 1008) is to stick to the assertion that, for example, Newton's theory _is_ falsifiable: "there is an infinite number of simple and quite different sets of possible observations (or potential falsifiers) which, if accepted, would refute Newtonian theory" (op. cit., 1004).

But surely this kind of reply does not take us to the heart of the matter, which is not, I think, a question of denying the formal point that it is possible to enumerate a long list of potential falsifiers for a theory like Newton's, but rather that workers in the field may systematically circumvent potential falsifiers through a series of respectable moves, opening thus the possibility of rather perennial protections before putative counter-examples.

It seems convenient to remark, at this point, that for any given generalization or theory, the nature of the sources, from which scientists may draw and build acceptable auxiliary hypotheses or adjustments, may vary widely, thus greatly increasing potential resources for facing the attacks. Taking the famous Mercury case to illustrate the point: it is not only through the postulation of, say, an unknown celestial body that one might be able to save the theory; but appeal to things like, for example, the shape of the sun, might lead to similar results (immunization before potential falsifiers). A hypothesis of this sort has in fact been worked out by the physicist R. H. Dicke, who was able to show that by postulating some elongations for the Solar equator, it would be possible to derive the apparent "irregularities" observed in the perihelion of Mercury (see R. Adler et al., Introduction to General Relativity Theory. New York: McGraw Hill, 1965; p. 202). One might conjecture what would have happened in the evolution of post-Newtonian physics if in due time some alternative solution to the problematic Vulcan--like the one just mentioned--had established itself in the field. It seems pertinent, at any rate, to underline the variety of sources usually at hand for scientists to manage problematic situations from the point of view of potential refutations.

Incidentally, and mutatis mutandi, rather similar considerations might eventually apply with respect to "background knowledge," even if room were provided for the difference in status between what comes under this heading, on the one hand, and auxiliary hypotheses, on the other, the former having the character of "tacit" or implicit assumptions held by scientific communities at given times (see, CR, 240; Schilpp II, 1080). Assumptions for example about the "size" of the universe might prove decisive in the evaluation of potential falsifiers for some astronomical hypotheses, like the stellar parallax.

But, then, if such is the case, what would be the real force of the Popperean reminder that, say, Newtonian physics is full of potential falsifiers? Rhetorical appeal to methodological rules would not lead us very far, unless quite specific decision procedures were provided so as to allow us to take definite sides in the face of dilemmas like the ones we have been discussing. That is, how to weigh the force of apparent observational "irregularities" from the viewpoint of a certain theory, or the methodological legitimacy of some auxiliary hypotheses, or the questionability of the "background knowledge" behind the theory, etc. Before difficulties and questions like these have been satisfactorily solved, for example, through methodological rules of the second kind, plus the corresponding decision procedures just mentioned, the _via negativa_ of falsificationism simply totters.

Once more, Popper cannot be accused of ignoring some of these difficulties. Thus, for example, his talk about the nature of the so-called "_ad hoc_" hypotheses might throw some light and point in some interesting directions. Nevertheless, the content seems so far very crude (see, for example, Schilpp pp. 32 ff.). Do we have, for instance, some methodological rule or procedure which would allow us to decide, say, whether or not a hypothesis like the Solar equator elongation would have been "_ad hoc_" one hundred and fifty years ago?

Let us just add here that, insofar as historical evidence may bear upon certain aspects of a methodology, the legitimacy of practices which might involve the conscious adoption of _false_ hypotheses from which interesting and fruitful results might be derived, could be properly certified by the scientific community. I think that Descartes was well aware of such procedures when he wrote that, "people are satisfied if the author's (Archimedes, Ptolemy, _et al._) hypotheses are not obviously contrary to experience and if their discussion is coherent and free from logical error, even though their hypotheses may not be true." ("Letter to Mersenne," May 17, 1638; trans. A. Kenny, Descartes, Philosophical Letters. Oxford: Clarendon Press, 1970; p. 55).

There might yet be, from a Popperean perspective, another type of attempt to get rid of the difficulties described, this time through a distinction between (formal) falsification and (factual) rejection of a theory or a generalization, thus providing room for the real evolution of science. But here again, rather similar difficulties seem to crop up, since, given first the systematic possibility of protecting theories through several, respectable strategies; and second, even providing room for formal falsifications, how would one know when it is legitimate "to reject" a theory presumably falsified? The lack of a systematic discussion about rules of the third type, that is to say, those having to do with the choice between competing theories, once more prevents a satisfactory answer from a falsificationist perspective.

The third line of criticism to be discussed here has to do with the subject of "basic statements" and it seems convenient to bear in mind that Popper himself, in his earliest discussion on the subject, opened the gate to a sea of problems related to the nature and status of those statements. In a widely quoted passage he wrote: "The empirical basis of objective science has thus nothing 'absolute' about it. Science does not rest upon solid bedrock... It is like a building erected on piles... if we stop driving the piles deeper, it is not because we have reached firm ground. We simply stop when we are satisfied that the piles are firm enough to carry the structure, at least for the time being" (LSD, 111).

Now, this is not an isolated statement since one finds throughout Popper's work several consistent references to the decisional elements involved in the scientific work, and the lack of ultimate foundations for the edifice of science as well; the implications for "basic statements" being then inescapable. Thus he says that "From a logical point of view, the testing of a theory depends upon basic statements whose acceptance or rejection, in its turn, depends upon our decisions. Thus it is decisions which settle the fate of theories" (LSD, 108); or again, "our 'basic statements' are anything but 'basic' in the sense of 'final' " (CR, 388).

And, at a more general level: "The distinguishing characteristic of empirical statements (is) their susceptibility to revision" (LSD, 49).

In other words, he who refuses to admit the possibility that scientific statements are immune to revision simply retires from the game (LSD, 53). But, then, a rather high price from the point of view of the efficiency of falsificationist tests is to be paid, since this makes "basic statements," from a theoretical perspective, as liable to damage as the theories they are supposed to damage. The more immune to damage they are, the less scientific they become.

Now, if "basic statements" are supposed to play the role expected of them in Popper's methodology, they ought to be more immune than the theories they test. But how would one know in every case that this is actually so? Through the "instinct" of the scientific community? To begin with, as it has been suggested, the language of "basic statements" is already immersed in the language of the very theories one is trying to refute through them. And, moreover, keeping track with Popperean falsificationist ethics equally promotes in every case the attempt to falsify those statements.

It has been said that Popper (_malgré lui_?) was opening here a Pandora's box, and I think this is right. His criticism of the purported "empirical basis" of science, if taken seriously, seems quite damaging, and, I agree with those who think that, on no small scale, an important and somewhat unorthodox branch in contemporary philosophy of science has just exploited and developed some of the insights contained in several of Popper's (and surely Duhem's) observations on the empirical basis of scientific theories. We have, on the one hand, the series of criticisms on the possibility of providing a purely observational language firmly anchored on the firm ground of sensorial experience, and, on the other, the role assigned to "decisions" that have to be reached at crucial stages in scientific work.

In the latter case, the stage would have been set for the entrance of _sociological_ considerations _via_ the role that scientists, as a "community," ought to be playing;

while in the former, the question of the relativization of the empirical basis of theories becomes prominent. When, for example, one reads in a recent paper by the British scientist Sir Herman Bondi, that what may have passed for a test at a given moment, might not be acceptable at a later time (see, L. John, ed., Cosmology Now. New York: Taplinger, 1975; p. 12), the relativization of scientific "tests" comes clearly into view. And, if one tries to fit considerations of this kind within Popperean coordinates, one would have to say that, for example, a theory T which has been presumably falsified through a test carried out at time t^1 might not (and this time without necessarily having recourse to some extra hypothesis) be falsified by the very same type of test at a later time t^2. Only a sort of inductive relapse, like "what has been falsified today will be falsified in the future" would guarantee the survival of the falsificationist effect.

One might try to go further and approach this problem of the empirical basis of scientific theories from a perspective clearly neglected or minimized by Popper, given his refusal to include it within the field of the philosophy of science proper (see, for ex., LSD, 43-4). I am alluding here to the whole array of problems associated with the question of the relationship between language and experience. I think, on the contrary, that one might focus here--and at a deeper level--on the types of considerations that go around the subject of theory dependence of observation statements.

Popper, as is well known, has persistently maintained the unavoidable theory dependence of observation statements. Now, if one tries to explore the way such a dependence might take place at a linguistic level, through, for example, the analysis of the way that several putatively, purely observational predicates function in language, one may discover that, say, the ascription of some key predicates is fundamentally determined by some theory. But, then, the vulnerability of "basic statements" would be shown to have a deeper source, increasing consequently their problematic character as potential destroyers of the very theories from which they originate.

In short, the areas of criticism just delineated establish a host of genuine difficulties for Popper's methodology in general, as well as for his attempted solution to the traditional problem of induction. We remember that his affirmative answer to the claim that an explanatory universal theory might be shown to be true or false through empirical reasons, rests on the possibility that falsificationist tests be established. To the extent that the efficacy of such tests may in general be questioned, his putative "answer" to Hume might be equally questioned. And it is not mere coincidence that somehow the three types of criticisms above described—having to do with logical, historical and empirical aspects--seem to have some degree of interconnection, thus possibly pointing to a common source from which the main difficulties may spring. I would like to end the present paper with some considerations which point to one of the very fundamental sources of the difficulties.

The passage from Bertrand Russell heading the present paper illustrates quite well a central motivation underlying the attitude of generations of philosophers--Popper included--concerning the problem of induction. We are told there that Hume's sceptical answer brought with it no less and no more than the bankruptcy of eighteenth century rationality. But one could well ask: what kind of rationality?

Kant, who, by his own confession, deserted the garden of dogmatic visions just by reading Hume (presumably in translation) could be an excellent guide for this purpose. We hear for example, in the Prefaces of his first _Critique_ about a "tribunal" which "will assure to reason its lawful claims" (A xi, trans. N. K. Smith. London: Macmillan, 1958) and learn that "a new light flashed upon the mind of the first man who demonstrated the properties of the isosceles triangle" (B xii). It was a "new light" because a _method_, the "true method," had been found which would allow us to discover _necessary_ connections between the elements, say, of a geometrical figure. Physics after Galileo followed the same path and achieved comparable success, whereas metaphysics, on the contrary, "has hitherto been a merely random groping" (B xv). Kant's task was then signaled by the effort to push metaphysics into the "secure path of science" (B x) in accordance with "the example set by geometers and physicists" (B xxii).

At the beginning of the present paper the reminder was made that Hume himself was closely associated with the consolidation of such motivations. His appeal to a missing premise which would guarantee the legitimacy of inductive steps is anchored in a tacit presupposition about rationality, namely the association of "rationality" with "formal demonstrability." Such association is, of course, not new. The novelty, however, of modern attempts from Bacon and Descartes onwards has to be linked, I think, with the search for a _method_ through which, if properly applied, infallible results can as a rule be achieved.

Now, it seems clear that if "rationality" be equated with "apodicticity," a demarcation is unequivocally established through which the non-apodictic and non-demonstrable falls on the side of the non-rational. Popper, for example, pointing out that there is no _logical_ method which may guide us in the acquisition of new ideas, concludes that every discovery has an _irrational_ element in it (LSD, 32).

In Popper, by the way, familiar components of such a _Weltanschauung_ are readily present: science characterized by its methods; the scientific model as a paradigm of rationality (OK, 27); the apodictic ideal pervading the method of testing theories (LSD, 32 ff.); the equation of "rationality" with "scientificity" (Schilpp I, 70). It is no wonder that his proposed "solution" to the problem of induction falls "entirely within the scope of deductive Logic" (OK, 12).

There is, unfortunately, no comparable discussion of _alternative_ notions of rationality. In this respect, Popper's position looks very much like a consummated fact around which no further questions can be raised. The trouble is that alternative answers to fundamental questions that may arise in respect to the concept of "rationality" may lead to a radical turn in the approach to the problem under discussion, namely, the justification of inductive inferences. And once more, a certain tension can be detected in the writings of Popper, since his not infrequent talk about "decisions," about "guesses," and about the "opinions" of scientists may invite reflection on alternative models of rationality, which may in fact be operating within the game of science itself. Moreover, on

Popper's own recognition, the very choice of "critical rationalism" cannot be rationally argued for. It is in fact, as he has expressly admitted, an "irrational decision" (see The Open Society and Its Enemies. London: Routledge and Kegan Paul, 1945; Ch. 24, I, II).

Nevertheless, and insofar as the Popperean logic of falsificationism is taken seriously, the former notion of rationality commands the background. It must be added in this respect that if such is the presupposition, then Popper's via negativa seems perhaps the only remaining alternative to answer Hume in Hume's own terms, that is to say, preserving the condition of apodicticity as the canon of rationality.

Popper's attempt, it must be recognized, has opened the door to some of the most important discussions of these issues since Kant's time. Failure, however, to answer Hume, under the above-mentioned presupposition, does not necessarily mean failure to provide a rational basis for the sort of reasoning that prompted Hume's problem. I am convinced that a fruitful discussion on the subject should start from an exploration of the related subject of rationality; a task which, to my mind, suggests itself as one of the most attractive and stimulating areas for philosophical reflection.

* This is a shortened and revised version of a paper read at a symposium on Popper held in Caracas, September 1980.

Francisco Miró Quesada

ON THE CONCEPT OF REASON

When empiricism began to decline, because it could in no way explain the nature of logic and mathematics while keeping its principles with respect to empirical knowledge, it evolved toward pragmatism in order to account for the *a priori* aspect of logical deduction and mathematical truth. For the Vienna Circle, logic was a kind of shorthand, whose rules were only the rules of an arbitrary syntax, justified because they could express in a concise and precise way the content of empirical science. This alliance between empiricism and pragmatism led, in a natural way, to relativism. This kind of relativism ruled the philosophical scene for many years as far as logic and mathematics are concerned, and even today one can say without exaggeration that a vast number of philosophers of science abide by it.

Let us consider, for instance, reason. With a few exceptions like Gödel, who thinks of himself as a Platonist,[1] and Pap, who defends the existence of synthetic *a priori* propositions,[2] the great majority of *scientific* or *rigorous* philosophers are convinced that it is not possible to talk properly about a faculty called "reason" by means of which objective knowledge is constituted.

The common argument of many philosophers concerns the evolution of logic and the results of metatheoretical research. Logic, they say, considered during more than two thousand years a perfect discipline in which the supreme principles of human reason were disclosed with unswerving evidence, has lost its unity and has transformed itself into a set of theories that do not seem to be systematically interrelated. Every different type of logic employs different principles so that the belief that our reason functions in a unitary, coherent manner through principles of necessary and universal value, cannot be sustained.

If we consider the results of metatheoretical research, the stated thesis is reinforced to some extent. The

J. J. E. Gracia et al. (eds.), Philosophical Analysis in Latin America, 397–411.

existence of logical antinomies in set theory shows that the safe procedures to validate the principles and the rules of logic are a myth. Systems of logic with "evident" principles were refuted by their own theorems, which in a way is the most infamous manner for a system to succumb: not so much because of contradiction with facts, but because of inconsistence within itself.[3] Moreover, the existence of radically different mathematics, comparatively speaking, like the classical and the intuitionist systems, show that it is a futile task to try to find a system of principles that could be considered as a manifestation of what has been traditionally called "reason." Wherever one looks one finds difference of systems, anomalies, contradictory results, in short, a situation wholly incompatible with the traditionally conceived unity of rational knowledge.

No doubt, if one follows the way of diversity, the mentioned results seem to impose a relativist and skeptical view of the matter. When differences are compared, as they are by a majority of "rigorous" philosophers, many of whom call themselves "analytic," then it is difficult to confer any kind of meaning to the word "reason." However, as it is always possible in philosophy, when a road is travelled in one direction, it can also be transited in the opposite direction. Instead of focusing our interest on the differences among logical systems, why not focus it on the coincidences? After all, to look for the differences is not a moral duty. I should say, rather, that it is the desire to reinforce a dogma: <u>there are no laws of reason</u>. But wishful thinking in philosophy, although a most common fact, is not, alas!, a criterion of truth. Nothing impedes, for this reason (my excuses for the redundance), our proceeding in exactly the contrary way to what, through so many years, has been done: in lieu of looking for the differences among logical systems, let us look for the coincidences; in lieu of enhancing the contradictory and incompatible results of metatheoretical research, let us show their deep relation to the principles of reason.

If we follow this <u>new</u> approach, the panorama offers unexpected horizons and wonderful vistas. We are amazed with the vast quantity of coincidences, convergences, compatibilities, common principles and analogous tendencies of the different logical systems, even of those which are farthest from being traditional; and the awe is reinforced by the systematic and unitary character of metatheoretical results. Contrary to what happens in astronomy, in this case the other face of the Moon reveals to us quite a different world.

Let us begin with logic. The system that must function as a paradigm for comparison is, of course, classical logic, because it is this system which presented for the first time rational principles that were necessary for the constitution of objective knowledge. I think most authors agree that classical logic is the system developed in the _Principia Mathematica_ of Whitehead and Russell or any of its equivalent systems; that is, a system of bivalent, non-modalized propositions that includes the three traditional principles: identity, contradiction and excluded middle. The first thing to observe is that the Aristotelian theory of the syllogism, the theory of systems of the Stoics and most of Apuleian, Boethian and medieval logic, are included in classical logic.

Let us now analyze, in a rather loose historical order, the systems which, in the framework of the prevailing conception, are considered as irrefutable proof that it is not possible to speak of "reason" as the faculty of universal and necessary principles. The first non-classical logic to appear was the intuitionist. Its difference from classical logic is that the principle of excluded middle is not among its principles. This absence imposes to intuitionist logic some theorematic differences, for instance, the classical law of double negation cannot be derived in this system.

The first thing we observe, however, is that leaving aside this difference, the formal apparatus of both systems are identical, and that, formally speaking, intuitionist logic is only a subsystem of classical logic. That means that all the theorems of intuitionist logic can be formally derived in classical logic, but there are classical theorems that cannot be obtained in the intuitionist system. Intuitionism rejects a classical principle, the _tertium non datur_, but it accepts the other ones: identity and contradiction. Moreover, as Gödel and Lukasiewicz have demonstrated, defining the logical connectives in a certain way, the whole classical propositional logic can be derived as a formal subsystem of the intuitionist one. This is, at first view, disconcerting, but philosophically it means that there exist very deep connections between both types of logic and that the existence of intuitionist logic is not a proof at all that there are no universal and necessary principles.

It is only a proof that the way our reason functions in the logical level is more complicated and profound than had usually been thought. The fact that the excluded middle is not applicable in intuitionist logic is due to the fact that mathematical truth is conceived by the intuitionists in a radically different way from the classical. Mathematical truth is not, for them, coincidence with the facts (abstract or ideal mathematical facts), but rather it is _to be constructed_. And falsity is not non-coincidence with the facts, but rather it is the _impossibility of being constructed_. And naturally, if a proposition "p" is not constructed, that is, if it is not true, there can be cases in which nobody knows whether it is impossible to construct it. So, in the intuitionist terminology "p" is neither true nor false. But the classical _tertium non datur_ was conceived within the _adequatio_ conception of truth. So, if this principle is invalid, there ought to be a counter-example within this conception of truth.

For instance, there should be an arithmetical attribute F, and a number a, such that Fa and not-Fa ought to be false. But nobody has ever found such an example and it is evident that no one ever will. So, up to now, nobody has presented a counter-example to the principle of the excluded middle.

The second so called non classical system was modal logic. Modal logic includes more primitive symbols than classical logic and, of course, has a different syntax and a different semantics. But as far as the nature of logical deduction is concerned, modal logic has nothing strange or deviant. Quite the contrary. First of all, it includes classical logic in both its aspects, the syntactical and the semantical; and secondly, its new symbols are employed to make explicit a most important feature of deduction that was present, although not explicitly expressed, in the deductive dynamics of classical logic: _the necessity of the relation of logical consequence_. Everybody knew that, when a mathematical deduction is made, the premises necessarily entailed the conclusion. When classical logic formally systematized the principles and rules of deduction that were employed in the practice of mathematical science, this necessity was not represented in the formal system. If the necessity of entailment is explicitly expressed we obtain modal logic, but this logic does not violate the classical principles; it includes them all![4]

The third kind of non-classical systems to appear was that of polyvalent logics invented by Lukasiewicz and Post. Pragmatists and relativists and, in general, all those who deny the existence of universal and necessary principles of reason, are especially fond of these systems. No wonder: not only do they reject the principle of excluded middle, but they also introduce a multitude of logical values besides truth and falsity. But elementary considerations are more than sufficient to show that polyvalent logics do not mean that there is no sense in speaking of the principles of reason; the situation is quite different: they rather allow better to understand its richness and complexity. For all of them have common principles. The best known systems, created by Rosser and Turquette, include for n values the principles of identity and contradiction, of universal instantiation, of the transitivity of implication, and many others. And instead of the excluded middle principle they have what could be called the _enesimum non datur_ principle. For n=2, this principle becomes the _tertium non datur_.

Our reason seems to impose in this way a kind of exclusion of values when the set of values is finite. Instead of the principle of the excluded middle one could speak of a _principle of exclusion_. This generalization is not surprising because the _tertium non datur_ never had the same status as the principles of identity and contradiction. Aristotle himself raises important critical observations concerning the contingency of the future and, even though the classical rationalists, for instance, Leibniz, put it at the same level of the other two, the evidence adduced in its support was never considered very strong.

If we think about probabilistic logic (the system of Reichenbach) we see that it is the final step in the process of generalization that constitutes the polyvalent logics. All that has been said about these logics is applicable to probabilistic logic, for instance, the principles of identity and contradiction are preserved, and also the transitivity of implication and universal instantiation.

One may conclude from the existence of polyvalent logics that the concept of proposition such as it was conceived in classical logic was too restricted. A proposition was classically an expression that is true or false. But, as the new logical systems show, a proposition is rather an expression <u>liable to be true or false</u>. Propositions can be perhaps, within quite determined conditions, non-true without being false and vice versa, but they must have the possibility of being true and of being false. This relation to truth and falsity imposes many common properties and principles to the different systems we have mentioned.

Temporal logics have also been considered a proof that the universal principles of logic cannot be maintained. From a philosophical point of view, temporal logics have a special interest: they lead to the problem of the relation between the ideal and intemporal structure of logical principles, and the temporal nature of empirical facts. But the existence of temporal logics does not show that there are two kinds of unconnected logical principles. Classical principles are valid if the truth of propositions is dated. Of course, the dating may present difficulties, as we know from physics. But, theoretically, if truth is dated, then the three classical principles are valid. To say that the principle of identity or of contradiction are not valid because a proposition "p" is true at date X, and false at date Y, is to talk nonsense. What is interesting, however, is that, when the change of propositional truth is introduced into the logical system, it is possible to discover evident deductive principles that can be applied to propositions with viable truth (falsity). It is also important to point out that the relations among the primitive temporal operators are parallel to the relations of the primitive modal operators. This shows that there are deep structural relations between logical systems that at first glance seem to be quite heterogeneous.[5]

When first Jaskowsky, and then da Costa in a deeper and more elaborated way, developed the systems that are known today as "paraconsistent logic," those who negate the possibility of employing the word "reason" in a coherent way greeted the event with joy. Paraconsistent logic allows perfect deduction applied to inconsistent systems. The supreme logical principle, the one that had resisted all the innovations of heterodox systems, the principle of contradiction, seemed at last destroyed.[6]

But, in spite of the deep interest of paraconsistent systems, their rejection of the contradiction principle is only apparent. What these systems make possible is the derivation of logical conclusions within inconsistent theories. These derivations do no trivalize the theories, contrary to what happens when the deductions are made through the application of classical logic. But this possibility does not mean that when a theorem is derived, its contradictory can always be derived. In order to function, paraconsistent logic must not trivialize the inconsistent theory to which it is applied; and this means that there must be certain theorems whose contradictories cannot be derived within the system. These theorems are called good theorems (when a theorem and its contradictory can be derived, then the theorem is a bad theorem). If every theorem were a bad theorem, the theory would collapse. That is precisely what happened in classical logic when it was applied to an inconsistent theory, because of the principle of Scottus.

This shows that our reason functions in a wider way than was thought by the classics. For them a single contradiction was sufficient to trivialize a theory; we know now that things are different. A theory can include many contradictions and maintain its value with the condition that contradictions cannot be derived with respect to certain theorems which have a special cognitive meaning.[7] So, the principle of contradiction is not eliminated from paraconsistent logic; quite the contrary. But it is applied in a more subtle way. The contribution of paraconsistent logic to the understanding of the way reason functions at the logical level is that a contradiction within a theory must not necessarily trivialize it. But it has shown that without some sort of guaranty of consistency (absolute consistency) logic cannot function.

In free logics the three classical principles are invalid. But if, instead of the values true, false and

<u>indeterminate</u>, we put <u>true</u> and <u>non-true</u>, the principles of identity, contradiction and excluded middle are formally derived and semantically valid. This shows that there are very close relations among the principles of free logics and of classical logic. The introduction into the system of propositions without any referent widens the scope of deduction, and the classical principles, due to this widening, appear to be invalid. But, as a matter of fact, they continue to be valid with respect to a subsystem of the free system. Classical logic was elaborated, after all, having in mind referential propositions. However, when non-referential propositions are added to the system the consequences of the broadening of content are quite intuitive; our reason not only keeps functioning on the logical level, but is able to grasp the principles that rule deduction in the new system. Instead of being a proof that human reason has no universal principles, free logics disclose a deep rational dynamics that is intuitively apprehended.

The recent development of fuzzy logic strengthens the former conclusions. One could think that fuzzy logic is a perfect example of a logic in which none of the three classical logical principles is valid. However, this would be a grave mistake, because the system of fuzzy logic created by Zadeh presupposes, for its development, the theorems of classical arithmetic, and even some theorems of integral and differential calculus; and these mathematical disciplines, such as Zadeh employs them, are developed within the framework of classical logic.[8] So fuzziness presents its own logical principles, different from the classical, but it also presupposes these last ones. The situation seems, then, to be the following: reason is able to discover principles that are necessary and universal with respect to fuzziness, and this validity depends on the validity of the classical logical principles which are applied to nitid sets.[9]

Nowadays fuzzy set theories and fuzzy logics are proliferating. There are systems that are different from Zadeh's conception. For instance, Peña has developed quite interesting systems from a point of view that is more philosophically centered than Zadeh's, and which are axiomatically developed (the Zadeh systems are far from being axiomatic). And it is striking to observe that in

Peña's system there is a value of truth according to which the classical principles are applicable.[10] So, in both conceptions of fuzziness, Zadeh's and Peña's, there is a clear and deep relation between the principles of classical logic and those of fuzzy logic.

Another type of logic that has been frequently considered as a proof that there are not universal logical principles is what can be called the athetic or non-propositional logic, for instance imperative logic, normative or deontic logic, erotematic logic (logic of questions), etc. In these systems there are, in fact, principles that are not applicable to propositional systems. Some classically minded logicians and philosophers affirm that these systems are not logical because deduction is only a propositional matter. Others accept the logical nature of athetic systems, but believe that their existence shows that there are no universal logical principles.

But when one analyzes without prejudice the structure of these systems and the way they function, one immediately sees two fundamental things: 1) in every one of these systems there is a large number of formal coincidences with propositional systems; their axioms and rules of inference and the form of the theorems are very much like those of classical and modal logic. So, there is no doubt that there are principles that are so universal that they must be applied, at least formally, even in the case of non-propositional deduction; 2) as for the idiosyncratic principles, if their meaning is duly analyzed, then they are apprehended with full fledged evidence: human reason is able to develop deductive non-propositional systems because there is non-propositional evidence about deductive relations of orders, norms, questions, etc. So, the existence of athetical logic, instead of being a proof that there are no universal logical principles, is a proof that human reason is able to discover universal principles applicable to non-propositional deduction. There are not only universal propositional logical principles, there are also universal non-propositional logical principles. The universality is limited by the realm within which the principles are valid: the propositional realm, the normative realm, the imperative realm, etc. Traditional logic was conceived with respect to propositions, and the fact that there are principles applicable to non-propositional deduction does not mean that the classical principles have lost their universality, because this universality was limited in its very origin to the realm of propositions.

We cannot, within the limited scope of an article, analyze in detail the situation with respect to metatheoretical research. But we can say, in a general way, that the results of this research, in spite of their spectacular and disconcerting aspects, are in no way a proof that we can no longer speak of general and universal principles of reason. The famous paradoxes of set theory, so frequently alluded to as a proof that we can no longer rely on our intuitive evidence, could never have produced such tremendous impression if the principle of contradiction had not been considered to be absolutely evident. On the other hand, the modification of Cantor's comprehension principle, that has been made to avoid the paradoxes, does not show at all that this principle was evident before the paradoxes were known and ceased to be evident after them. Quite the contrary. The modification was introduced on the most evident basis because, although it is true that the evidence of Cantor's principle was jeopardized by the paradoxes, the only change was that there were some cases to which it could not be applied. But within these limits nobody can doubt that the comprehension principle is as evident as it is unshakable.

The existence of intuitionist mathematics has been considered to be a proof that mathematical truth is relative and that it is impossible to trust intellectual intuition. But the real situation is completely different. Intuitionist mathematics is based on very strong intuitions and, on certain elementary levels, intuitionist arithmetic fully coincides with classical arithmetic. What the intuitionists try to do is to develop mathematics free from contradiction and based on absolute evidence. But this does not mean that classical mathematics is not based on rational evidence. There are cases in which evidence has failed and it seems that the introduction of the actual infinite in set theory is the cause of the failure. But we must not think that all the concepts related to the actual infinite are vague and unevident. There are many that are quite evident, for instance, many axioms of classical set theory, like the axiom of pairing, of powers, of the sum, of replacement and others. Even the axiom of choice is, in our opinion, evident. Perhaps we could concede that its evidence, as it happens with respect to the principle of comprehension, is not unlimited. But it is evident as far as finite sets are considered, and this is much to say.

The former analysis enables us to mention two results that are deeply significant. In the first place, the famous proliferation of different logical systems does not mean that all rational logical principles are relative. In all known logical systems we can find common principles, even in those which appear are completely different from classical logic; and this leads us to think that there is a deep common structure revealed, in one way or the other, in all of them. <u>This structure can only be called rational</u>. The lesson to be learned from the comparative study of the different logical systems is that one cannot summarily dismiss rational principles without a deep analysis of the formal and semantical inter-relations among these systems.

In the second place, the evolution of modern logic and the results of metatheoretical research show that the concept of reason has a much wider scope than what was classically thought. Our reason works through a system of necessary, universal principles; but it is a dynamic system. At the logical level, the most universal and transparent of all, there can certainly be tricky and deceiving intuitions. But the only way to detect them is through evidence about which it is impossible to have doubts (for example, in intuionist logic the limitations that are imposed on the excluded middle are compensated through the certainty of consistency, and in the first trivalent system of Lukasiewicz, although the principle of contradiction and of the excluded middle are not valid, it is impossible that two contradictory propositions be both true or both false; so, in a very clear way, there is an essential aspect of the classical principles that is preserved).

Since our reason constantly conquers new knowledge, we become more and more conscious of its principles. Departing from an original nucleus of principles, we begin to distinguish the really universal and necessary from those that are less universal (that have more restricted applications) and from those which are only a product of habit, generalized through ill-founded extrapolations. So, the original nucleus is readjusted in accordance with the advance of our rational knowledge. But it is also increased when we continue to advance and are able to discover new principles (for instance, the principle of constructivity that is today accepted by all logicians and mathematicians),[11] and to proceed without end in our effort to find a good foundation for our scientific knowledge. Through this process we find many coincidences, many connections among the new principles and the old ones; we find a coherent dynamic that, as it advances, develops itself not by chance but by a sort of internal rationality, through closely related ways according to definite vectors. In a word, there exists an advance, a historical expansion of reason. This expansion is made according to some patterns that are typical of rationality.

To understand deeply these patterns and their respective relations within the dynamics of reason, to understand what is the unity within a diversity which advances without end, is the most important goal of a philosophy of knowledge which pursues the understanding of what is really happening in modern scientific knowledge. If the importance of this goal is not understood, if the unity of logic within diversity is considered only a matter of convention or of habit, then logic and, in consequence, all mathematics and empirical science (that cannot be duly constituted without mathematics) are reduced to non-sensical jargon.

NOTES

* <u>Revista Latinoamericana de Filosofía</u> 1, 3 (1975), 183-191.

1. Gödel, "Russell's Mathematical Logic," in <u>Philosophy of Mathematics</u>, ed. Paul Benacerraf and Hilary Putnam, (Oxford: Basil Blackwell, 1964).
2. Pap, <u>Semantics and Necessary Truth</u>, (New Haven: Yale University Press, 1958).
3. Klimovsky, G. <u>El método hipotético-deductivo y la lógica</u>. La Plata: Instituto de Lógica y de Filosofía de las Ciencias de la Universidad Nacional de la Plata, SIF, p. 14.
4. As a matter of fact, any logical system can be, at least in principle, modalized, because one of the essential

traits of logic is the necessity of deduction. If this necessity is formally expressed, then the system becomes modal. So, modality never means a change of logical principles, it only means that the necessity of entailment is explicitly represented. Of course, a modal logic, due to the principles that govern the relations among modalities and among modal and categorical propositions, can have more principles than a classical system. But this fact does not mean that the classical principles must be changed.

5. It is very interesting to observe that most temporal logics include propositional classical logic as a subsystem.
6. Before da Costa, Lukasiewicz had developed a system in which the contradiction principle was not valid. But he did not apply it to inconsistent systems. And Jaskowsky's system includes the contradiction principle in spite of the fact that his system can be applied to inconsistent theories.
7. This does not mean that contradictory theorems are not important. For example, in the inconsistent theory of sets developed by da Costa and his school, although the principal theorems of classical set theory are derived and it is not possible to derive their contradictories, there are highly interesting contradictory theorems about Russellean sets (da Costa 74).
8. Zadeh, 65, 79.
9. "Nitid" is an English archaic word recorded in the Oxford English Dictionary, that expresses very precisely the nature of sets as it was conceived in classical set theory. "Nitid" is the antonym of "fuzzy."
10. Peña, 72, 80.
11. This principle does not mean that only a constructed proposition can be true, but only that if a proposition is constructed, then it is true (only the intuitionists think that constructivity is a necessary condition of truth).

BIBLIOGRAPHY

Arruda, Ayda. A Survey of Paraconsistent Logic. Relatorio interno N 106. Universidade Estadual de Campinas. Instituto de matemática, Estadística e Ciencia da Computação, 1979.

DaCosta, N. C. A. "On the Theory of Inconsistent Formal

Systems," Notre Dame Journal of Formal Logic, 15 (1974), 497, 510.

DaCosta, N. C. A. "Logique Mathématique-Remarques sur les calculs C , C* , C= et D ," C. R. Acad SC, Paris 278 (18 mars 1974), Serie A-819.

DaCosta N. C. A. and Wolf, R. G. Studies in Paraconsistent Logic I: The Dialectical Principle of the Unity of Contraries. Unpublished.

Goddard, Leonard and Routley, Richard. The Logic of Significance and Context. Edinburgh and London: Scottish Academic Press, 1973.

Gödel, K. "Russell's Mathematical Logic," in Philosophy of Mathematics, ed. Paul Banacerraf and Hilary Putnam. Oxford: Basil Blackwell, 1964.

Haack, Susan. Deviant Logic - Some Philosophical Issues. Cambridge:Cambridge University Press, 1974.

Heyting, A. Intuitionism--An Introduction. Amsterdam: North-Holland Publishing Company, 1956.

Jaskowski, S. "A Calculus of Propositions for Contradictory Deductive Systems." Studio. Soc. Scen. Torunensis, section A.1 (1948), 27-77.

Klimovsky, G. El método hipotético-deductivo y la lógica. Instituto de Lógica y de Filosofía de las Ciencias. Universidad Nacional de La Plata, SIF. p. 14.

Kneale, William and Kneale, Martha. The Development of Logic. Oxford: Clarendon Press, 1962.

Lukasiewicz, J. Selected Works. Amsterdam: North Holland Publishing Company, 1974.

Miró Quesada, Francisco. "Las lógicas heterodoxas y el problema de la unidad de la lógica." in Lógica - Aspectos Formales y Filosóficos. Lima: Pontificia Universidad-Católica del Perú, 1978.

Miró Quesada, Francisco. "Lógica y Razón." Ponencia presentada a la Sociedad Peruana de Filosofía, 1979.

Pap, Semantics and Necessary Truth. New Haven: Yale University Press, 1958.

Peña, Lorenzo. Contradiction et verité-Étude sur les fondements et la porté épistémologique d'une logique contradictorielle. Liege, 1979.

Peña, Lorenzo. Apuntes introductorios a la lógica matemática elemental. Quito: Pontificia Universidad Catolica del Ecuador, 1980.

Rosser, J. B. and Turquette, A. R. Many-valued Logics. Amsterdam: North-Holland Publishing Company, 1952.

Zadeh, L. A. "Fuzzy Sets," Information and Control 8

(1965), 338-365.
Zadeh, L. A. "Fuzzy Logic and Approximate Reasoning," Synthese 30 (1975).
Zadeh, L. A. "Inference in Fuzzy Logic." Research supported by the National Science Foundation, 1978.

Jorge J. E. Gracia

APPENDIX

LATIN AMERICANS RESIDING IN THE UNITED STATES AND CANADA

The number of Latin American philosophers who work in the United States and Canada is not large. While there are comparatively large numbers of physicians, physicists, mathematicians, and other Latin American professionals and scientists dedicated to teaching and research, the number dedicated to the study and teaching of philosophy is very small. Moreover, the Spanish speaking minorities that live in the United States have produced no important philosophical figures. There are no major Puerto Rican or Mexican American philosophers to date. The few philosophers of Latin origin that work in the United States and Canada are immigrants, having moved to these countries for various personal or political reasons. Their philosophical orientation reflects to a large extent the predominant currents of Latin American philosophy. Most of them fall within the continental traditions. Nevertheless, the most distinguished among them work within the analytic tradition or at least are sympathetic to it. In fact, the two most important Latin American philosophers to date who presently live in Anglo-America are Mario Bunge and Héctor-Neri Castañeda; both work within the analytic tradition.

Bunge (b. 1919) was educated in Argentina (see section on Argentina) and settled in Canada permanently in 1966, after having visited a number of universities in Europe and the United States. He is a very prolific writer, having authored or edited more than twenty books and two hundred articles. His main interests are in the philosophy of science. His works have had a substantial impact both in North and South America as well as in Continental Europe. He founded the Society for Exact Philosophy, the Asociación Mexicana de Epistemología and is very active in editorial work.

J. J. E. Gracia et al. (eds.), Philosophical Analysis in Latin America, 413–415.

The case of Castañeda (b. 1924) is somewhat different, even though he is Bunge's contemporary. Castañeda's college education took place in the United States and Oxford, and although he returned to Guatemala, his place of birth, in 1954 as a professor at the Universidad de San Carlos, he went back to the United States the following year (see section on Central America). Since then he has taught philosophy at various American universities. He is, like Bunge, a prolific author who has produced seven books and more than one hundred articles. His impact on American philosophy has been greater than Bunge's, but his influence on Latin America has been lesser. Nevertheless, he has maintained a presence in Latin America through publications, attendance at congresses, and as consulting editor of Latin America's first completely analytic journal, Crítica. He is the founding editor of Nous, one of the most prestigious journals of philosophy published in the United States. Castañeda's main interests center on ethics, metaphysics, the philosophy of mind, and the philosophy of language.

In addition to Bunge and Castañeda there are four other, younger philosophers who merit mention: Alberto Coffa, Ernesto Sosa, Ignacio Angelelli, and Jorge J. E. Gracia. Of these the first two are clearly analytic philosophers. Angelelli and Gracia, on the other hand, are historians of philosophy who have strong sympathies for philosophical analysis. They typify the extent to which analysis has spread in the United States.

Alberto Coffa (b. 1935) was born and partly educated in Argentina. He received M.A. and Ph.D. degrees in philosophy from the University of Pittsburgh in 1970 and 1972 respectively. He had previously worked on philosophy and the history of science in the Universidad de Buenos Aires and the Universidad de La Plata. Also, he had taught at the Universidad de La Plata. Since 1970, however, he has been teaching at Indiana University in the Department of the History and Philosophy of Science. He has written more than a dozen articles on the philosophy of science and its history.

Ernesto Sosa (b. 1940) was born in Cuba, but most of his education took place in the United States. He received a Ph.D. from the University of Pittsburgh in 1964. He has taught in Canada and, since 1964, at Brown University, where he was chairman from 1970 to 1976. Over the last few summers he has been a regular visitor at the Instituto de Investigaciones Filosóficas at UNAM for lectures and seminars. Sosa's main interests are in epistemology and metaphysics. He has edited four books and has written a large number of articles, some of which have appeared in Latin America.

Ignacio Angelelli (b. 1933) was born in Italy, but was educated in Argentina. In 1957 he went to Fribourg, where he studied philosophy and history of logic with Bochenski, receiving the doctorate in 1965. At present he teaches at the Univesity of Texas. He has taught also at the universities of Buenos Aires and Notre Dame. His field of specialization is the history of logic and particularly Frege. He has edited several books and is the author of a substantial number of articles.

Jorge J. E. Gracia (b. 1942) was, like Sosa, born in Cuba. He came to the United States as a result of the political changes that took place in the island in 1959. His philosophical education, of classical and scholastic orientation, took place in Chicago and Toronto. He specializes in the history of metaphysics, particularly in the medieval period. He maintains a strong interest in Latin American philosophy and has written some non-historical articles from an analytic perspective. He is presently Chairman of the Department of Philosophy at the State University of New York at Buffalo.

Although none of these philosophers live in Latin America and most of them were at least in part trained in the United States, they are mentioned here for three reasons: First that they were born in Latin America; second that some of their work is written in Spanish; and third, that through publications and professional contacts they have maintained an important presence in Latin America. The history of analysis in Latin America could not be complete without reference to them. The Spanish version of this Anthology contains selections from their work. However, since they are well known in the English speaking world, and we were pressed for space, a decision was made to exclude their articles from this edition.

BIOGRAPHICAL NOTES

CARLOS EDUARDO ALCHOURRON (b. 1931, Argentina). Educated in Argentina. Professor of Philosophy of Law at the University of Buenos Aires. Coauthor of _Normative Systems_ (1971, with E. Bulygin), and author of many articles on logic, philosophy of logic and the methodology of law.

EUGENIO BULYGIN (b. 1931, Argentina). Educated in Argentina. Independent Researcher of the Consejo Nacional de Investigaciones Científicas and Associate Professor of Philosophy of Law of the University of Buenos Aires. Co-author of _Normative Systems_ (1971, with C. E. Alchourrón), and author of several articles on logic and the philosophy and methodology of law.

GENARO R. CARRIO (b. 1922, Argentina). Educated in Argentina, the U.S.A. and England. Former Professor of General Theory of Law at the University of Buenos Aires. Author of _Recurso de amparo y técnica judicial_, _Notas sobre derecho y lenguaje_, _Recurso extraordinario por sentencia arbitraria_, _Sobre el concepto del deber jurídico_, _Principios jurídicos y positivismo jurídico_, _Algunas palabras sobre las palabras de la ley_, _Sobre los límites del lenguaje normativo_, _Dworkin y el positivismo jurídico_, and of many articles on philosophy of law and constitutional law.

MARCELO DASCAL (b. 1940, Brazil). Educated in Brazil, France and Israel. Professor of the University of Campinas and of the University of Tel-Aviv. Editor of _Lo racional y lo irracional_ (1974), _Lo justo y lo injusto_ (1977), _Fundamentos metodológicos da linguística_ (1978), _Baruch de Spinoza_ (1979), and others. Author of _La semiologie de Leibniz_ and of many articles on semantics, Leibniz, ethics and the history of philosophy (XVIIth and XVIIIth centuries).

JORGE J. E. GRACIA (b. 1942, Cuba). Educated in Cuba, the United States, Spain and Canada. Professor and Chair of the Department of Philosophy of the State University of New York at Buffalo. Co-editor of _El hombre y los valores en la filosofía latinoamericana del siglo XX_ (1975, 1980), and editor _Man and His Conduct: Philosophical Essays in Honor of_

Risieri Frondizi (1980). Author of Suárez on Individuation (1982), Introduction to the Problem of Individuation in the Early Middle Ages (1984), and of many articles on ontology and the history of medieval philosophy.

GREGORIO KLIMOVSKY (b. 1922, Argentina). Educated in Argentina. Researcher of the Instituto Di Tella, Professor of Epistemology of the University of Belgrano, and Professor of Mathematics of Caese. Author of several articles on logic, set theory and epistemology.

ZELJKO LOPARIC (b. 1939, Yugoslavia). Educated in Yugoslavia and Belgium. Professor of the University of Campinas and Coordinator in the area of epistemology and natural sciences for the Centro de História e Filosofia da Ciência. Author of several articles on epistemology, the philosophy of science and the history of philosophy.

PEDRO LLUBERES (b. 1928, Venezuela). Educated in Venezuela, the United States and England. Director of the Instituto de Urbanismo of the University of Simón Bolivar and member of the Instituto de Filosofía of the Central University of Venezuela. Author of Platón y la evolución de los establecimientos humanos en el mundo helénico (1966), Ciencia y escepticismo: aproximación a Descartes (1976), and several articles in the philosophy of language and the philosophy of science.

FRANCISCO MIRO QUESADA (b. 1918, Peru). Educated in Peru. Professor of the University of Lima and the University Cayetano Heredia and Professor Emeritus of the University of San Marcos. Author of many books, such as Apuntes para una teoría de la razón (1939), Filosofía de las matemáticas (1980), and Problemas funcionales de la lógica jurídíca (1956), and also of many articles on logic, epistemology and the philosophy of science.

JOÃO PAULO MONTEIRO (b. 1938, Portugal). Educated in Portugal and Brazil. Director of the Department of Philosophy of the University of São Paulo. Author of Teoria, retorica, ideologia (1975) and of several articles on epistemology, political philosophy, and Hume.

CARLOS ULISES MOULINES (b. 1946, Venezuela). Educated in Venezuela, Spain, and Germany. Researcher of the Instituto de Investigaciones Filosóficas of the National Autonomous University of Mexico. Author of La estructura del mundo sensible (1973), Exploraciones en filosofía de la ciencia (1981), and several articles on epistemology and the philosophy of science.

CARLOS SANTIAGO NINO (b. 1943, Argentina). Educated in Argentina. Adjunct Researcher of the Consejo Nacional de Investigaciones Científicas and Associate Professor of Law of the University of Buenos Aires. Author of Introdución al análisis del derecho (1980), Consideraciones sobre la dogmática juridíca (1976), and several articles on the theory of law, ethics and political philosophy.

RAUL ORAYEN (b. 1942, Argentina). Educated in Argentina. Full Professor of logic at the Universities of Comahue and La Plata. Editor of two volumes: Ensayos actuales sobre Adam Smith y David Hume (1978) and La ontología de Frege (1971). Author of articles in semantics, philosophy of logic and ontology.

EDUARDO RABOSSI (b. 1930, Argentina). Educated in Argentina. Independent Researcher of the Consejo Nacional de Investigaciones Científicas and Professor of the Philosophy of Language of the University of La Plata. Author of Análisis filosófico, lenguaje y metafísica (1977), La justificación moral del castigo (1978), Estudios éticos (1980), and several articles on ethics, philosophy of language, modern philosophy and theory of knowledge.

JOSE A. ROBLES (b. 1938, Mexico). Educated in Mexico and the United States. Professor of Logic of the National Autonomous University of Mexico and Researcher of the Instituto de Investigaciones Filosóficas of the same university. Editor of El problema de los universales: el realismo y sus críticos. Author of articles on ontology and the history of modern philosophy.

ALEJANDRO ROSSI (b. 1932, Venezuela). Educated in Venezuela and Mexico. Professor of the Autonomous National University of Mexico and Researcher of the Instituto de Investigaciones Filosóficas of the same university. Author of books and articles on varios philosophical topics.

FERNANDO SALMERON (b. in Mexico). Full Research Fellow in philosophy at the National Autonomous University of Mexico. Author of *Las mocedades de Ortega y Gasset*, *La doctrina del "Ser Ideal" en tres filósofos contemporáneos: Husserl, Hartmann y Heidegger*, *La filosofía y las actitudes morales*, among others, and of more than thirty articles.

ENRIQUE VILLANUEVA (b. 1938, Mexico). Educated in Mexico and England. Director of the Instituto de Investigaciones Filosóficas of the National Autonomous University of Mexico. Author of *El argumento del lenguaje privado* and of several articles on the history of philosophy, philosophy of mind and the metaphysics of persons.

INDEX OF PROPER NAMES